Cryobiology for South American Neotropical Fish Species

Editors

Danilo P. Streit Jr.

Department of Animal Science
Faculty of Agronomy
Federal University of Rio Grande do Sul
Porto Alegre, RS
Brazil

Tiantian Zhang

Faculty of Science and Technology
Bournemouth University
Poole
United Kingdom

Estefania Paredes

Centro de Investigación Mariña (CIM)-Universidade de Vigo
Ecoloxía e Bioloxía Animal
Facultade de Ciencias do Mar
Vigo (Pontevedra)
Spain

CRC Press
Taylor & Francis Group
Boca Raton London New York

CRC Press is an imprint of the
Taylor & Francis Group, an **informa** business
A SCIENCE PUBLISHERS BOOK

Cover credit: Life cycle of Piracanjuba (*Brycon orbignyanus*). Image designed by Itamar Cossina Gomes, student of Dr. Danilo P. Streit Jr.

First edition published 2025
by CRC Press
2385 NW Executive Center Drive, Suite 320, Boca Raton FL 33431

and by CRC Press
4 Park Square, Milton Park, Abingdon, Oxon, OX14 4RN

CRC Press is an imprint of Taylor & Francis Group, LLC

Library of Congress Cataloging-in-Publication Data (applied for)

ISBN: 978-1-032-21658-4 (hbk)
ISBN: 978-1-032-21659-1 (pbk)
ISBN: 978-1-003-26944-1 (ebk)

DOI: 10.1201/9781003269441

Typeset in Times New Roman
by Prime Publishing Services

Foreword

Aquatic species represent an incredible diversity of forms and challenges. Aquaculture (e.g., farming of fish and shellfish) by itself involves dozens of commercially relevant species, and when trying to respond to needs for wild fisheries, conservation, and biomedical models the number of lines and species can be counted in the thousands. In some cases, such as the U.S. catfish industry, production of a hybrid between two different species (channel catfish female x blue catfish male) which requires artificial spawning—controlled mixing of sperm and eggs—is used to produce most of the fingerlings needed by farms. Other industries take different approaches to supply animals for farms such as same-sex production (e.g., all-female trout), or the use of polyploid animals with extra sets of chromosomes (e.g., triploid oysters). Currently most genetic resources of aquatic species are maintained as live populations which is unnecessarily risky and expensive.

The lack of ability to protect, distribute, and commercialize the genetic resources of aquatic species is a rapidly expanding global problem. Unlike livestock and other sectors, we have been unable to build the infrastructure necessary to safeguard the genetic wealth of fish, shellfish, and amphibians. The need is great and cannot be accomplished through the actions of individuals alone; it must be addressed through larger cooperative efforts. This book is a wonderful example of the broad approach necessary to address this problem, and it could not be centered at a more appropriate time or place than now in South America, especially Brazil.

As this volume is being finalized (early in 2024) we can see there are multiple threats to genetic diversity due to habitat change, for example, caused by artisanal gold mining and Amazonian deforestation. Reductions in habitat also affect population and ecosystem health and resiliency. At present there is a drought in the Amazon River Basin, the likes of which have not been seen for 100 years. Such droughts can become more frequent given the perils of climate change. This affects wild aquatic species and is reflected in the deaths of Amazon river dolphins (boto, *Inia geoffrensis*). Another dimension seen in South America is expansion of the range of the non-native lionfish (genus *Pterois*) into Southern coastal waters. Introduced species can transform aquatic ecosystems through predation and increased competition, and this expanded distribution can also be linked to climate change.

With respect to the aquaculture sector, we see increased expansion in South America and elsewhere. Any modern livestock industry is based on the interaction of production systems with factors such as nutrition and genetics. The genetic component is often undervalued at first, with more attention initially going to production systems. Genetic resources, however, become increasingly important as industries mature. It is imperative to collect genetic material from the wild populations that provide source material for aquaculture before selection has reduced the diversity. Cryopreservation and development of storage facilities (repositories) are not going to solve such problems, but the do offer time to identify and develop other options. Research activities such as protocol development represent a foundational step, but these protocols need to be converted into applied practical methods (which eventually may be viewed as community-level pathways) by taking into consideration commercial-level factors such as throughput, efficiency, cost, personnel involved, and space.

Other important concerns for repository development are data management, incorporation of genotypic and phenotypic information, inventory status, and quality management. Well-preserved and curated samples are only as valuable as they are available. As such, strong user interface mechanisms need to be developed. At this time, deployment of genetic resources for aquatic animals would primarily consist of the use of frozen sperm. This is due to a general lack of technical means to cryopreserve eggs, embryos, and larvae (with a few exceptions such as oysters). Deployment of genetic resources could be carried out at national fish hatcheries that have the expertise necessary to reproductively condition broodstock, collect eggs, and perform artificial spawning with frozen germplasm. Genetic resource sharing could also be carried out through partnerships at state, private, and academic hatchery facilities. Major considerations would include the relative lack of cryopreserved samples overall in collections, biosecurity infrastructure to prevent transfer of pathogens and parasites, and a current lack of goals and planning for deployment.

As indicated above, germplasm cryopreservation in aquatic species has been actively studied at a benchtop-research level, but little effort has been placed in scaling of protocols into stable applied platforms for use at the repository level. Because most work is research-oriented there is low reproducibility, and no standardization (of methods) or harmonization (of results) within individual species, or among them. This causes problems with processing of samples and prevents standardized or systematic approaches to deployment, for example due to the free-ranging diversity of freezing containers, cryopreservation protocols, and labeling methods used. These problems could be addressed by establishing and strengthening central facilities with high levels of capability and expertise for in-house research, processing, storage, and distribution. These central facilities could also operate mobile capabilities to bring high-level capabilities to the locations where animals can be accessed. This is important because transportation of fish or samples can substantially reduce gamete quality.

A new mechanism would be to develop distributed capability, where user communities could perform reproducible work by use of standardized devices shared over the internet as computer design files. Such devices can take advantage of consumer-level technologies, such as microprocessors and 3-D printing, to produce open-hardware devices cheaply and easily. Continued reliance on basic research will not address these large global problems and lack of large-scale infrastructure. Attention at an inter-agency level or among countries to establish national or regional frameworks for conserving and utilizing aquatic genetic resources could be used for developing expanded capability and integration of collection and deployment strategies, as well as addressing biosecurity concerns. This would also enhance resiliency in response to new diseases, changes in market structures, and climate change.

Overall, going forward we must reach outside traditional approaches contained within reproductive biology and cryopreservation. We need to bring in other expertise and link that with existing activities in other spheres such as habitat restoration, conservation management, and agriculture. Interdisciplinary and cooperative actions will be key, for example bringing together expertise in science, engineering, economics, public administration, sociology, education, and communications. We must ensure continuity in linking the past and future, especially with regard to including indigenous peoples and their connections to the natural world. All in all, the South American perspective has much to offer the world for aquatic genetic resources, and the contents of this book can provide utility there and beyond. With forethought and appropriate actions, we can sustainably protect, distribute, and commercialize the genetic resources of the residents of our lakes, rivers, estuaries, and oceans with the genetic material that we pass onto the future.

Terrence R. Tiersch
ttiersch@agcenter.lsu.edu
http://aggrc.com/

AGGRC

Preface

Dr. Danilo Streit is a Full Professor at the Department of Animal Science at the Federal University of Rio Grande do Sul in Brazil, he works in two Postgraduate Programs and is a Research Fellow at the Brazilian National Research Agency (CNPq). His research focuses on developing solutions using cryobiology, for example, to apply a strategy to reconstitute a population or to make a certain species viable for aquaculture. Projects applied in the electricity generation sector and solid partnerships with a research collaboration network with groups from the USA and Europe guarantee the quality of the researcher's publications. Over a sixteen-year career, his research with cryobiology applied to gametes, embryos and reproductive tissues of native South American freshwater species generated more than 100 scientific articles and trained more than 40 postgraduate students. Currently focusing on developing a strategic plan for the conservation of aquatic genetic resources of South American fish fauna.

Dr. Estefania Paredes is a tenure-track professor at the Department of Ecology and Animal Biology at the University of Vigo in Spain. Her research is focused on marine invertebrates and very recently fish both from the most teorical Cryobiology research to applied and translated research for aquaculture enterprises and government providers in Europe and United States. Most interested about preserving fertility and genomic biodiversity of marine and brackish organisms has published over 50 scientific articles and trained more than 20 postgraduate students. Dr. Paredes was elected member of the board at the Society for Cryobiology twice from 2019 to 2024 and is currently the President of the Spanish Society for Cryobiology since 2023.

Dr. Tiantia Zhang is now the Deputy Dean of Research and Professional Practice in the Faculty of Science and Technology. Professor Tiantian Zhang is a leading figure internationally in research on cryopreservation of gametes and embryos of fish species and her research interests have been in the areas of cryopreservation of reproductive cells and the effect of cryopreservation on genome integrity and cellular metabolism with applications in biomedicine, conservation and aquaculture. Her research activity has led to over 150 publications and over 100 presentations at international conferences and workshops and she supervised more than 30 PhDs. Professor Tiantian Zhang has obtained substantial funding from funding bodies such as the UK, the Wellcome Trust and the EU Research Councils. She is currently a member of the Board of Governors of the Society for Cryobiology and was elected President of the Society for Low Temperature Biology (SLTB) from 2005 to 2008, after serving as General Secretary, Treasurer and Committee Member of the SLTB for 10 years. She received an award from the Society of Cryobiology for her services in 2016.

This book aims to offer the reader information about the use of cryobiology for genetic conservation of South American freshwater ichthyofauna. By sharing knowledge and perspectives, we hope to inspire concrete actions to protect and preserve South American fish fauna for future generations.

As we explore the South American fish fauna, we are taken on a journey of lush aquatic ecosystems, where a multitude of life forms coexist in a complex and delicate balance. The global relevance of South American biodiversity cannot be underestimated, as many species are unique and play fundamental roles in maintaining global ecosystems. In this context, the germplasm conservation strategy emerges as an essential tool to protect the genetic diversity of South American species, ensuring their survival and adaptation to environmental changes.

Over the last forty years, numerous researches applying cryobiology methods have been and are being developed. And this book is the result of a vast experience that brought together global stakeholders interested in South American freshwater fish species, as well as expertise in conservation and management techniques for these populations using cryobiology.

In this context, the book's editors have organized a sequence of chapters so that the reader initially understands the application of cryobiology to fish germplasm banks and subsequently has access to a compilation of research already carried out with species of South American freshwater ichthyofauna. It is evident that also in the chapters, the reader will find future perspectives on research and application of cryobiology to South American ichthyofauna. The Editors' long experience with research in cryobiology applied to cells and reproductive tissues as well as fish embryos, made it possible to bring together a team of young and experienced collaborating researchers, recognized in different areas of activity on the subject of this book.

Finally, we hope that the document will contribute to the conservation of genetic resources of aquatic species of South American freshwater fish fauna, consolidating cryobiology as a fundamental *ex-situ* strategy in this process.

Contents

Foreword iii

Preface vii

1. **Integrating the Components Necessary for Germplasm Repository Program Development to Safeguard the Genetic Resources of Aquatic Species** 1

   Jack C. Koch, Lucía Arregui, Sarah Bodenstein, Teresa Gutierrez-Wing, Yue Liu, Rose Upton and *Terrence R. Tiersch*

2. **Commercial Application of Sperm Cryopreservation in Fish: Production of Salmonids in Chile** 27

   Iván Valdebenito Isler, Maritza Pérez-Atehortúa, Leydy Sandoval-Vargas and *Elías Figueroa Villalobos*

3. **Biobanks: Application in Environmental Conservation** 54

   Borys Dzyuba and *William V. Holt*

4. **Biobanks of Biological and Aquarium Model Species** 69

   Patrícia Diogo and *Rômulo B. Rodrigues*

5. **Gonads and Gametes of South American Neotropical Species** 96

   Diógenes H. Siqueira-Silva, Maira da Silva Rodrigues, Jeane Rodrigues, Daniel Fernandes Da Costa, Maria Clara Rosa-Silva and *Rafael Henrique Nóbrega*

6. **Quali-Quantitative Analysis of Semen of Neotropical South American Species** 118

   Eduardo A. Sanches and *Victor Gallego*

7. **Maintenance and Management of Breeders of Neotropical South American Species in Laboratories** 142

   Sergio Ricardo Batlouni, Ruy Alberto Caetano Corrêa Filho, Rafael Yutaka Kuradomi, Louise Nex Spica, Mariana Roza De Abreu, Lucas de Oliveira Brasileiro, Naiara Cristina Motta, Laíza Maria De Jesus-Silva, Susana Amaral Teixeira and *Jayme Aparecido Povh*

8. Embryos and Ovarian Tissues of South American Neotropical Fish: Current State of Development 171

Leandro Andrés Miranda, Thaíza Rodrigues de Freitas, Monica Rodrigues Ferreira Machado and *Lis Santos Marques*

9. Laboratory Techniques for the Analysis of Cells and Reproductive Tissues of Neotropical Species 183

Lauren Dorneles Dipp, Lis Santos Marques, Rômulo Batista Rodrigues, Thales de Souza França, Nathalia dos Santos Teixeira, Thaiza Rodrigues De Freitas, Raquel Santos Dos Santos, Larise Lima, Jhony Lisboa Benato and *Danilo P. Streit Jr.*

10. Sperm Cryopreservation Protocols for Neotropical South American Species 193

Danilo P. Streit Jr., Rômulo Batista Rodrigues, Eduardo Antônio Sanches, Jayme Aparecido Povh, Diógenes Henrique de Siqueira-Silva, Ana Carina Nogueira Vasconcelos and *Tiantian Zhang*

11. Stem Germ Cells in Fish, the Methods for Cryopreservation in Neotropical South American Species and Its Potential for Fish Conservation and Restoration 220

Saynara Souza dos Santos Miranda, Hadda Tercya, Taiju Saito and *Diógenes Henrique Siqueira-Silva*

12. Potential Use of Germplasm Banks of South American Neotropical Species in Environmental Conservation in Brazil 241

Evoy Zaniboni-Filho and *Josiane Ribolli*

13. Potential Use of Germplasm Banks of Neotropical South American Species for Fish Farming in Brazil 256

Danilo P. Streit Jr., Jayme A. Povh, Darci C. Fornari, Jonhy L. Benato, Fernanda De Mello, Eric A. B. Routledge, Estefania Paredes and *Tiantian Zhang*

Index **273**

CHAPTER 1

Integrating the Components Necessary for Germplasm Repository Program Development to Safeguard the Genetic Resources of Aquatic Species

Jack C. Koch,[1] *Lucía Arregui,*[1] *Sarah Bodenstein,*[1,2] *Teresa Gutierrez-Wing,*[1] *Yue Liu,*[1] *Rose Upton*[1,3] and *Terrence R. Tiersch*[1,*]

INTRODUCTION

1. Germplasm Repository Components

This section discusses the requirements and purposes for the development of germplasm repositories, as well as clarified terminology, components, process pathways, and equipment needs. This will provide a basic description of germplasm repositories and their role in support of wild fisheries, aquaculture, conservation, and biomedical research.

[1] Aquatic Germplasm and Genetic Resources Center (AGGRC) School of Renewable Natural Resources Louisiana State University Agricultural Center 2288 Gourrier Ave, Baton Rouge, LA 70820, USA.

[2] Current Address: Louisiana Sea Grant - Louisiana State University Sea Grant Bldg, Baton Rouge, LA 70808, USA.

[3] Current Address: The Conservation Biology Research Group School of Environmental and Life Sciences The University of Newcastle Callaghan, NSW 2308, Australia.

Emails: jkoch@agcenter.lsu.edu, lucia.arregui@gmail.com, mwing@agcenter.lsu.edu, yliu@agcenter.lsu.edu, sboden2@lsu.edu, rose.upton@newcastle.edu.au

* Corresponding author: ttiersch@agcenter.lsu.edu

1.1 Purpose and Needs of a Germplasm Repository

The climate-change crisis currently at play worldwide has led to a long list of stressors on aquatic biodiversity, which is sensitive to seemingly small changes in the environment. Germplasm repositories integrated with reproductive technologies have the potential to advance and complement current aquaculture and conservation efforts at a relatively low cost by preserving and allowing the management of the genetic diversity of aquatic species. As a tool, germplasm repositories are often overlooked and have experienced limited application in aquatic species despite the great potential to preserve and reintroduce genetic diversity at reduced costs. Examples of germplasm repository capabilities currently exist for aquaculture species, including blue catfish (*Ictalurus furcatus*), Atlantic salmon (*Salmo salar*) and Pacific oysters (*Magallana gigas*); these capabilities also extend to aquatic biomedical models like zebrafish (*Danio rerio*) and platyfish and swordtails (*Xiphophorus* spp.). Additionally, germplasm repositories are utilized for the conservation of small-bodied live-bearing fishes (e.g., family Goodeidae).

Germplasm repositories are most useful with active user communities (discussed further in Section 3). Germplasm repositories can have a broad range of applications; therefore, the purpose of a repository can vary and should be based on the needs identified by the user communities. Indeed, different communities may access the same repository for different purposes. Conservation specialists may be more interested in samples that can be available for decades or more to safeguard broad genetic diversity and act as protection from extinction. These groups might require a few samples collected from multiple locations (e.g., on-site cryopreservation, discussed further in Section 3). Biomedical researchers and educational entities may be interested in specific natural or genetically modified genetics, with uses ranging from short storage time to long-term preservation. These communities may process samples in centralized locations but require multiple batches of varied genetic materials. Commercial groups may require high throughput of samples with improved genetics to produce cohorts of organisms for grow-out, reducing the costs of maintaining broodstock year-round. Whatever the purpose, effective repository programs require a combination of components and capabilities beyond the foundational protocols needed for the production of frozen samples. Program developers should keep in mind that germplasm repositories must demonstrate impact by continual use of samples and remain accountable to their user communities and sources of funding. This requires samples in germplasm repositories to be identifiable and in continual linkage with relevant information, even with changes in staff. Samples and associated information should be collected and maintained, assuming that other unknown users will seek to use them, sometimes many years in the future.

By establishing repository capabilities beyond protocol development, generalizable and scalable cryopreservation pathways can be applied to programs and can be extended to aquaculture and conservation groups for commercial, imperiled, hobby, and biomedical aquatic species, including Neotropical fishes. Robust outreach programs can address these sectors and inform potential users of the capabilities of repositories. Training programs for the use of preserved genetic material, coupled with widespread collection of samples, can increase the impact of these repositories. These diverse purposes highlight the need for community input and interaction (discussed further in Section 3).

1.1.1 The Need for Consistent Terminology

With the list of imperiled aquatic species growing, clear and consistent communication surrounding the importance and necessity for germplasm storage is vital for a unified approach. Despite this, heterogeneous growth in aquatic cryopreservation has resulted in a lack of organized terminology, including terms that are used interchangeably. This problem is pervasive at the level of protocols and research and is seen at higher levels of organization. For instance, for the purpose of this book, the term "biobank" has been used, with the definition of any cells kept alive using some type of biotechnique, such as cryopreservation (D.P. Streit, personal communication). This definition is often extended to include the preservation of DNA, RNA, and somatic tissues. While the term biobanking historically refers to a collection of human materials, the term has gained popularity in recent years to also refer to collections for wildlife purposes. Similar terms, such as biorepository, cryo-bank, gene bank, frozen zoo, and genome resource bank, have been proposed to address such purposes and have been used interchangeably over the years, with multiple definitions and variations in usage. In the future, standardized usage of such terms will be essential for effective communication, policy development, and harmonization of approaches with networks.

For this chapter, the focus will be constrained to the cryopreservation of germplasm, such as sperm, oocytes, and early life stages, that can be used to reconstitute specific genetic resources, and thus the term "germplasm repository" will be used as it is most appropriate to describe these specific activities. However, the concepts and organizational principles described would be applicable to numerous other related activities.

1.2 Defining Repository Organizational Principles Using Process Mapping

How do we define a repository on a conceptual level? One tool that can guide this understanding comes from industrial engineering in the form of process mapping. This method is used to systemically define and document processes, such as those required to operate a manufacturing facility (Renger et al. 2016; Snyder et al. 2005). Process maps can be developed by combining information from discussions with individuals involved in the process and observations from outside parties using standardized approaches and symbology.

Visual representations of a process can be created that track the flow of information and materials into and out of the system (Figures 1A; 1B). Evaluation of process maps and creation of simulation models can assist in identifying requirements (e.g., time, cost, quality, personnel, and bottlenecks), areas of waste or redundancy, and specific points within a system where improvements can be implemented to achieve goals desired by community members (Renger et al. 2016; Snyder et al. 2005; Pegden and Sturrock 2014). Process maps have historically been used in commercial manufacturing settings, such as automobile factories. More recently,

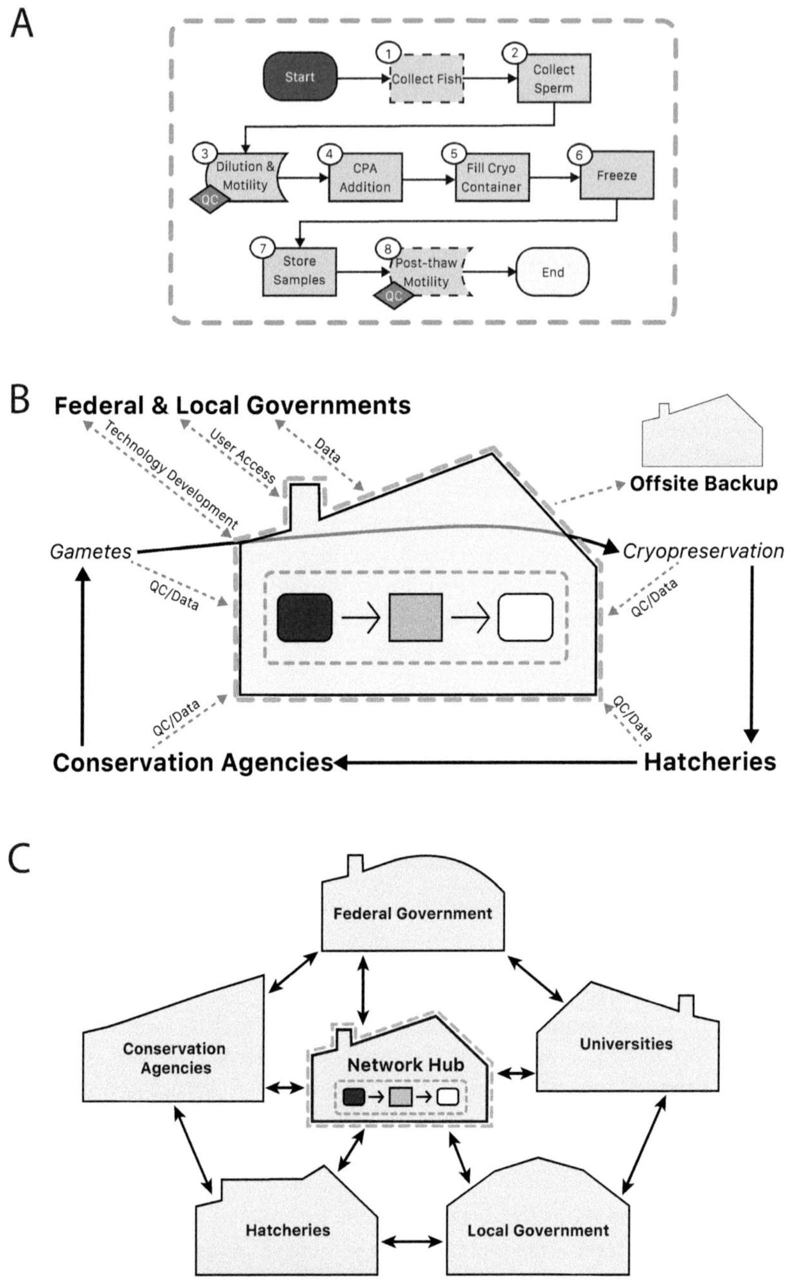

Figure 1. Organizational levels of a repository network. Pathway (A), repository (B), and network (C) level process flow diagrams. (A) Simplified cryopreservation pathway with integrated quality management steps. (B) Repository activities, community interaction, quality control, and sample processing are central to repository operations. (C) A centralized hub can manage the repository network that facilitates community interaction, addresses community needs, and integrates community input. The blue dashed line indicates a pathway, and the orange dashed line indicates a repository.

process maps have been used in agricultural and biomedical fields to outline the relationships among producers, customers, and suppliers (Apaiah and Hendrix 2005; Emelogu et al. 2019). Another recent application has mapped the cryopreservation processes for blue catfish (Hu et al. 2015) and Eastern oyster, *Crassostrea virginica* (Bodenstein et al. 2023). In the context of community integration into a repository framework, process maps can illustrate how communities currently interact with each other and the repository (Figure 1C). Such maps also allow for future interactions to be outlined if the repository framework expands.

Therefore, process maps and simulation models are useful tools when developing a repository framework and attempting to understand how communities and the repository will interact. Moreover, the practice of creating maps and models can drive interactions as it necessitates discussions between community members and the repository. With ample input during repository development and operations, a framework can be developed that serves all community members and is flexible enough to accommodate expansion to new members in the future.

1.3 Components of Germplasm Repositories

A germplasm repository is not limited to the physical facilities and the equipment used to process and store samples but includes additional components (Figure 2). The repository must be organized around standardized and repeatable procedures that will result in reliable recovery of samples months, years, or decades after their collection. The procedures must be performed by qualified and well-trained personnel who can evaluate and maintain the quality of the material being processed. At a minimum, these procedures must be developed as part of an integrated pathway that includes protocols for sample collection and transport, reliable cryopreservation, documentation and information management, secure storage (including backups in multiple locations), and training programs to ensure continuity with future staffing. A quality management program must also be integrated across the entire process to reduce waste in materials and workload, protect from loss of genetic resources, and ensure the utility of samples and their associated information.

In order to maximize the utility of samples in a repository, accurate and complete information should accompany each sample that is deposited. Specific information on the organisms, such as collection location, culture site and conditions, and unique identifiers, should be recorded in a centralized facility and an off-site backup location. Physiological data, measurements, processing conditions, including chemical cell protectants (cryoprotectants), exposure times (i.e., “equilibration time”), cooling rate, container type, and other pertinent data must also be recorded. This information must be maintained in a well-structured database that can be accessible to users through an outward-facing interface such as a website (discussed further in Section 3.1.1). Systematic updating of the database procedures ensures that technological advances will not render the information unrecoverable.

Centralized storage of cryopreserved samples is a foundational approach to safeguarding genetic resources. But even the most secure locations can fall victim to catastrophe. As with secure maintenance of any valuable resource, auxiliary locations and networked systems are needed where backup versions of material

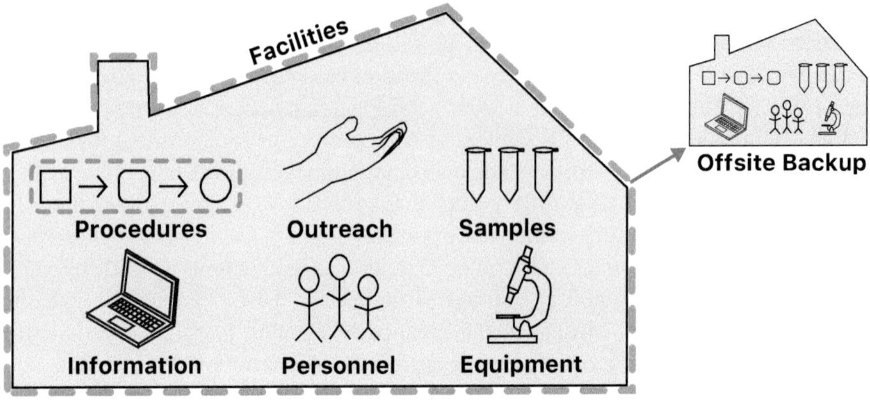

Figure 2. The major components of a germplasm repository. A germplasm repository should be comprised of eight major components: (1) facilities that the germplasm repository is built within; (2) procedures that dictate sample management; (3) an outreach program to connect with current and new repository communities as well as engage in educational outreach opportunities; (4) samples that are housed within the repository; (5) metadata associated with each sample, the information management system and database; (6) personnel who manage the repository and samples; (7) equipment used to process and store samples; (8) off-site backup where duplicate samples are stored in case of disaster.

can be stored. Such locations could, for example, be at the national level, such as the germplasm bank of the Embrapa Recursos Genéticos e Biotecnologia (Brazilian Agricultural Research Corporation 2024) of the Brazilian Agricultural Research Corporation (EMBRAPA). Similar institutions, such as the United States Department of Agriculture's National Animal Germplasm Program (USDA-NAGP), also house samples and information, including some with relevance to aquatic species of Brazil (e.g., United States Department of Agriculture, Agricultural Research Service, National Animal Germplasm Program and Brazilian Agricultural Research Corporation 2024). Smaller repositories distributed geographically can provide insurance against loss if they maintain the same quality management practices and information management systems.

1.4 Generalized Steps of Sample Processing

Processing samples for integration in a repository can be accomplished by central processing, mobile facilities, or aggregate collection. Although the scale of processing changes among these approaches, the basic processes and requirements are the same: to produce high-quality samples through standardized pathways that lead to reliable recovery of genetic material. In central processing, the organisms or germplasm are maintained or sent to a central facility where they are processed and cryopreserved, perhaps using automated industrial-scale high-throughput methods, typically overseen by experienced personnel (Childress 2017; Liu et al. 2021b).

Mobile facilities essentially deliver central processing capabilities to a worksite to take advantage of collecting high-quality samples (e.g., at a fish hatchery) without the shipping of animals or gametes that can reduce sample integrity (Childress 2017). This type of capability is often purpose-built and may include less expensive and portable forms of equipment or more expensive equipment capable of higher-throughput processing.

Community members collect and preserve samples with well-established procedures and quality management in aggregate collection and may send them to a central repository. The expensive equipment found in centralized facilities would not be cost-effective if only a few samples were processed by outside groups, and lower-cost standardized approaches are thus needed (see Sections 2.1.1 and 3.2.1 for discussion of open hardware). Nonetheless, no matter the sample processing type, it must produce high-quality material.

An essential component of cryopreservation pathways is a quality management (QM) program that includes quality evaluation (QE), quality control (QC), and quality assurance (QA). Quality management is the comprehensive set of methods and approaches (including training) put in place as an over-arching program to ensure that practices will yield reliably consistent outcomes in the present and future. At the base level, inexpensive, practical methods for evaluating sample quality are needed to ensure efficiency and identify problems. Researchers typically use QE methods to collect data (e.g., microscopic assessment of sperm motility or morphology) when studying gametes or developing cryopreservation protocols. When protocols move into practical application, QE can characterize samples and make QC decisions relevant to ensuring reliable final products, such as the disposal of inferior materials early in the process. While QC approaches are product-based, QA approaches are process-oriented. For example, programs to ensure that equipment is calibrated and properly functioning would be part of a QA system.

From a strategic overview, a cryopreservation pathway would include a range of activities, starting with collecting samples through storage in a repository and eventual thawing and use to produce offspring (Figure 1A). With planning, the entire process can incorporate QM practices at each step. For several biological and practical reasons, these steps can be organized into three main groupings: (1) fresh sample handling, (2) cryopreservation core processing, and (3) post-freeze management (Figure 3).

Since the most common cryopreserved material is currently sperm, we will present an example pathway based on these cells. Although other types of cells or tissues can have different specific variables, the overall pathway outlined below would be generalizable for all (Blackburn et al. 2022).

A. Fresh Sample Handling

1. *Sample Collection*: Initial sample collection is of utmost importance for successful cryopreservation and adding value to the accessions in a germplasm repository. Animal health, condition, method of collection, environmental or culture variables, time and method of sample collection, type of sample (tissue and germplasm), and other variables that may affect the quality of the material should be noted and included in the information that will be associated with the sample in the repository database.
2. *Initial Dilution and Quality Evaluation*: After a sample is collected, it can be diluted initially with an extender solution or buffer. This is sometimes necessary,

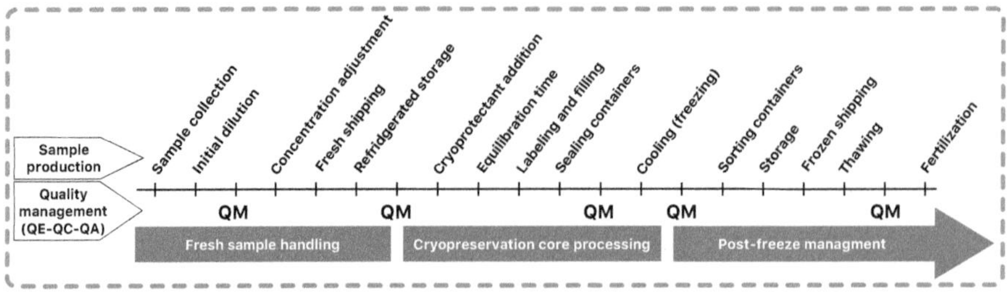

Figure 3. Sample production and quality management timeline pathways. The three main groupings of sample production are (1) fresh sample handling, (2) cryopreservation core processing, and (3) post-freeze management. Each sample production grouping is divided into major steps. This figure is an alternative visual representation of the cryopreservation pathway in Figure 1A. Quality management ensures quality samples and includes QE, QC, and QA. Indicated above the gray arrow are several example opportunities to integrate QM practices, and such integration could occur at each step in the pathway. (Figure adapted from Liu et al. 2021.)

for example, with species in which samples are routinely contaminated with urine during collection that can activate sperm motility and cause cells to become non-viable within minutes after collection (e.g., razorback suckers; Tiersch et al. 1998). Initial dilution can also ensure that viscous samples (e.g., produced by crushing of testis) are within a cell concentration range where they can be processed, and initial quality can be evaluated. For example, in the case of sperm samples, concentration, motility, or membrane integrity may be used for QC. If, at this point, samples are deemed to be below an established standard, the process would be stopped, and the sample discarded as part of a pre-established QC approach.

3. *Concentration Adjustment*: Samples that pass initial QC checks require measurement of cell concentration and, if necessary, adjustment with extender solutions or buffers. This step ensures that samples receive equivalent chemical and physical treatments for repeatable results. Lack of cell concentration adjustment can lead to differences in the effective concentration of cryoprotectants, affecting toxicity, freezing, and thawing conditions and resulting in unreliable performance. Adjustment of cell concentration also enables reliable control of the number of eggs that a cryopreserved sperm sample can fertilize. In fully practical pathways, this relationship (fertilization dose) should be adjusted to ensure efficient use of sperm numbers in relation to the desired number of eggs to be fertilized (e.g., one straw per female). Quality evaluation of the sample would be performed after adjustment as a step in the QC process.
4. *Fresh Sample Shipping*: At this point, if samples were collected on-site for preservation at a central facility, they would be shipped or transported. Transportation conditions should be controlled to prevent circumstances that could damage the cells. For example, high or low temperatures can lead to cell death, unprotected sample vials can break during transit, and long transport times can result in loss of viability. Specific forms of packaging (e.g., insulation, chill packs, or cushioning) can protect the samples. Monitoring of conditions (e.g., time, temperature, or accelerometry) can be used for QC purposes and

can be integrated into the sample information contained within the database. Open-hardware approaches can be extremely useful in addressing these problems (discussed further in Sections 2.1.1 and 3.2.1).

5. *Refrigerated Storage*: Upon arrival at the central processing facility, the samples should be adjusted to the processing temperature. Sample quality should be evaluated upon receipt and again after refrigerated storage (which may be for hours or days) when processing begins. If, at any of these QC points, samples are deemed unusable, the samples will be discarded.

B. Cryopreservation Core Processing

Freezing methods should be determined by research in advance to establish the most practical combinations of cryoprotectant and concentration, equilibration time, and cooling and warming rates. Two common forms of cryopreservation exist: equilibrium cooling (Mazur 1990) and vitrification. The most common approach, equilibrium cooling, uses cryoprotectants at lower concentrations (e.g., 5–15% v:v), longer exposure (equilibration) times (e.g., 15–30 minutes), and often controlled cooling rates (e.g., 5–40°C per min) for sample volumes of 0.25–5.0 mL within containers such as French straws or cryo-tubes. The equilibration time should be rigorously set, for example, from when cryoprotectant is added until the freezing process (which can be defined in various ways, but it should be explicit). For vitrification, mixtures of more concentrated cryoprotectants (e.g., 30–50%) are added for shorter exposure times (e.g., 3–5 minutes), and samples are frozen by ultra-rapid cooling (e.g., > 10,000°C per minutes) for volumes of 2–20 µL (often suspended in a thin film or droplet) by direct plunging in liquid nitrogen or other cryogenic liquids to produce an amorphous (non-crystalline) glass.

6. *Cryoprotectant Addition*: After the sample has been adjusted to the desired cell concentration, the selected cryoprotectant is added. Cryoprotectants protect cells during freezing in several ways. For example, penetrating cryoprotectants can enter the cell and reduce intracellular ice crystal formation, and non-penetrating cryoprotectants can act osmotically outside the cell to promote cell dehydration during freezing. Some cryopreservation protocols call for a mixture of both types of cryoprotectants. The cryoprotectant to be used should be selected based on previous toxicity screening to maximize protection while minimizing toxicity. This type of screening can be viewed as range-finding experiments during protocol development research to reduce the number of treatment variables during freezing experiments.
7. *Equilibration Time*: After the cryoprotectant is added, it is allowed to equilibrate for a set amount of time based on practical concerns. For example, a short time (e.g., 5 minutes) may not allow the cryoprotectant to diffuse properly throughout the sample, but a longer time may result in cell damage or death due to toxicity. It is also important to ensure that the selected equilibration time will be sufficient to allow filling, sealing, labeling, and racking operations to allow containers before the samples are placed into the freezing chamber. Depending on the time required by the cryoprotectant, the number of trained personnel, the

availability of automated equipment, and the number of samples to be frozen, 15–30 minutes would be a reasonable target for equilibration.

8. *Labeling, Filling, and Sealing Containers*: During cryoprotectant equilibration, samples must be packaged into containers that will be sealed and labeled. There are several different containers available for cryopreservation, including glass ampules, plastic cryo-vials (i.e., with a gasket to ensure proper sealing), glass capillary tubes, and plastic French straws of 0.25 mL or 0.5 Ml capacities, although larger straws are also available. The end of the equilibration period offers an extremely valuable QC point to evaluate the entire batch by subsampling. For example, if, for some reason, the samples exhibited no motility (perhaps due to cryoprotectant toxicity), the batch could be discarded before freezing or storage. Vitrified samples are still rare for aquatic species, and there are few practical systems for sealing and labeling (Tiersch et al. 2020; Liu et al. 2021a).
9. *Cooling (Freezing)*: When the equilibration time is completed, samples are cooled at a controlled rate or specific static temperature, as determined through previous experiments. Cooling can be done in multiple ways, including computer-controlled freezers, liquid nitrogen-based chambers, or dry ice, depending on the number of samples and the required conditions. Among the devices for small numbers of samples are commercial products (e.g., Thermo Fisher Mr. Frosty™) and open-hardware devices. Design files for open hardware can be distributed over the internet and produced inexpensively by users through fabrication methods such as 3-dimensional (3-D) printing. Such devices can offer a range of cooling rates by positioning horizontal samples at various heights above liquid nitrogen (Hu et al. 2017) or by vertical positioning within a standard shipping dewar or dry shipper (Childress et al. 2021) that can be used for sample storage and shipping when in the field. Thus, various options exist for cooling devices, and several may be used at a single facility based on available resources and scale of production. Ensuring the proper functionality of devices before freezing is an essential QA operation.

C. Post-Freeze Management

10. *Sorting Containers*: After the samples are frozen, they typically must be sorted and placed in containers (e.g., plastic goblets) for organization, inventory, and retrieval in long-term storage units (dewars). At this point, a subset of samples can be used for QC, and if satisfactory, the rest would be placed into storage.
11. *Storage*: Long-term cryogenic storage can be done on location; samples can be sent to a backup location or ideally both. Some facilities use mechanical –80°C freezers for storage, but the preferred method is to place samples in long-term storage dewars with liquid-phase or vapor-phase nitrogen. Although dewars require monitoring of temperatures and nitrogen levels, they are not affected by electrical power outages, and the loss of nitrogen is relatively predictable, particularly if samples are not being continuously moved, or the dewar opened.
12. *Frozen Shipping*: Samples can be sent to other locations, such as germplasm repositories, stock centers, or user facilities. Before shipping, QE of a subset of samples can be performed as part of a QC program. The shipping can be best

done in shipping dewars (dry shippers) with specialized internal construction that prevents liquid nitrogen spills if the dewar is tipped. Sometimes, samples are shipped on dry ice, which is cheaper but less reliable.

13. *Thawing*: Evaluation of thawing procedures should be performed prior to working with important samples, as the process of deciding this procedure may involve trial and error. This evaluation could be performed at the repository or at the site of sample use. The final temperature goal of the sample is a vital consideration and could be set based on the biology of the species or based on experimental evaluation. For example, if the species of interest lives in cold water (i.e., 10°C), the final temperature goal should be close to 10°C. With this final temperature in mind, it is important not to overshoot this goal, especially when thawing samples using higher temperatures (e.g., 40°C).

    The final consideration is the method of thawing: water bath or air. Water-bath thawing is a common method. Samples held under liquid nitrogen are near –196°C, and thus, in general, the temperature at which they are thawed changes the thawing rate relatively little (i.e., the thaw rate of a –196°C sample in a 20°C water bath is not likely to have thawing consequences compared to samples in a 40°C water bath). This lack of difference in thawing rate can be used as a temperature-protective measure in the thawing process, as samples in a water bath will never warm above that temperature. During water-bath thawing, samples should be continuously moved to facilitate convective heating. After storage or shipping, a subset of samples must be thawed for QC. This assessment should be done before the use of the samples to avoid waste of valuable eggs or other materials and ensure that samples shipped or used in fertilization are reliably recovered.

14. *Fertilization*: A common evaluation of cryopreservation and recovery is fertilization. Although this is often the actual purpose of the samples, relying only on this parameter to assess sample quality can lead to loss of material, defective samples, unreliable recovery of genetic material, and waste of resources because it occurs at the end of the process. Quality control evaluations made early and at multiple steps of the process are more useful. It should also be noted that fertilization is not solely dependent on the cryopreserved sample. Viable thawed sperm may fail to fertilize if the eggs have low quality or if conditions are not conducive to gamete activation or interaction. For these reasons, control groups of eggs without sperm added and fertilized with high-quality (e.g., fresh) sperm should be included for each female.

1.4.1 The Role of Protocols

As indicated above, creating and maintaining a germplasm repository is impossible when based on protocol development alone. Germplasm repositories encompass multiple factors and activities that must work together to preserve recoverable genetic material reliably. This does not mean that protocols are not important. They are the first step of a cryopreservation process and the foundation of a repository process pathway. A well-developed protocol will optimize steps at the level of the cryopreservation process, while a well-developed pathway ensures that the protocol will be useful for application by considering factors such as cost, reliability, efficiency, automation, safety, facility design, information management, and ease of training.

Seven decades and significant resources and effort have been devoted to the development of protocols for individual steps of cryopreservation in aquatic species. There are established protocols for animal maintenance, sample collection, and considerable work on "optimum" extenders, cryoprotectants, equilibration times, and cooling rates (Yang et al. 2022). However, simply combining a series of independently optimized steps does not address the practical needs of a repository. A view of the bigger picture shows the need for a framework that combines a production pathway, where samples are collected and processed within a QM program, with multiple evaluation points to enable QC and process integrity is provided by QA. These production and QM pathways must be well integrated with data collection and management to ensure that necessary information is being updated and will be available in the future.

1.4.2 High-Throughput Processing

With the increasing number of populations of aquatic organisms that are being generated for research (e.g., by gene editing), aquaculture (e.g., by selective breeding), or collected for conservation, the creation and maintenance of germplasm repositories is of increasing importance. The impossibility of maintaining thousands of live populations of organisms has shifted perception from the cost of germplasm repositories to their potential value. The biomedical animal model community is a clear example of the pressure of maintaining live animals with continual production of transgenic, gene-edited, and mutated lines, with billions of USD in research investment. The Zebrafish International Resource Center currently lists more than 44,000 lines of zebrafish (Zebrafish International Resource Center 2024), and there are many other aquatic animal models.

As the number of lines increases, processing samples manually in small batches can become impractical. Thus, it is important to ensure that protocols used initially to develop programs can be scaled up for higher levels of production. Each step in a process should be evaluated not just for maximizing quality but also for factors such as cost, reliability, ease of training, and value at the level of real-world practical application. For example, research-oriented evaluation would tend to favor approaches leading to diminishing returns, such as improving post-thaw motility from 75% to 85%, while practically speaking, post-thaw motility of 50% might be sufficient to accomplish the goals of producing satisfactory fertilization to reconstitute lines. Expensive ingredients or complicated protocols may improve the likelihood of publishing research results, but they are of less value when trying to operate a repository.

Container choice is another example. If work is performed in French straws, it can be easily elevated in scale by the use of semi-automated or automated equipment, whereas other container types may represent a "dead-end" in terms of increasing production scale. The same holds true for storage and shipping. Standardized containers such as straws have well-developed systems to support their use. At higher levels of production, labeling becomes extremely problematic, especially with regard to integration with information systems. Features such as barcodes enable data management systems to interface directly with databases.

Practically speaking, the most important considerations for high throughput are cost and quality. Mass production is the reduction of per-unit cost through increased scale of production (i.e., the economy of scale; Caffey and Tiersch 2000). If done properly, processing more samples should offer economic benefits to a repository. Also, QM is essential. It is not useful to produce thousands of inferior-quality samples that need to be stored before removing them from the system after they are found to provide poor fertilization.

1.5 Equipment Needs

Equipment is needed that allows for processing and quality control at each step of the process, from sample collection to final usage. Developing a general list of equipment needed for these tasks is difficult, as the scale and characteristics of each facility, application, and taxa can differ. Despite these differences, the generalized pathway and processing steps described above can provide a basis for almost all species. In order to be most useful, generalized models for processing and QM should be able to account for a range of features.

For example, collecting sperm from zebrafish will require equipment with precision to evaluate, dilute, adjust concentration, and package samples with initial volumes of < 2 µL, while processing sperm of blue catfish may yield > 150 mL (Hu et al. 2011). Some taxa, such as amphibians, have extremely large sperm (e.g., *Ambystoma* with a 30 pg haploid genome size), while some fishes have extremely small sperm (e.g., *Xiphophorus* with a 0.75 pg haploid genome size). Organisms have different reproductive strategies, such as internal or external fertilization; thus, the collection and fertilization methods can be different. Animals reside in freshwater, estuarine, and marine environments, which can greatly affect sperm activation and extender solutions. In addition, organisms such as invertebrates (e.g., *Aplysia* sea hares) may need to be preserved as larvae that are produced in tangled compartmented egg strands. However, if we can look past these differences and instead focus on the similarities, we can see many areas of convergence with respect to sample processing and QM. As such, the expensive equipment found in central facilities for cryopreservation could be used to accommodate a wide range of species and applications, better justifying their utility as a generalized service center rather than addressing the needs of a single species or activity (e.g., conservation or aquaculture).

2. Germplasm Repository Operations

The previous section addressed the necessary components of a repository, including pathway-level activities. This section looks at how these components can be assembled into the next hierarchical level by addressing higher operational levels, including models for sample acquisition, sample export, sample storage, backup planning, information management, and the concepts of standardization and reproducibility. To do this, we will utilize three primary models of germplasm repositories:

central processing, mobile laboratories, and aggregate throughput (introduced in Section 1.4).

2.1 Sample Acquisition

The primary model (Figure 4) to transform raw germplasm, such as sperm samples, into repository accessions is processing through central facilities. Materials can be sent to the central facilities as live animals, dissected gonads, or diluted sperm. These facilities must have consistent and adequate funding to support core capabilities such as advanced equipment, experienced staff, dedicated space, established sample processing and quality management pathways, and high throughput. These facilities require at least one designated technician or research assistant with years of training to handle repository operations. In addition, standard processes and procedures for sample production and QM would be continuously improved after initial establishment and be strictly executed. With these resources, a central facility could achieve sustained high throughput (e.g., hundreds or thousands of samples per day) with strong QM. Facilities of this type could reliably process hundreds of thousands to millions of high-quality samples annually. Given the specialized nature and high expense of aquatic species repository development, there is an overall lack of capability to address current and future needs.

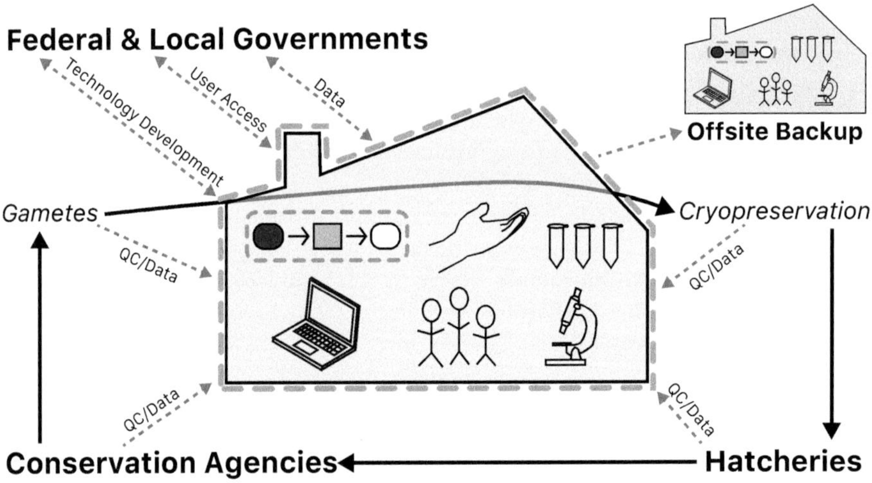

Figure 4. Repository operations. A generalized germplasm repository operation model based around central processing and off-site backup facilities. This model includes interactions with federal and local governments, hatcheries, and conservation agencies, but the communities can be interchanged and expanded based on specific situations and needs. Quality control (QC) information and sample or processing data are input into the repository information management system at different points throughout the operation pathway.

2.1.1 Mobile Facilities and Aggregate Throughput as Complementary Strategies to Central Facilities

Another strategy to address these limitations of centralized processing is through the development of mobile cryopreservation capabilities. A self-contained customizable mobile laboratory was established and evaluated for application with aquatic user groups to provide biosecurity and QC levels similar to, if not superior to, those of most centralized facilities (Childress et al. 2018; Childress et al. 2019). This mobile laboratory demonstrated the feasibility of high-throughput capability and QM by on-site cryopreservation at several remote locations for various user groups (i.e., federal hatcheries, biomedical research institutes, and fisheries research institutes), various species (i.e., saltwater and freshwater, egg-laying, and live-bearing), and different locations (i.e., Louisiana, Mississippi, and Texas).

With mobile laboratories, samples could be processed and cryopreserved within hours of collection, reducing variability due to the effects of refrigerated storage and transportation (Childress et al. 2018). This strategy can provide access to genetic resources and user groups that would be difficult or impossible to accommodate with typical centralized processing, facilitating repository development. Mobile laboratories are most suitable for destinations of < 1,000 km, but despite their advantages, they still present the requirements of expensive equipment and highly trained personnel (Childress et al. 2018).

An emerging strategy to address the limitations of traditional models is "aggregate throughput" (Liu et al. 2021b), which has the potential to achieve large-scale production that is independent of centralized processing. With this strategy, samples are processed by community members themselves, with relatively low throughput individually, yet combining for high throughput when aggregated in a unified repository system (Hu et al. 2017). For example, it would be relatively easy for an individual laboratory to cryopreserve several dozen samples if simple, reliable, and inexpensive options existed for the processing and submitting samples and associated information to a central location. If thousands of laboratories or users could submit high-quality samples to repositories, the aggregate throughput of tens of thousands of accessions would be comparable to that of a single central facility.

In recent years, the advent of open scientific hardware has fueled a new movement (Pearce 2017, 2012). The term "open hardware" comes from terminology originally used in the "open-source" software movement in the 1980s and 1990s. Open hardware, used in this context, refers to devices designed to be accessible and fabricated by a wide range of community members. The accessibility features of open hardware include low manufacturing costs and minimal skill requirements. The increased availability of fabrication methods such as 3-D printing makes it possible for rapid, inexpensive prototyping of devices and fabrication of hardware. This type of approach has been introduced to sperm cryopreservation of aquatic species (Tiersch and Monroe 2016; Hu et al. 2017). Open technologies can allow global accessibility of customized, standardized devices (Liu et al. 2021b, 2020), enabling an aggregated high-throughput model for repository development.

2.2 Sample Export

Cryopreserved germplasm should be available for research and commercial applications to improve the management of species and ensure that sufficient genetic diversity is available to reconstitute lost populations. Sample transport requires consideration of biosecurity (discussed further in Section 2.3.1), donor requirements, user needs, and user intent (Hiemstra (Ed), 2004). Germplasm repositories must have policies in place for the release of samples and may implement material transfer agreements (e.g., Zebrafish International Resource Center 2024) to help inform this process.

2.3 Sample Storage

Repositories must ensure the proper quality of samples and data over time. Germplasm samples should be stored in liquid nitrogen or vapor-phase nitrogen to avoid metabolic activity and maintain high quality during long-term storage. For cryopreserved samples, it is essential that the samples are properly labeled to link them with specific information and inventory (localization inside the repository). At the minimum, sample containers should be permanently labeled to indicate species, donor identification number, date of freezing, and cryopreservation protocol used.

Plastic French straws are commonly used containers for sample storage because they present advantages over other methods of packaging. French straws allow standardization of freezing and thawing rates due to their high surface-area-to-volume ratio, can be sealed by ultrasound, heat, polyvinyl alcohol (PVA) powder, or open-hardware devices, and come in two common volumes (0.25 mL and 0.5 mL) that allow efficient use of space for storage and shipment and enable scalable manual and automatic sample processing. These containers can also be permanently labeled (handwriting, printing, or with cryogenic adhesive labels) and are manufactured in different colors that aid in sample identification, organization, and processing. French straws are stored in plastic tubes called goblets that are available in different sizes, colors, and shapes and allow further organization of samples within dewar storage units.

2.3.1 Biosecurity

Many microorganisms (viruses, bacteria, archaea, and fungi) can survive cryogenic temperatures and are a potential source of cross-contamination among samples kept in storage units. Contamination of gametes with pathogenic or commensal microorganisms in a repository presents a significant risk as microorganisms could compromise the quality of biological materials after thawing, impair development of embryos or larvae after *in vitro* fertilization (IVF), and spread disease to other organisms in a facility or external populations (Sooksawat et al. 2020; Christensen and Tiersch 1996; Jenkins and Tiersch 1997). Liquid nitrogen and vapor-phase nitrogen can each be vectors for microorganism transfers (Grout and Morris 2009). Microorganisms can enter liquid nitrogen via contaminated surfaces, leaking samples, or directly from the atmosphere when airborne water droplets freeze above an open liquid nitrogen container. The liquid nitrogen can also be contaminated during transport and storage before reaching the repository facility. However, contaminants originating from stored samples are most likely to be sources of contamination and show larger effects on stored biomaterials (Bajerski et al. 2021b; Bajerski et al. 2021a).

Thus, germplasm containers must be correctly sealed. It should be considered that microorganisms colonize all parts of the animal body and, therefore, neither the sample nor the process of sampling can be considered sterile. In essence, all samples should be considered contaminated, and precautions should always be taken to avoid the spread of microorganisms. This would be especially important when transporting samples across regions and through time (e.g., decades). Special care should be taken when sampling wildlife where a disease-free status cannot be guaranteed.

During vitrification, samples are in direct contact with the liquid nitrogen, presenting the risk of sample or liquid nitrogen contamination (Criado et al. 2011). Under adequate handling and storage conditions, the load of microorganisms in liquid nitrogen has been shown to be small, and contaminants were most likely to be found in accumulated ice and debris at the bottom of the storage unit and underneath the lid (Bajerski et al. 2020). The microbial load could increase with use via the incorporation of new samples or by opening the storage unit; consequently, the risk of contamination could rise with time (Bajerski et al. 2020), so it would be recommended to periodically clean storage units.

2.4 Sample Information Management

The information associated with archived samples is among the most valuable assets in a repository, as it can be used even if the samples are lost. Therefore, information technology is a critical component, and efforts should ensure that this capability is effective and secure. Efficient management and tracking of samples as they are received, processed, stored, and shipped is a key aspect of a repository. This can only be performed with a well-designed computerized database. A database is a matrix of information that can be searched or evaluated based on specific queries, allowing retrieval of desired information (discussed further in Section 3.1.1).

Many commercial applications and open-source software options offer sample information management specifically designed for inventory control of cryopreserved samples. These systems avoid potential errors that derive from manual data entry, allow sample traceability, and accommodate small and large scales of production. However, they are sometimes expensive and often developed for human reproduction. Low-cost, modifiable, and user-friendly databases can be developed with open-source database software, such as MySQL, MongoDB, and SQL Lite, or commercially available database software, such as Microsoft Access, Oracle, Microsoft SQL Server, and DB2.

2.4.1 Information Types and Needs

Information associated with the samples can be classified into three groups: (1) data related to the sample donor at the levels of individual (e.g., sex, age, phenotypic data) and population (e.g., GPS coordinates of capture site, environmental data, genetic characteristics); (2) data related to the sample itself (e.g., method of collection, method of processing, quality, and disease screening); (3) inventory (e.g., number of samples and current location in the repository). Repository managers should establish minimum data submission requirements in relation to specific data fields and the structure of the database.

Essential information kept in the database can be used to make decisions on management or research applications. Therefore, the information needed will depend on the specific objectives, applications, and species in the repository, and it should consider the potential for currently unknown uses of samples and information in the future. Although it is typically understood that it is imperative to maintain a current and accurate inventory, staffing plans often underestimate the required workload.

2.5 Backups and Contingency Plans

Germplasm repositories should be implemented to be operational for at least several decades. Samples or associated data could be threatened by internal or external risks (e.g., power outages, internet outages, lack of liquid nitrogen supply, and flooding). Internal risks can be further classified according to the source: human mistakes (e.g., mislabeled data), equipment malfunction (e.g., dewar failure, errors in LN probes), or infrastructure problems (e.g., fire or reductions in base funding). In order to mitigate the risk of losing samples or data at the storage facility, contingency plans should be developed. Such plans should contain a comprehensive risk management approach that covers assessment and mitigation. These must be considered "living" documents that undergo annual review and updates. Common risks can be minimized by considering the establishment of minimum precautions at the central facility. For example, staff should be properly trained, and storage dewars could be equipped with a continuous automatic monitoring system with alarms linked to phone notifications at predetermined liquid nitrogen levels.

Risk management often involves trade-offs. Germplasm collections can be split into multiple storage locations for safety, further complicating recordkeeping and inventory. It is worth noting that the rarest risks are sometimes the most damaging, and assessments must balance the frequency of events against their severity. In case of equipment failure, at least one completely empty storage unit could be held available for sample relocation. Duplicate collections at off-site storage locations in different geographical regions could be used to diminish losses caused by devastating disasters (e.g., building collapse in an earthquake-prone area). Data security is integral to the reliable and efficient management of a repository. Information loss can render biological material useless. Therefore, on-site and off-site data backup should be implemented, and policies instituted to maintain data integrity and security.

2.6 Standardization and Reproducibility

The key to maintaining and operating repositories is the reliable production of samples with defined quality. This is challenging because of the pervasive lack of reproducibility among aquatic cryopreservation communities (Torres and Tiersch 2018). Overall, this lack of reliability and efficiency creates substantial doubt about the quantities of material needed to maintain and recover lines. The response to this is often to build in a safety factor using over-collection of animals and the processing and storage of multiple repeated samples. After thawing, non-reproducible fertility leads to increased use of females and eggs to ensure the production of offspring. All told, this leads to routine waste of supplies, storage space, time, and personnel, which considerably decreases efficiency and increases costs.

Why is reproducibility a common problem despite many hundreds, if not thousands, of published protocols on sperm cryopreservation? Most efforts thus far have addressed single-entity approaches (e.g., within a laboratory) rather than community-recognizable standardization to address this problem. The concept of "standardization" is often interpreted as a unified or improved protocol with "optimal" cryoprotectants and cooling rates. However, even optimized protocols will not guarantee good outcomes without strict adherence to control of important factors, such as sperm concentration and equilibration time for a given concentration of cryoprotectant in a specified container type (e.g., 0.5 mL French straw).

Therefore, we recognize four essential elements for standardization and reproducibility of an integrated pathway to address reproducibility issues within and among communities: (1) process recognition, (2) documentation and reporting, (3) training, and (4) hardware for sample production and quality management. In the development of pathways, attention should be paid to standardization and scalability at the level of the entire process instead of optimizing rates and reagents as separate entities.

As indicated above, measurement and adjustment of sperm concentration, timing control, and container choice are extremely important at the applied pathway level but highly variable, unreported, or poorly controlled at the research protocol level (Torres and Tiersch 2018; Tiersch 2011). An emergent property of recognizing the applied nature of pathways is that they can be protocol-independent; for example, the same pathway can be adjusted and standardized (generalized) for different species or cryoprotectants (Liu et al. 2019).

Problems of reporting are prevalent among publications and documentation. For example, for most aquatic species cryopreservation research, much of the relevant information is typically not recorded, controlled, or reported, which produces fundamental challenges for reproducibility (Hu et al. 2017). This results in persistent challenges in reproducing research outcomes, identifying problems in sample processing, and utilizing frozen samples. Standard checklists and worksheets would assist in reporting procedures in research and documentation of information for samples destined for repositories. Achieving reproducibility will be difficult without the widespread availability of appropriate training to ensure the proper execution of protocols, processes, and reporting to support standardization efforts (Torres and Tiersch 2018).

The hardware used in aquatic species cryopreservation is extremely variable, perhaps due to its continued residence at the research level, and it is difficult to find any two groups in aquatic species work that utilize the same tools and equipment. Indeed, this problem is so ubiquitous standardization (based on the similarity of process) may not be an appropriate goal for existing hardware. Instead, harmonization, based on the similarity of results (i.e., equipment independent), maybe the best approach.

Efforts such as these must shift the focus toward engagement with broad categories of users as overlapping communities rather than continuing the current isolating trend of working as individual laboratories.

3. Germplasm Repository Community Interactions

The following sections highlight the necessity for community establishment and involvement in developing an effective germplasm repository system. We focus on community interaction, technology development, and communication as key repository considerations.

3.1 Community Interaction

Community interaction is vital for building and operating a successful and relevant germplasm repository (Figure 5). Repositories exist to serve their surrounding communities by providing services and a central location to access and store germplasm material and corresponding data. Too often, protocol development takes place in a "vacuum" that is disconnected from the facilities and people who will

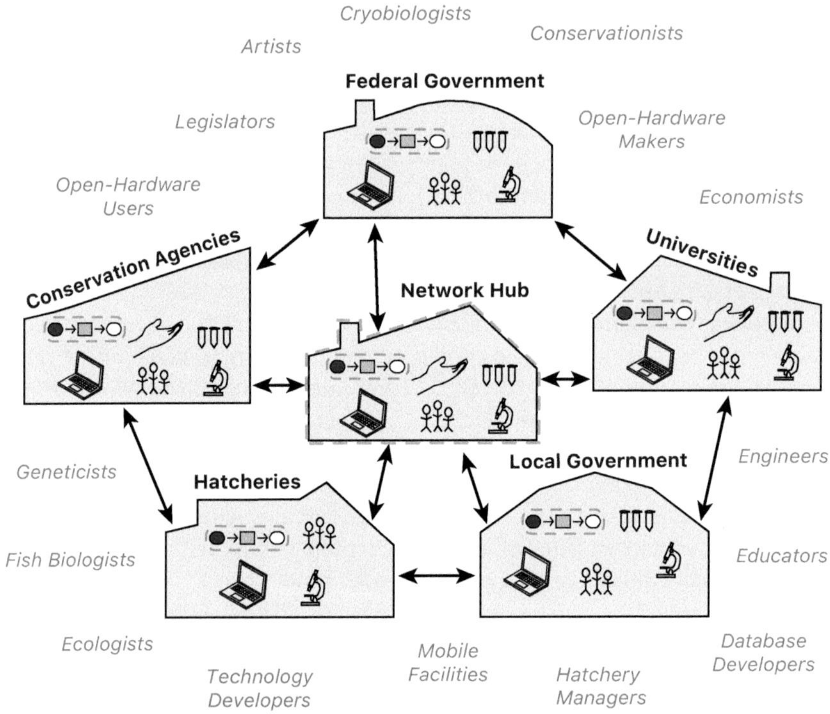

Figure 5. A repository interaction network. Repository network activities and interactions are coordinated by the network hub facility. There are core network interactions between satellite repositories and edge interactions between repositories and other community members (light gray text). Some satellite repositories have modified capabilities (e.g., hatcheries do not necessarily need to store samples on-site) that support the network.

use these protocols and processes daily. Continued dialogue with the community is paramount to ensure that protocols, processes, and samples are usefully managed. While the necessity for community interaction with a repository has been recognized (e.g., Torres and Tiersch 2018), tools such as websites, public databases, on-site cryopreservation, and mobile facilities are needed to facilitate community integration.

3.1.1 Website and Public Databases are Critical for Germplasm Repository Communities

A website and public database are essential community interaction tools for germplasm repositories. Both raise community awareness and provide information about the purpose and goals of the repository. The initial purpose of a germplasm repository website would be to communicate the goals and objectives, benefits, and resources (samples) of the repository to its community and other interested parties. The community can be strengthened and work toward standardizing and harmonizing cryopreservation efforts by expanding the website to include other features such as a learning management system, training materials, protocols, videos, open-source files, and a forum or chat. There is also the opportunity to consolidate communication efforts into a single place by providing a space for community information dissemination. This space could be a blog, social media feeds, or an advertisement board for upcoming events like workshops, courses, or conferences.

There are many websites and databases that can be starting points for the development of a germplasm repository information system. The Animal Genetic Resources Information Network (A-GRIN; United States Department of Agriculture, Agricultural Research Service, Animal Germplasm Program; Brazilian Agricultural Research Corporation; Agriculture and Agri-Food Canada 2024) of the USDA-NAGP is an excellent example of an outward-facing community interface that provides powerful tools for users to evaluate repository samples and associated information. The A-GRIN website also provides information relevant to samples held by the Brazilian Agricultural Research Corporation (Brazilian Agricultural Research Corporation 2024) and the Animal Genetic Resources of Canada (Agriculture and Agri-Food Canada 2024), which partnered in the website and database development. Websites such as this can be used to determine what genetic resources are available and to develop plans for material that needs to be added to the collection. Integration of website and repository databases can provide additional benefits.

Spreadsheets, while often used at the initial stages of repository development, lack many important features of databases, which are powerful tools that can be used for sharing data across studies, increasing reproducibility, and driving inter-field collaboration. The data contained within a database can also provide opportunities to perform additional studies, such as meta-analyses, to generate new insights without needing new data. Meta-analyses are often limited by inconsistencies in data collection and reporting. With a strong community, website, and database, these limitations can be overcome with process standardization and harmonization.

A repository is fully functional when there is a strong interaction with its community; otherwise, it is simply a collection of samples. Inputs and uses by the community are critical to the mission of germplasm repositories. Websites and databases are most powerful when the data and information are accessible, organized, and complete. These decisions require community input from the inception of these tools, and these decisions should continue to evolve with their communities. Evaluation by an external policy coordination committee or internal advisory board can also be useful. However, although there are numerous advantages of internet-driven interaction with a community, including low cost and the ability to cover wide geographic distances, in-person interactions are also needed.

3.1.2 On-Site Cryopreservation and Mobile Facilities

On-site cryopreservation provides an important option to cryopreserve imperiled, wild, or commercial species and engage in community interaction at the same time. The practice of on-site cryopreservation is not new to fisheries or conservation (e.g., Wayman et al. 1997). Between 1990 and 2006, the World Fisheries Trust (WFT) invested considerable resources into field cryopreservation for fish in North and South America (Harvey 2011) with simple tools such as shipping dewars. On-site cryopreservation can be further enhanced by adding mobile facilities (discussed previously in Section 2.1.1). In combination, mobile facilities and on-site cryopreservation benefit the community by providing opportunities for training, research, and outreach. An example of the effectiveness and potential scope of mobile facilities and on-site cryopreservation in research was documented when researchers from the AGGRC traveled to three field sites (covering over 4,000 km), including a fish farm, a national biomedical stock center, and a coastal laboratory to cryopreserve species from three genera (Childress et al. 2019). Thus, mobile facilities and on-site cryopreservation can complement and extend existing tools for research.

Outreach can be implemented on its own or incorporated into training and research. With the addition of mobile facilities and on-site cryopreservation, outreach activities can bring hands-on experience to user communities. They can also be used for early education programs, such as bringing mobile facilities to primary or secondary schools, much like the local fire department might bring a fire truck. Mobile facilities can also provide a mechanism to advance topics such as raising awareness about wildlife and habitat conservation or emerging opportunities such as 3-D printing and open technologies.

3.2 Technology Development Can Assist Standardization and Reproducibility

Technology is developed through design, innovation, and distribution. Traditionally, this process is closely guarded so that companies can monetize products and protect design and manufacturing techniques. In this form, technology development is separated from the communities that the technology serves, and there are often few options for input or feedback. Open-source technologies (introduced in Section 2.1.1) represent a growing field that acts as a counterpoint to traditional proprietary development and, though not new in principle, is an important consideration and powerful approach in the context of repository community interaction. Open-source technologies can drive initiatives for standardization and reproducibility through communication and aggregate throughput.

3.2.1 Open-Source Hardware as a Means of Communication

Technology facilitates communication. Consider the internet and the advent of electronic mail (e-mail). The internet and e-mail increased the frequency and speed of communication. In the context of germplasm repositories, open-source technology provides a new platform to expand communication with communities. Open-source technology can be broadly split into two categories: open-source software and open-source hardware (OSH).

The most recognizable form of this is open-source software, an established example being the Linux operating system, which expanded rapidly in the 1990s with currently more than 250 distributions ("flavors") actively maintained (Loli 2022). It is clear how open-source software could facilitate communication and support community development of thousands of computer programmers and users motivated by the same mission. Recall the previous discussion on websites and databases, which are examples of communication and some of which are built using open-source platforms. Broadly, open-source community members can be described as users, makers, and developers, with each level typically gaining knowledge of open source and increasing technical skills and autonomy. We refer to this development as the user-maker-developer (UMD) progression which is driven by the communication networks that open-source communities and technology bring forth.

Like open-source software, OSH should be accessible to a wide range of community members by ensuring low cost and minimal skill for manufacturing. Community communication, enhanced by OSH, can assist members to progress in understanding and enable participation in OSH initiatives (e.g., by UMD progression). Communication via OSH is not limited to single communities and can lead to community expansion. The broadening of communication driven by OSH can be useful as a teaching tool and practical application. In these forms, OSH can be applied in different scenarios, many of which overlap in purpose, much like mobile facilities and the multiple roles they can play in research, processing of accessions, and outreach. This is a new area, but there is a specific example of OSH as a mode of communication, demonstrated by the shipping dewar positional cooling device (SDPCD) or "Cajun Ejector" (Aquatic Germplasm and Genetic Resources Center 2023, Childress, Liu and Tiersch 2021).

As part of the development of the Cajun Ejector, the AGGRC conducted a beta-testing workshop (Childress et al. 2021) that simultaneously provided user testing of the device and opened portals for potential community members to begin their UDM progression. Beta testers were invited to assemble and test the Cajun Ejector and to engage in dialogue about open hardware devices. The activity drove communication among testers and device developers, creating an improved design and a broader understanding of open hardware. Thus, an open hardware device has enabled new modalities of communication across communities.

3.2.2 Aggregate Throughput Promotes Communication

One method of ensuring community interaction with a repository is developing an aggregate throughput model (introduced in Section 2). In aggregate throughput, community members receive cryopreservation training to produce standardized samples of set quality. However, this presents a challenge because standardized and reproducible cryopreservation protocols and equipment are necessary for producing samples of consistent quality in an aggregate throughput model (Liu et al. 2021b; Torres and Tiersch 2018). Open hardware offers a solution to this problem as it can be made widely available to diverse communities much more easily than traditionally manufactured pieces of equipment (Liu et al. 2021b).

Open hardware takes advantage of recent advancements in 3-D manufacturing technologies such as fused filament fabrication (FFF) and resin printing. These technologies are cost-efficient, and they can be used with minimal training that is readily available from multiple online sources. With FFF or resin printing, open-hardware devices for cryopreservation can be designed, and the files distributed to community members who use them to fabricate cryopreservation devices to freeze samples in a standardized and reproducible way. An aggregate throughput model combined with open hardware ensures continuous community interaction with the repository because community members are actively involved in contributing samples and modifying or creating devices.

4. Conclusions

There is a critical need for the preservation and conservation of South American neotropical fishes and other aquatic animals. This chapter presented existing examples from fisheries, engineering, and biomedical research to illustrate how these technologies can help the future of South American Neotropical fishes. Germplasm repositories can provide technologies that will assist in maintaining the genetic resources of these species for decades into the future. It is important to identify and explore the components, operations, and interactions of cryopreservation-based germplasm repositories to help guide current and future research and application efforts. Germplasm repositories cannot be created or maintained in a vacuum; the planning, implementation, and maintenance must include local, regional, and national communities, the private sector, and government input and support. Together, people and repository technology can help preserve and distribute biologically and economically valuable genetic resources for the future of the world.

Acknowledgments

This work was supported in part by funding from the National Institutes of Health, Office of Research Infrastructure Programs (R24-OD010441 and R24-OD028443), with additional support provided by the National Institute of Food and Agriculture, United States Department of Agriculture (Hatch project LAB94420), and LSU-ACRES (Audubon Center for Research of Endangered Species) Collaborative Program.

References

Agriculture and Agri-Food Canada. Agriculture and Agri-Food Canada. Available online: https://agriculture.canada.ca/en (accessed on 4 January 2024).

Apaiah, R.K. and E.M.T. Hendrix. 2005. Design of a supply chain network for pea-based novel protein foods. J. Food Eng. 70: 383–391.

Aquatic Germplasm and Genetic Resources Center. Sperm Cryopreservation - Shipping Dewar Positional Cooling Device (SDPCD, Known as Cajun Ejector) for V3.10.45. Available online: https://3d.nih.gov/entries/17520/3 (accessed on 4 January 2024).

Bajerski, F., A. Bürger, B. Glasmacher, E.R.J. Keller, K. Müller, K. Mühldorfer et al. 2020. Factors determining microbial colonization of liquid nitrogen storage tanks used for archiving biological samples. Appl. Microbiol. Biotechnol. 104: 131–144.

Bajerski, F., B. Glasmacher, E.R.J. Keller, K. Müller, K. Mühldorfer, M. Nagel et al. 2021a. What factors affect the presence of microorganisms in cryotanks? - A culture-independent approach to assess potential microbial colonization of liquid nitrogen storage tanks. Cryobiology 103: 161.

Bajerski, F., M. Nagel and J. Overmann. 2021b. Microbial occurrence in liquid nitrogen storage tanks: a challenge for cryobanking? Appl. Microbiol. Biotechnol. 105: 7635–7650.

Blackburn, H., L. Torres, Y. Liu and T.R. Tiersch. 2022. The need for a framework addressing the temporal aspects of fish sperm motility leading to community-level standardization. Zebrafish 19: 119–130.

Bodenstein, S., I. Nahmens, B. Callam and T. Tiersch. 2023. Simulation analysis of high-throughput oyster cryopreservation at three scales of production. Aquac Int. doi.org/10.1007/s10499-023-01139-y.

Brazilian Agricultural Research Corporation. Research and Innovation for Brazilian Agriculture. Available online: https://www.embrapa.br/en/international (accessed on 4 January 2024).

Caffey, R.H. and T.R. Tiersch. 2000. Economics and marketing of cryopreserved fish sperm. pp. 388–408. *In*: Tiersch T.R. and P.M. Mazik [eds.]. Cryopreservation in Aquatic Species - World Aquaculture Society, Baton Rouge, LA, USA.

Childress, W.M. 2017. On-site Sperm Cryopreservation for Aquatic Repository Development and User Group Linkage. Ph.D. Thesis, Louisiana State University, USA.

Childress, W.M., R.H. Caffey and T.R. Tiersch. 2018. Design and cost analysis of a self-contained mobile laboratory for commercial-scale aquatic species cryopreservation. J. World Aquac. Soc. 49: 805–826.

Childress, W.M., B. Bosworth, E. Chesney, R. Walter and T.R. Tiersch. 2019. On-site capabilities of a mobile laboratory for aquatic germplasm cryopreservation. N. Am. J. Aquac. 81: 349–363.

Childress, W.M., Y. Liu and T.R. Tiersch. 2021. Design, alpha testing, and beta testing of a 3-D printed open-hardware portable cryopreservation device for aquatic species. J. Appl. Aquac. 1–24.

Christensen, J.M. and T.R. Tiersch. 1996. Refrigerated storage of channel catfish sperm. J. World Aquac. Soc. 27: 340–346.

Criado, E., F. Moalli, N. Polentarutti, E. Albani, G. Morreale, F. Menduni et al. 2011. Experimental contamination assessment of a novel closed ultravitrification device. Fertl. Steril. 95: 1777–1779.

Emelogu, A., S. Chowdhury, M. Marufuzzaman and L. Bian. 2019. Distributed or centralized? A novel supply chain configuration of additively manufactured biomedical implants for southeastern US States. CIRP J. Manuf. Sci. Technol. 24: 17–34.

Grout, B.W.W. and G.J. Morris. 2009. Contaminated liquid nitrogen vapour as a risk factor in pathogen transfer. Theriogenology 71: 1079–1082.

Harvey, B. 2011. The application of cryopreservation in fish genetic conservation in North and South America. pp. 783–788. *In*: Tiersch T.R. and C.C. Green [eds.]. Cryopreservation in Aquatic Species. World Aquaculture Society. Baton Rouge, LA, USA.

Hiemstra, S.J. 2004. Guidelines for the Constitution of National Cryopreservation Programmes for Farm Animals. ERFP, Paris.

Hu, E., H. Yang and T.R. Tiersch. 2011. High-throughput cryopreservation of spermatozoa of blue catfish (Ictalurus furcatus): Establishment of an approach for commercial-scale processing. Cryobiology 62: 74–82.

Hu, E., T.W. Liao and T.R. Tiersch. 2015. Simulation modeling of high-throughput cryopreservation of aquatic germplasm: a case study of blue catfish sperm processing. Aquac. Res. 46: 432–445.

Hu, E., W. Childress and T.R. Tiersch. 2017. 3-D printing provides a novel approach for standardization and reproducibility of freezing devices. Cryobiology 76: 34–40.

Jenkins, J.A. and T.R. Tiersch. 1997. A preliminary bacteriological study of refrigerated channel catfish sperm. J. World Aquac. Soc. 28: 282–288.

Liu, Y., H. Blackburn, S.S. Taylor and T.R. Tiersch. 2019. Development of germplasm repositories to assist conservation of endangered fishes: Examples from small-bodied livebearing fishes. Theriogenology 135: 138–151.

Liu, Y., M. Chesnut, A. Guitreau, J. Beckham, A. Melvin, J. Eades et al. 2020. Microfabrication of low-cost customisable counting chambers for standardised estimation of sperm concentration. Reprod. Fertil. Dev. 32: 873–878.

Liu, Y., A. Lin, T.R. Tiersch and W.T. Monroe. 2021a. A 3D printed vitrification device for storage in cryopreservation vials. NATO Adv Sci Inst E. 11: 7977.

Liu, Y., W.T. Monroe, J.A. Belgodere, J.W. Choi, M.T.G. Wing and T.R. Tiersch. 2021b. The emerging role of open technologies for community-based improvement of cryopreservation and quality management for repository development in aquatic species. Anim. Reprod. Sci. 106871.

Loli, F. 2024. Linux Distribution Timeline. Available online: https://github.com/FabioLolix/LinuxTimeline (accessed on 4 January 2024).

Mazur, P. 1990. Equilibrium, quasi-equilibrium, and nonequilibrium freezing of mammalian embryos. Cell Biophys. 17: 53–92.

Pearce, J.M. 2012. Building research equipment with free, open-source hardware. Science 337: 1303–1304.

Pearce, J.M. 2017. Emerging business models for open source hardware. J. Open Hardw. 1: 1–14.

Pegden, C.D. and D.T. Sturrock. 2014. Rapid Modeling Solutions: Introduction to Simulation and Simio. Simio LLC, Sewickley.

Renger, R., M. McPherson, T.K. Bartels and K.L. Becker. 2016. Process flow mapping for systems improvement: lessons learned. Can J. Program Eval. 31: 109–121.

Snyder, K.D., P. Paulson and P. McGrath. 2005. Improving processes in a small health-care network. Bus. Process Manaj. J. 11: 87–99.

Sooksawat, T., T. Boonthai, S. Nimrat and V. Vuthiphandchai. 2020. Cryostorage of silver barb (Barbodes gonionotus) semen in dry shipper: Efficacy, risk of bacterial cross-contamination and effects of Aeromonas hydrophila on post-thaw sperm quality. Cryobiology 96: 184–196.

Tiersch, C.J., Y. Liu, T.R. Tierch and W.T. Monroe. 2020. 3-D printed customizable vitrification devices for preservation of genetic resources of aquatic species. Aquac. Eng. 90: 102097.

Tiersch, T.R., C. Figiel Jr., W.R. Wayman and J.H. Williamson. 1998. Cryopreservation of sperm of the endangered razorback sucker. Trans. Am. Fish. 127: 95–104.

Tiersch, T.R. 2011. Process pathways for cryopreservation research, application and commercialization. *In*: Tiersch T.R. and C.C. Green [eds.]. Cryopreservation in Aquatic Species. World Aquaculture Society, Baton Rouge, LA, USA.

Tiersch, T.R. and W.T. Monroe. 2016. Three-dimensional printing with polylactic acid (PLA) thermoplastic offers new opportunities for cryobiology. Cryobiology 73: 396–398.

Torres, L. and T.R. Tiersch. 2018. Addressing reproducibility in cryopreservation, and considerations necessary for commercialization and community development in support of genetic resources of aquatic species. J. World Aquac. Soc. 49: 644–663.

United States Department of Agriculture Agricultural Research Service National Animal Germplasm Program; Brazilian Agricultural Research Corporation. Explore the Inventory. Available online: https://agrin.ars.usda.gov/tax_inv_drilldown_page_dev?language=EN&record_source=BR (accessed on 4 January 2024).

United States Department of Agriculture, Agricultural Research Service, Animal Germplasm Program; Brazilian Agricultural Research Corporation; Agriculture and Agri-Food Canada. Information to Bring the World Together. Available online: agrin.ars.usda.gov (accessed on 4 January 2024).

Wayman, W.R., R.G. Thomas and T.R. Tiersch. 1997. Refrigerated storage and cryopreservation of black drum (Pogonias cromis) spermatozoa. Theriogenology 47: 1519–1529.

Yang, H., E. Hu, J.L. Matthews, Z.M. Varga and T.R. Tiersch. 2022. Is catalase an effective additive to alleviate oxidative stress during cryopreservation of zebrafish sperm at the repository level? Cryobiology 104: 70–78.

Zebrafish International Resource Center. Zebrafish Lines at ZIRC. Available online: http://zebrafish.org/fish/lineAll.php (accessed on 4 January 2024).

CHAPTER 2

Commercial Application of Sperm Cryopreservation in Fish

Production of Salmonids in Chile

Iván Valdebenito Isler,[1,*] *Maritza Pérez-Atehortúa,*[2] *Leydy Sandoval-Vargas*[3] and *Elías Figueroa Villalobos*[1]

INTRODUCTION

1. Context

Although salmonids are not Neotropical species, their commercial importance in this region has increased progressively. Chile, in particular, has become the second largest producer in the world, with about 27% of the world's production of farmed salmon (after Norway with over 50%). In 2021, Chile produced 995,158 tons of salmonids, of which 72.9% were Atlantic salmon (*Salmo salar*), 21.4% coho (Pacific) salmon (*Oncorhynchus kisutch*), and 5.7% rainbow trout (*Oncorhynchus mykiss*) (Sernapesca 2021). To reach this production, slightly over 1 billion oocytes and about 800 million pigmented eye embryos are required; consequently, approximately 100,000 females and 20,000 males are needed for reproduction each year. Currently, fewer than 1% of embryos are imported, and the national production of oocytes and milt is an essential factor for the industry.

In rainbow trout, the most frequently used reproductive biotechnology is to produce all-female populations, which is a normal production practice in this species; small numbers of triploid specimens are produced to improve the quality of

[1] Núcleo de Investigación en Producción Alimentaria, Facultad de Recursos Naturales, Universidad Católica de Temuco, Temuco, Chile.

[2] Ph.D. Program in Agricultural Sciences. Catholic University of Temuco. Temuco, Chile.

[3] Postdoctoral Researcher. Catholic University of Temuco. Temuco, Chile.

Emails: efigueroa@uct.cl, maritza.perez2021@alu.uct.cl, lsandoval@posgrado.uct.cl

* Corresponding author: ivisler@uct.cl

the meat since the sterility of female triploid specimens allows optimum productive performance, thanks to their low gonadosomatic index (GSI), resulting in high quality of the fillet. In salmon, these biotechnologies have only been carried out experimentally but are not applied in mass production.

Sperm cryopreservation is widely carried out in a variety of species:

i. *Marine Species*: In some marine species, such as golden cusk eel (*Genypterus blacodes* Figure 1), sea bass (*Eleginops maclovinus*, Figure 2), Patagonian toothfish (*Disostichus eleginoides*), and southern hake (*Merluccius australis* Figure 3), milt cryopreservation has been used experimentally and with very good results in terms of sperm motility, but further experiments are required to evaluate the fertilizing capacity of thawed sperm.
ii. *Salmonids*: In salmonids, the spermatozoa are immobile in the male genital tract, and flagellar activity is only triggered by contact with an aqueous medium – fish farms usually use fresh water, coelomic fluid and/or an artificial sperm activator. These aqueous mediums reduce potassium levels inside the spermatozoa and incorporate Ca^{2+} ions, activating the cascade of intracellular signals that leads, in fractions, to the triggering of flagellar activity; this activity lasts for less than one minute in freshwater but can be prolonged by sperm activators used in the mass production of these species (Valdebenito et al. 2009, 2010). The first experiment in the cryopreservation of salmonid milt (rainbow trout) in Chile was carried out by Almendras (1993), who compared different cryopreservation protocols using the cryoprotectants of Mounib (1978) and Alderson and MacNeil (1984), DMSO as penetrating cryoprotectant, and 0.5 mL straws and 5 mL cryotubes as containers. The results of this study were highly variable, with survival rates in the incubation phase between 1% and 94.9%. In Chile, protocols currently exist for large-scale cryopreservation of sperm from rainbow trout, coho salmon, and Atlantic salmon (Figure 4). However, due to the large milt produced by male rainbow trout and coho salmon, its use in these species has not generally been necessary; moreover, the results do not exceed 50% fertility. In Atlantic salmon, on the other hand, cryopreserved sperm is used on a large scale with successful results; in fact, the fertility rate using cryopreserved sperm usually does not present statistically significant differences compared with fresh sperm. It should be noted that this species produces a low volume of milt when abdominal massage is applied; fish farmers are, therefore, forced to sacrifice specimens and macerate the testicle to obtain cells for cryopreservation. The result is that milt is available in greater volumes and for extended periods. The protocols applied have not been developed by research carried out in Chile but by foreign companies that offer cryopreservation services and sell the reagents, dilution medium, and equipment for cryopreservation.

This chapter describes the species in which the process is applied, the cryopreservation medium, freezing rates, the types of containers used, freezing systems and equipment, and the results obtained with this biotechnology. It also provides images and videos that will facilitate an understanding of the importance of cryopreservation in Chilean salmonid farming.

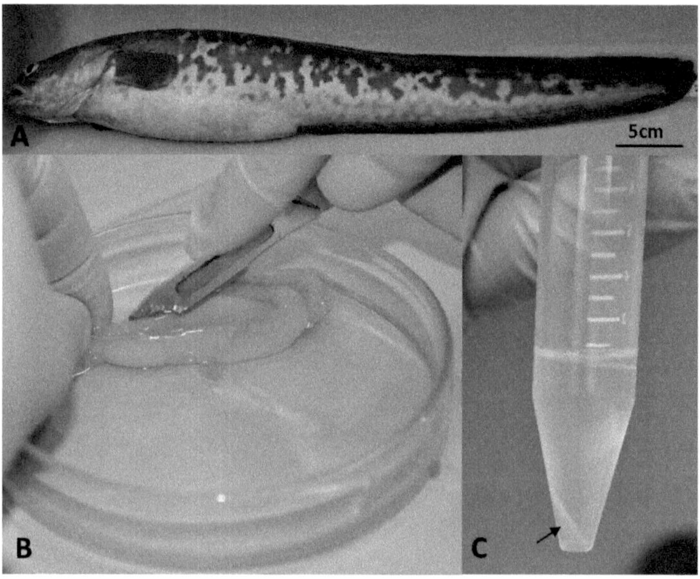

Figure 1. Immature male specimen of golden cusk eel (*Genypterus blacodes*) during artificial extraction of gametes. A: Immature specimen. B: Preparation of the testicular macerate. C: Obtaining the sperm pellet by centrifugation.

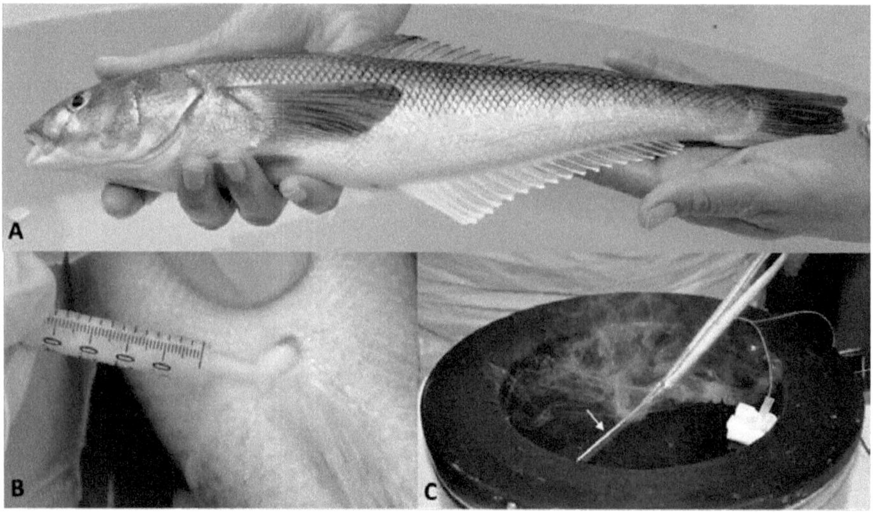

Figure 2. Male sea bass specimen during the artificial extraction of gametes. A: Mature specimen. B: Milt extraction by abdominal massage. C: Straw thawing (arrow) in a thermoregulated bath.

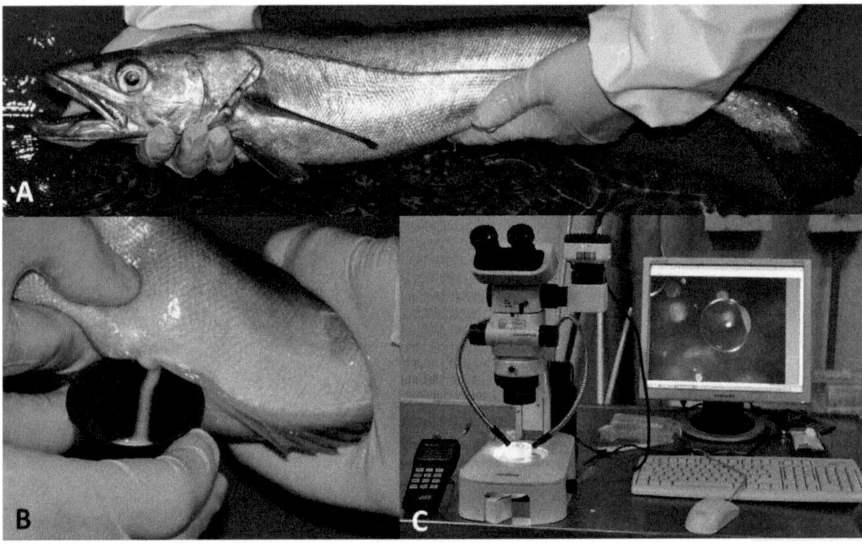

Figure 3. Male specimen of southern hake during the artificial extraction of gametes. A: Mature specimen. B: Milt extraction by abdominal massage. C: Optical equipment for quality evaluation.

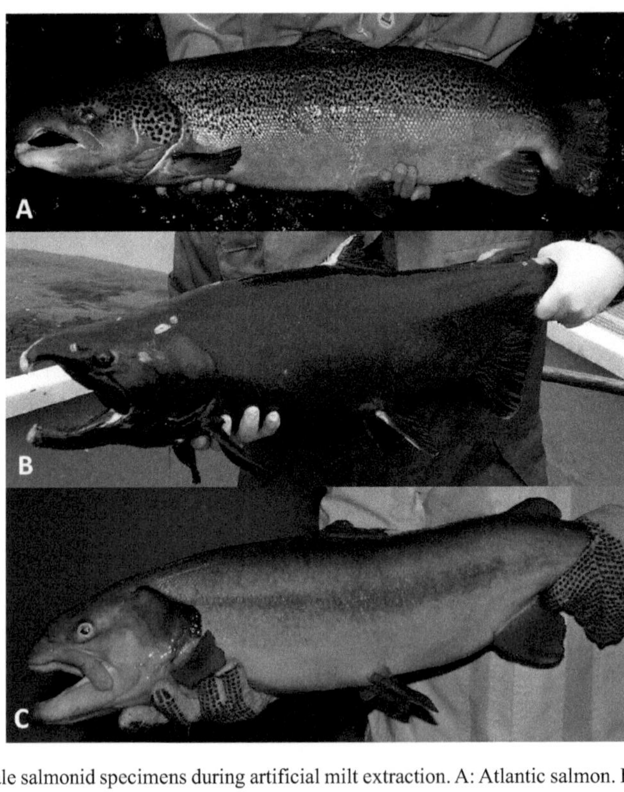

Figure 4. Male salmonid specimens during artificial milt extraction. A: Atlantic salmon. B: Coho salmon. C: Rainbow trout.

1.1 Results

1.1.1 Species

Rainbow Trout (*Oncorhynchus mykiss*)

Rainbow trout is a species native to the west coast of North America, living in the water courses of the coastal chain from Alaska to Baja California; local varieties or subspecies can be distinguished.

In nature, rainbow trout are sedentary fish that occupy aquatic spaces of pure, crystalline waters flowing through channels with marked topographic accidents; these give rise to rapids, jumps, and waterfalls typical of high mountain rivers (Bromage and Roberts 1995). The species requires cold water for reproduction (less than 10°C); however, the adults can tolerate a wide range of temperatures, subsisting in temperatures of 25°C for several days. As a result, rainbow trout have been introduced into a large part of the world's fresh waters, and intensive production has been developed in tropical countries, such as Colombia, Brazil, and Peru, based on the importation of embryos from countries with cold waters, such as Chile or USA.

The rainbow trout was introduced into Chile at the beginning of the twentieth century to develop sport fishing and is now distributed throughout practically the entire country, generating a great environmental impact. The first industrial-scale project started in 1970; since then, production has increased and currently represents around 6% (931,000 tons) of national salmonid production (Sernapesca 2021). It is an iteroparous species with annual spawning, although Estay et al. (1994, 1998) describe a cycling population in southern Chile. Females regularly reach sexual maturity in the second year, while a variable percentage of precocious males in a population do so in their first year. Adult males produce an average of 30 mL of milt, while the females have a relative fecundity of 1,825 oocytes/kg. Currently, the production of rainbow trout in Chile is based on all-female populations obtained from the testicular macerate of neomales. These frequently contain portions of gonad with the appearance of testicles and others with the presence of oocytes (hermaphrodites) (Figures 7 and 8). Cryopreservation in this species is only applied to sperm with variable results. In view of the easy adaptation of the species to the environmental conditions of southern Chile, cryopreservation is still little used; basically, this biotechnology is applied in genetic improvement programs and prolonged storage of "female milt" for the production of all-female populations.

Coho salmon (*Oncorhynchus kisutch*)

Coho salmon is a diadromous species endemic to the entire North Pacific basin. It is distributed from northern Japan through Kamchatka, across the Bering Sea to Alaska, and south through the California coastal area (Groot and Margolis 1991; Pennell and Barton 1996). It is an anadromous, semelparous fish that can only reproduce once in its life; therefore, expends all its energy and effort in its reproduction cycle.

This species was introduced into Chile at the beginning of the twentieth century to encourage sport fishing. However, the introductions were not successful at that time, and it was only in 1982 that the first populations were introduced for the development of fish farming in open systems (ocean ranching) in the extreme south of the country. Currently (2021), coho salmon production represents 21.4%

of national salmonid production, with 219,959 tons. In Chile, this species matures during the summer, often after one winter at sea, and spawning occurs in late summer and fall, while natural Northern Hemisphere populations may mature after several years at sea.

In fish farming, two-year-old broodstock reaches a total weight of between 4 kg and 8 kg. The males can produce a large amount of milt (55.8 mL mean) compared with other salmonids (Sandoval-Vargas et al. 2021); the relative fecundity can be around 986 oocytes/kg and survival during incubation can reach 74.3% (Estay et al. 1995). Cryopreservation is used only for male gametes, and the results are significantly lower than with fresh milt; similarly, fertility using cryopreserved sperm usually does not exceed 50%. Despite the low fertility results, several companies continue to use and improve this technique, mainly for genetic improvement programs.

Atlantic salmon (*Salmo salar*)

The Atlantic salmon is a diadromous fish native to the temperate and subarctic regions of the North Atlantic Ocean, inhabiting both the United States and Canada, as well as Portugal, Russia, the United Kingdom, Norway, Iceland, Ireland, and Greenland (Aas et al. 2010; Pennell and Barton 1996). It is an anadromous, iteroparous species, the adults typically return from the ocean to the rivers where they were born to reproduce. In Chile, this species was introduced for the development of sport fishing at the beginning of the twentieth century and developed for industrial purposes from the end of the 1980s. The first harvests were recorded in 1997 in small productions based on embryo imports; however, it now presents the highest volumes in Chile, accounting for over 70% of national production with less than 1% of embryos imported. In breeding systems, this species reaches maturity after two or three winters at sea. The males can reach a total weight of over 30 kg, while the females are regularly smaller and do not exceed 20 kg. Estay et al. (1999) report a relative fecundity for the species in Chile of 946 eggs/kg and survival in the incubation stage of 76.1%. Despite the large size of the males, the quantity of milt obtained by abdominal massage is very small (no more than 2 mL).

For this reason, the males are sacrificed to remove and macerate the testes; the intratesticular sperm is then used to fertilize oocytes (Figures 5 and 6). Cryopreservation in this species is only applied to milt, with good results; in fact, these do not differ significantly from the results with fresh sperm. As a result, several companies use milt cryopreservation in a large part of their production. In this case, the main benefits are that milt is available throughout the year, and companies can carry out directed crossing in genetic selection programs.

Experimental work has also been carried out on other native species of interest for fish farming, such as golden cusk eel (*G. blacodes*), sea bass (*E. maclovinus*), Patagonian toothfish (*D. eleginoides*), and southern hake (*M. australis*). The effect of cryopreservation on sperm motility has been evaluated in these species (Magnotti et al. 2022).

Below, we describe the main aspects of the sperm cryopreservation protocols applied in the salmon industry, such as cryopreservation medium, freezing and thawing curves, as well as the prospects for development.

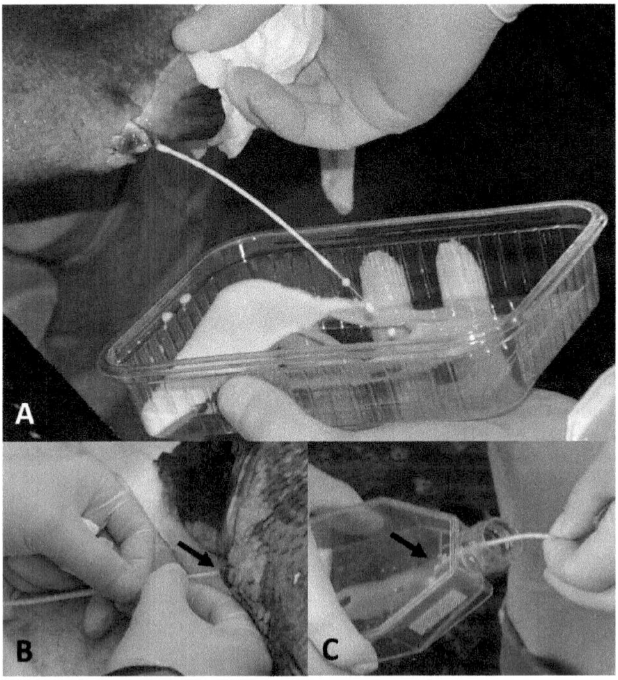

Figure 5. Artificial extraction of "fresh" milt from salmonids. A: Abundant milt extraction by abdominal massage in coho salmon. B: Extraction of scarce fresh milt by cannula in Atlantic salmon; the cannula is introduced into the genital pore of the male (arrow), and the milt is deposited in a container (C).

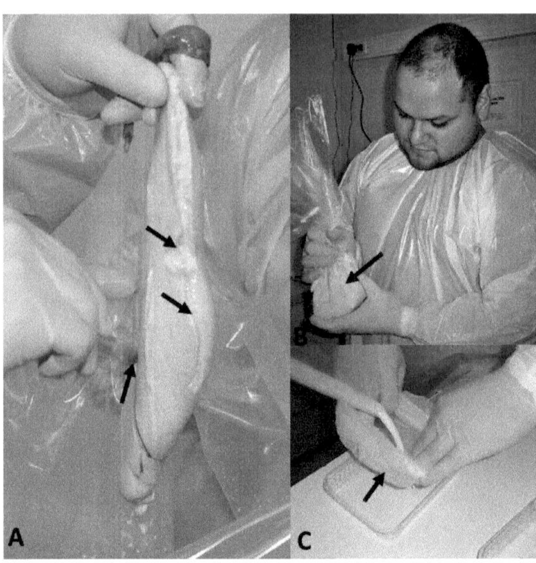

Figure 6. Atlantic salmon. A: Application of longitudinal cuts in the testis (Arrows). B: Extraction of intratesticular sperm (arrow) inside an aseptic bag using a sperm diluent. C: Separation by filtering testicular tissue from intratesticular spermatozoa using a sterile pleated cap as a filter (Arrow).

Figure 7. Phenotypic characteristics of "neomale" specimen of rainbow trout in advanced sexual maturity.

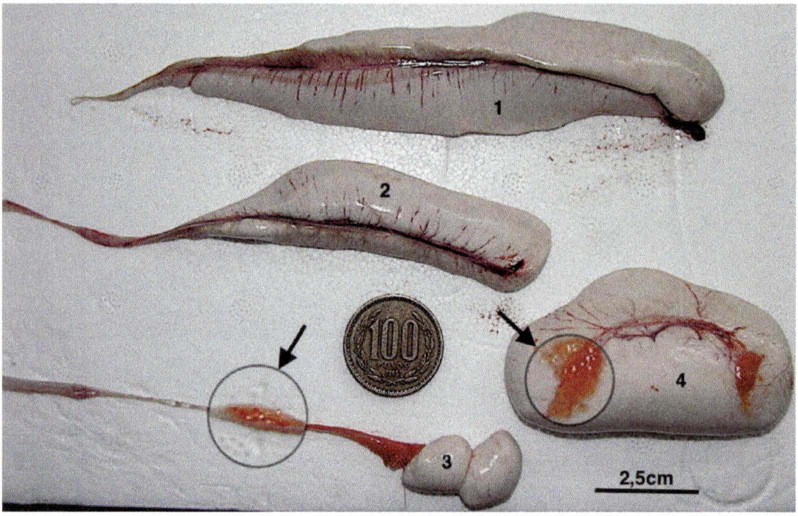

Figure 8. Testicles of male rainbow trout in advanced maturity: 1 and 2 testicles of normal males with vas deferens. 3 and 4 testicles of "neomales" with the presence of oocytes in some portions of the "testicle."

2. Cryopreservation Medium

The storage of genetic material in liquid nitrogen (LN_2), also called cryostorage, generally requires a specific cryopreservation medium. The medium consists of cryoprotectants, distilled water and, in some cases, other chemical substances. In the cryopreservation literature, the medium is often described as "extender." However, it should be understood that an extender is a solution of salts with a certain osmolality and pH used to (i) obtain a larger quantity of milt, keeping the spermatozoa immotile for artificial management (Torres et al. 2016) and (ii) protect the sperm from cold and heat shock and cryo-toxicity during cryopreservation (Muchlisin 2004). Cryoprotectant, in contrast, is a chemical that is added to an extended sperm dilution to help lower the freezing point of water (Devismita and Kumar 2015), thus reducing cell damage by ice crystal formation (Muchlisin 2005). In the present work, we will use the extender to refer to the salt or chemical solution used to dilute the cryoprotectants, while the cryopreservation medium will be used to refer to the solution containing both the diluent and the cryoprotectant.

Generally, cryoprotectants are classified into permeating and non-permeating agents (Meryman 1971). Permeating cryoprotectants were used for the first time by Polge et al. (1949) in fowl spermatozoa, testing different concentrations of glycerol. Since then, other compounds, such as methanol, ethylene glycol, 1,2-propanediol, butanediol, dimethyl sulfoxide, and dimethylacetamide, have been tested in several species, including fish (Suquet et al. 2000; Magnotti et al. 2018). These cryoprotectants have a low molecular weight, which allows them to enter cells and displace part of the intracellular water to prevent the increase in water volume during freezing (Meryman 1971; Weng et al. 2011). Non-permeating agents (e.g., sugars such as glucose or sucrose and polymers such as polyvinylpyrrolidone, dextran, polyethylene glycols, egg protein, milk, and anti-freeze proteins) increase the concentration of extracellular solute, activating water loss by the cell before freezing to allow the entry of permeating cryoprotectants. This reduces ice crystal formation during freezing and helps to balance sudden rehydration after thawing by providing stabilization of cell membranes and cell metabolism (Babin et al. 2007; Tiersch and Green 2011). Although several of these permeating cryoprotectants have been tested in fish sperm cryopreservation (Balamurugan and Munuswamy 2017; Lee et al. 2021), the most frequently reported are methanol (MeOH) and dimethylsulfoxide (Me_2SO) due to their cryoprotective characteristics and lower toxicity compared with the other agents. Interestingly, they are the most suitable for salmonids and are used worldwide for these species. As can be seen in Table 1, the cryopreservation protocols established in Poland (Ciereszko et al. 2014; Nynca et al. 2015, 2016; Dietrich et al. 2016; Judycka et al. 2016, 2018) for Atlantic salmon and other salmonids are based on simple medium with constant final concentrations of cryoprotectants. They have been used successfully for cryopreserving the spermatozoa of several salmonid species with 0.15 M glucose and 7.5% MeOH.

At laboratory scale, the cryopreservation protocol standardized for Atlantic salmon in Chile includes the use of 1.3 M Me_2SO, 0.3 M glucose, 2% bovine serum albumin (BSA) diluted in Cortland solution (Figueroa et al. 2016, 2018, 2019; Díaz et al. 2019). There are few studies for coho salmon, but a recent study evaluating different cryopreservation mediums reported better results when 8% and 10% Me_2SO were supplemented with 0.18 M glucose and 10% hen egg yolk (Sandoval-Vargas et al. 2021). It is important to pinpoint that the use of whole hen egg yolk in the cryopreservation process has been questioned due to the sanitary risk posed by pathogens (Yildiz et al. 2013). Nevertheless, this compound is essential to protect the sperm cells in some species due not only to the presence of specific protein or lipid molecules, such as low-density lipoprotein (LDL), but also to the presence of other components, such as antioxidants, vitamins, and amino acids (Dong et al. 2011). The exact protection mechanism is still unknown, some studies suggest that the cryoprotective effect of LDL is related to its gelling capacity and that the phospholipids and triglycerides contained in the lipoprotein may form a film around the membrane or become inserted in it, thus protecting the cell against ice formation (Pérez-Cerezales et al. 2010).

In general, the post-thaw motility, fertility, and hatching rates reported in salmonid sperm treated with these cryoprotectants and protocols are variable

Table 1. Cryopreservation protocols described for different species of the Salmonidae family.

Species	Cryo	Extender	Dilution milt: Medium	Volume/ Device	Equilibration Time	Thawing Conditions	Parameters			Reference
							Mot	Fert	Hatc	
			(v:v)				(%)			
Atlantic salmon (*Salmo salar*)	10% M_2SO 10 and 12.5% PG	Similar to the seminal plasma of *S. salar*	1:4; 1:9	1 mL	3 hours	At 4°C/as quickly as possible	NR	5.0–19.0	ND	Truscott and Idler 1969
Rainbow trout (*Salmo gairdneri*), now known as (*Oncorhynchus mykiss*)	10% M_2SO	Different protein sources	1:3	Pellet	NR	At 9°C/?	NR	31.0–77.5	29.8–73.8	Stoss and Holtz 1983
Brown trout (*Salmo trutta m. lacustris* L.)	10% M_2SO	0.3 M glucose, 20% egg yolk, extender of Stein (1984)	1:3	Pellet of 0.05 mL	NR	30°C/?	NR	31–85% of the controls (fresh sperm)	ND	Piironen 1993
Salvelinus alpinus L. (*Salvelinus alpinus*)	5–25% M_2SO 5–25% Glycerol	0.3 M glucose, 10% Egg yolk, Basic extender of Mounib (1978)	1:4	0.05 mL	NR	25 and 30°C/?	NR	25–75% of the controls (fresh sperm)	ND	Piironen 1993

Rainbow trout (*Oncorhynchus mykiss*) Lake trout (*Salmo trutta f. lacustris*) Brown trout (*Salmo trutta f. fario*) Charr (*Salvelinus alpinus*)	10% MeOH 5% M_2SO+ 1% glycerol	103 mM NaCl, 40 mM KCl; 1 mM CaCl2, 0.8 mM MgS04, 20 mM Hepes, 1.5 BSA, 7% egg yolk, 0.5% sucrose	1:3	0.5; 1.2 and 5 mL	10–15 minutes	25 and 30°C/30s	NR	64.0–89.6	ND	Lahnsteiner et al. 1997
Rainbow trout (*Oncorhynchus mykiss*)	7% M_2SO	Extender #6 from Erdhal and Graham (1980) 4 mg/mL BSA, 10% egg yolk, 7.5 mg/mL DanPro S760	1:3	0.5 mL	15 minutes	25°C/30 s	17.7– 42.2	38.4–73.9	ND	Cabrita et al. 2001
Rainbow trout (*Oncorhynchus mykiss*)	10% MeOH	103 mM NaCl, 40 mM KCl, 1 mM $CaCl_2$, 0.8 mM $MgSO_4$, 20 mM hepes,1.5% BSA, 7% egg yolk, 0.5% sucrose, pH 7.8	1:3 or 1:5	0.5 mL	10 minutes at 4°C	25°C/30 s	15–56	64.8 –74.1	ND	Salte et al. 2004
Caspian brown trout (*Salmo trutta caspius*)	10% M_2SO 10% MeOH	0.3 M glucose 0.6 M sucrose 10% egg yolk	1:3	0.5 mL	10 minutes: 2cm above the surface of LN_2	5°C/90s 15°C/45s 25°C/30s	NR	25.3–66.6	24.3–66.2	Sarvi et al. 2006
Rainbow trout (*Oncorhynchus mykiss*)	7.5 MeOH	0.075 M, 0.15 and 0.25 M glucose	1:1; 1.3; 3:1	0.25 mL	3 minutes: 3 cm above the surface of LN_2	40°C /5s	0–49.9	> 60 and < 85	67.1–84.4	Ciereszko et al. 2014

Table 1 cont. ...

...Table 1 cont.

Species	Cryo	Extender	Dilution milt: Medium	Volume/ Device	Equilibration Time	Thawing Conditions	Parameters			Reference
							Mot	Fert	Hatc	
			(v:v)				(%)			
Caspian brown trout (*Salmo trutta caspius*)	10% MeOH	0.3 M glucose	1:3	0.5 mL	15 minutes on ice	40°C/5s	> 60	~ 57 – ~ 73.5	~60– ~75	Golshahi et al. 2015
Adriatic grayling (*Thymallus thymallus*)	10% MeOH (final concentration)	200 mM glucose, 40 mM KCl, 30 mM Tris (pH 8.0)	1:1	0.5 mL	3 minutes above surface of LN_2	40°C/13s	33	27.4–91.3	25.4–82.3	Horváth et al. 2012, 2015
European huchen (*Hucho hucho* L.) European grayling (*Thymallus thymallus* L.)	9% MeOH (7.5% final concentration)	0.18 M glucose (0.15M final concentration)	1:5	0.25 mL	15 minutes on ice	40°C/5 s	45 and 68	87–88 and 50	75–80 and ~ 45	Nynca et al. 2015
Atlantic salmon (*Salmo salar*) Brook trout (*Salvelinus fontinalis*) Brown trout (*Salmo trutta*) *Whitefish* (*Coregonus lavaretus*) Rainbow trout (*Oncorhynchus mykiss*)	9% MeOH	0.18 M glucose 0.18 M sucrose 0.18 M trehalose	1:5; 1:9	0.25 mL	3 cm above ice for 15 minutes.	40°C/5s	38–73 (depending on the species)	ND	ND	Nynca et al. 2016

Whitefish (*Coregonus lavaretus* L.)	9% MeOH	0.18 M glucose	1:5	0.25 mL	3 cm above ice for 15 minutes plus 3 cm above LN for 5 minutes	40°C/5s	53	54–56	ND	Dietrich et al. 2016
Rainbow trout (*Oncorhynchus mykiss*) *Sex-reversed female Whitefish* (*Coregonus lavaretus*) Brown trout (*Salmo trutta*) Brook trout (*Salvelinus fontinalis*)	9% MeOH	0.18 M glucose supplemented with 20 or 40 mM KCl	1: 5; 1:9	0.25 mL	15 minutes on ice	40°C/5s	~20– ~75 (exact values not described)	ND	ND	Judycka et al. 2016
Atlantic salmon (*Salmo salar*)	1.3 M Me_2SO	0.3 M glucose 2% BSA, Cortland's medium	1: 3	0.5 mL	7 to 10 minutes at room temperature plus 10 minutes 2 cm above LN	35°C/9s	58.5	81.6	ND	Figueroa et al. 2016
Atlantic salmon (*Salmo salar*)	1.3 M Me_2SO	0.3 M glucose 2% BSA, Cortland's medium, supplemented with antioxidants	1: 3	0.5 mL	7–10 minutes at room temperature plus 10 minutes 2 cm above LN	40°C/9s	58.5–74.4	80.3–85	ND	Figueroa et al. 2018, 2019

Table 1 cont. ...

...Table 1 cont.

Species	**Cryo**	**Extender**	**Dilution milt: Medium**	**Volume/ Device**	**Equilibration Time**	**Thawing Conditions**	**Parameters**			**Reference**
							Mot	**Fert**	**Hatc**	
			(v:v)				**(%)**			
Brown trout (*Salmo trutta m. fario*) Sea trout (*Salmo trutta m. trutta*), Brook trout (*Salvelinus fontinalis*) Atlantic salmon (*Salmo salar*)	7.5% MeOH (final concentration)	0.15 M glucose (final concentration)	NA	0.25 and 0.5 mL	15 minutes on ice	40°C/10s for 0.5 ml straws 40°C/5s for 0.25 ml straws	58–76 (depending on the species)	ND	ND	Judycka et al. 2018
Atlantic salmon (*Salmo salar*)	5, 10% Me_2SO 5, 10, 15% MeOH (final concentrations)		1:1, 1:3, 1:19	0.5 mL	30 minutes	40°C/12s	0–36	0–70	ND	Yang et al. 2018
Coho salmon (*Oncorhynchus kisutch*)	8, 10% Me2SO 9, 12% MeOH	0.14 M trehalose 0.18 M glucose 10% egg yolk, 10% StorFish® solution	1:3	0.25 mL	4◦C for 8 minutes	40°C/10s	20.5–35.8	12–31%	ND	Sandoval-Vargas et al. 2021
Atlantic salmon (*Salmo salar* L.)	10% MeOH (final concentration)	200 mM glucose, 40 mM KCl, 30 mM Tris (pH 8.0)	1:6	0.5 mL	3 minutes above the surface of LN_2	40°C/15s	19.6	33.7	31.43	Erraud et al. 2022

Cryo: Cryoprotectant; Mot: Motility; Fert: Fertility; Hatc: Hatchery
NR: not reported; ND: Not determined.

(Table 1), depending on other aspects of the protocol. For example, in Chile, the motility of post-thaw samples of Atlantic salmon milt was between 58.5% and 74.4%, better than the results found for coho salmon (20.5–35.8%).

3. General Principles of Cryopreservation: Freezing Rates

Cryopreservation of cells for storage at a low temperature is desirable for biological and commercial reasons. Keeping cells at extremely low temperatures (–196°C) allows enzymatic activity, cell respiration, metabolism, growth, multiplication, etc., to be halted completely. This allows cells to be stored for a long period without affecting their viability or causing genetic changes (Schneider and Mazur 1984). However, most mammalian cells die when exposed to low temperatures unless they have previously been exposed to a protective solution and are subjected to specific cooling and heating ranges (Shaw et al. 2000).

Cryopreservation of gametes implies contact with the cryoprotective medium, subsequent freezing, and storage at sub-zero temperatures. Once thawed, the cryoprotectants are diluted and eliminated, and the sperm are then transferred to a physiologically optimal environment that allows them to develop properly. These processes involve great stress (mechanical, thermal, and chemical) for the sperm cells.

The greatest challenge that the cell must endure during cryopreservation is not caused by low-temperature storage but by the lethality of an intermediate temperature interval (+15°C to –5°C) through which the cells must pass during the freezing and thawing processes (Mazur 1963).

Cryopreservation protocols have been classified as slow or fast according to the cooling speed, type, and concentration of cryoprotectants used. However, the principles and objectives of cryopreservation apply to both types: (a) protect from the effects of cooling and freezing; (b) avoid the formation of intracellular ice; and (c) protect from the toxic effects of cryoprotectants at both low and high temperatures (Parks and Ruffing 1992; Critser et al. 1997; Paynter et al. 1999).

3.1 Slow Freezing

Slow freezing is a cryopreservation technique in which there is a balance between the rates of cooling, dehydration, and the formation of ice nuclei (Figure 9). The main objective of this type of cryopreservation is to control the cooling rate in such a way that, as the temperature drops, the cryoprotectant penetrates the cell, producing an osmotic equilibrium and reducing the probability of the formation of intracellular ice crystals.

To prevent intracellular ice formation or minimize the damage it can cause, all freezing protocols are aimed at dehydrating the cell. For slow freezing protocols, this process is achieved by placing cells in a solution containing 10–11% (v/v) cryoprotectant (approximately 1.5 M). The temperature then decreases, and ice seeding is caused within this solution. As the ice crystals grow, the water in the solution changes from a liquid to a solid, and the extracellular concentration of the solute increases, causing water to escape from the cell. The lower the temperature,

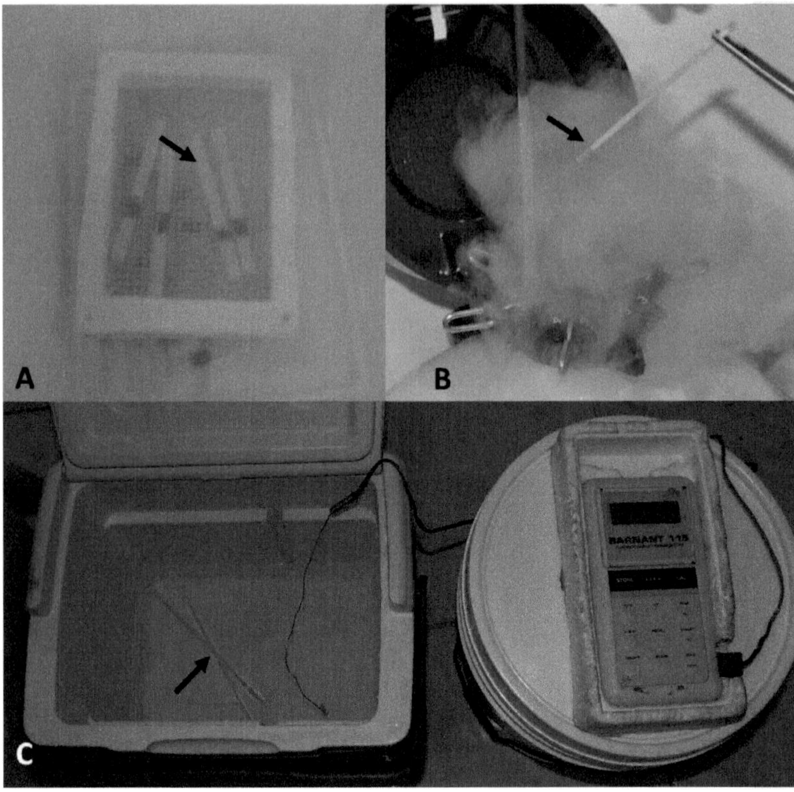

Figure 9. *In situ* protocol for cryopreservation of salmonid sperm. A: Straws (arrow) in nitrogen vapor 2 cm above the nitrogen level. B: Immersion of straw in LN_2 (Arrow). C: Thawing process; placing straws in water at 37°C (Arrow).

the more water can be converted to ice, but the cell's ability to remove intracellular water also decreases with the temperature. Therefore, the success of a slow freezing protocol is based on reaching a balance between the rate at which water leaves the cell and the rate at which this water turns to ice.

When a cell is cooled to –5°C, the cell and the medium remain in a supercold state (supercooled). Between –5 and –10°C, the formation of ice nuclei begins (either spontaneously or induced by seeding) in the external medium, while the intracellular medium remains liquid. This super cold extracellular water has a high chemical potential, causing water to flow from inside the cell to the external medium, where it is captured by the extracellular ice nuclei (Gao et al. 1997).

When the cell loses water, it becomes dehydrated; consequently, penetration of the cryoprotectant occurs to maintain the osmotic balance. After the ice formation temperature is passed (usually between –5 and –9°C), the temperature must drop to –33°C and –40°C, at which point the cells can be immersed in LN_2. Cells frozen using this methodology should be thawed following the rapid thawing protocol (Shaw et al. 2000).

The cooling rate for this technique ranges from 0.2 to 0.3°C/minute. However, to achieve a correct cooling rate, expensive equipment is needed to control cooling automatically at all times. If a cell is cooled too fast, the water does not have time to maintain osmotic equilibrium; in other words, the intracellular water does not have enough time to come out and form ice nuclei outside the cell; as a result, intracellular ice nuclei form, injuring the cell (Mazur 1992). Conversely, if the cooling rate is too slow, the efflux of water from the cell is so slow that the cell is exposed for too long to a high concentration of salts and other solutes. This can reduce cell survival since a high concentration of salts and solutes can cause changes in intracellular pH, lipid peroxidation, etc. (Gao et al. 1997). An optimal cooling range must, therefore, be achieved, in which there is a balance between the drop in temperature and the removal of water from the cell.

There are several slow-freezing protocols used for fish spermatozoa (Ekici et al. 2022). When freezing in 0.6 M sucrose solution + 10% dimethyl sulfoxide (DMSO), these authors obtained good fertilization percentages with *O. mykiss* spermatozoa (81% and 83%) but lower percentages with other species. Variants have, therefore, been introduced to improve the fertilizing capacity of frozen spermatozoa, resulting in different protocols using different cryoprotective agents, such as propylene glycol, ethylene glycol, DMSO, sucrose, methanol, and glycerol (Ciereszko and Dabrowski 1993; Ciereszko 1999; Fabbrocini 2000; Tiersch 2000; Linhart et al. 2000; Babiak et al. 2002; Cabrita et al. 2005, 2009) accompanied by different cooling rates. For example, 40 seconds between 0°C and 2°C followed by 5 minutes at –79°C and then cooling to –196°C in LN_2 (Babiak et al. 2001). The addition of albumin (BSA fraction V) to the cryoprotective solution (Lahnsteiner et al. 2000) allows very high fertilization percentage results (95.3%) to be obtained in some salmonid species; however, these protocols require rigorous control of cooling times and a series of components necessary for freezing, as well as special cryobiological equipment.

There is, therefore, a need to develop a technique based on quick or ultra-quick freezing to ensure favorable results for the salmon industry. It must be simple and easily implemented in the laboratory, giving fast results at a low cost (Figure 9).

3.2 Quick and Ultra-Quick Freezing

This technique prevents the formation of intracellular ice by dehydrating the cell. In order to achieve this, the cell is exposed to high concentrations of a permeable cryoprotectant and subsequently to quick or ultra-quick cooling (Figure 10).

Quick-freezing protocols use high concentrations of solute (permeating and not permeating cryoprotectant) that rapidly remove water from cells. In these solutions, the cells dehydrate rapidly, favoring penetration of the cryoprotectants; they can then be immersed directly in LN_2 or liquid nitrogen vapor. The freezing rate achieved with quick-freezing techniques (11,000°C/min to 14,000°C/min) drastically decreases chilling damage, allowing the use of less concentrated (less toxic) cryoprotectant solutions and reducing the exposure time of the cell to the cryoprotectant (Martino et al. 1996).

This technique was first used in mammalian oocytes and embryos; it is less effective in spermatozoa due to the high concentration of cryoprotectant required

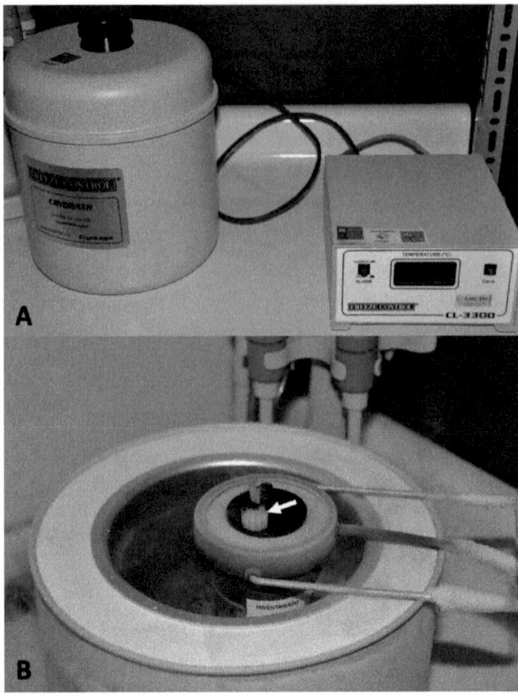

Figure 10. Programmable system for cryopreservation of salmonid sperm. A: Nitrogen container (left) and programming unit (right). B: Straws (Arrow) inside the nitrogen container.

and the osmotic and cytotoxic effects that it produces (Vanderzwalmen et al. 2003; Isachenko et al. 2008). Due to the low viability rate of spermatozoa treated in this way, quick-freezing procedures have been modified to use lower concentrations of cryoprotectant and avoid direct exposure to liquid nitrogen vapors, allowing a sperm preservation method that would provide more security than conventional freezing techniques (Cervera and García 2003).

Quick-freezing protocols are divided into two subcategories depending on whether ice formation occurs in the solution during freezing (quick freezing) or not (vitrification – ultra-quick freezing). Although the differences are not well established, vitrification should be used only for those techniques in which ice crystals do not form during freezing and thawing, either inside or outside the cell. If even trace amounts of ice are formed during these processes, quick freezing is the correct term (Shaw et al. 2000).

4. Containers

Three types of containers are used in cryopreservation processes:

a. *Straws*: For salmonids, 0.25 mL and 0.5 mL straws are regularly used experimentally, while on a large scale, companies use 0.5 mL straws and squarepacks. The company IMV Technologies recommends using 1 to 2 million sperm for each oocyte after thawing, which means fertilizing between 300 and 500 oocytes with each straw, depending on the sperm concentration used. The straws are made of an ionomeric resin that is biocompatible and inert. They are designed to be sealed at both ends, with colored cotton plugs for identification (CBS High security straws from Imv Technologies) (Figure 9).
b. *Squarepacks*: These containers are produced by the Norwegian company Cryogenetics and have a total capacity of 12 mL, allowing up to 6,000 oocytes to be fertilized with each squarepack (Figure 12).
c. *Cryotubes*: Bioassays of the vitrification of salmonid milt have been carried out experimentally (Merino et al. 2011). In this protocol, milt is dripped directly onto LN_2 (Figure 11), forming spheres that are stored in 1.5 mL cryotubes – the results in rainbow trout produced mitochondrial membrane potentials varying between 5.5% and 54.7%. Figueroa et al. (2013) determined fertility values between 15.8% and 31.0% for vitrified rainbow trout milt.

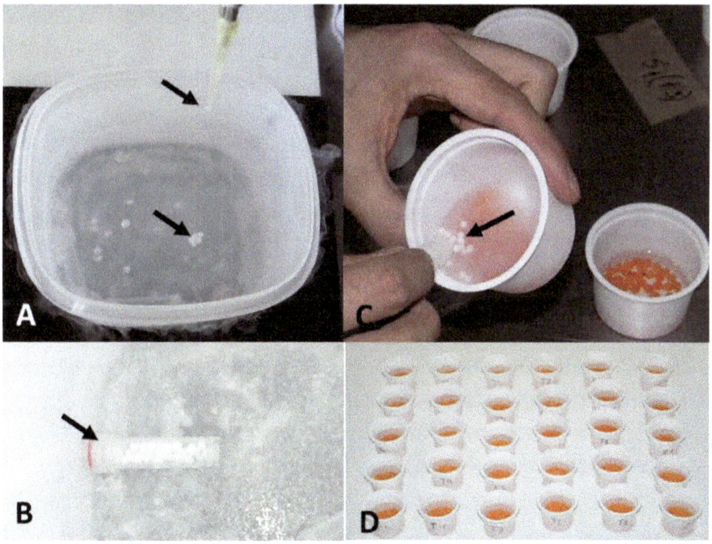

Figure 11. Salmonid sperm vitrification protocol. A: Direct dripping of milt (Arrows) onto LN_2. B: 5 mL cryotube containing vitrified milt drops (Arrow). C: Fertilization using spheres of vitrified sperm (Arrow) directly on the oocytes. D: Experimental design for evaluating vitrified sperm (5 treatments) and a control (fresh sperm) with five replicates per treatment.

5. Freezing Systems

Different strategies and protocols have been developed for freezing fish sperm. First, we will consider conventional (*in situ*) methods that do not require complex equipment and are, therefore, cheap and simple (Figure 9). In this type, control of the freezing parameters is generally limited to the freezing agent, for example, dry ice, alcohol, and LN_2. The liquid nitrogen vapor method is widely used, as it offers some advantages over methods based on dry ice and alcohol. Since LN_2 boils at –196°C, its vapor produces faster freezing rates and lower final temperatures (–160°C at 2.5 cm above the LN_2 surface) (Cabrita et al. 2009; Merino et al. 2011). The cooling profiles for the sperm of different fish species are established empirically by placing sealed straws on a freezing rack at various heights above LN_2 to determine the relation between height and cooling rate. Once the optimum height (corresponding to the desired cooling rate) above the surface of the LN_2 for freezing the sperm of the candidate species is established, the straws with milt are placed on a freezing rack at this height (most frequently 2 cm), and chilled for 10 minutes. During this period, they are slowly supercooled from +4 to –130°C; they are then immersed directly in LN_2 (–196°C) in a cryogenic container for long-term storage (Figueroa et al. 2019; Betsy and Kumar 2020). Another conventional cryopreservation protocol is the use of LN_2 dry shippers; the temperature at the bottom of a fully loaded dry container (without free LN_2) is usually below –150°C and, therefore, suitable for most applications. Freezing can be carried out simply by dropping the straws to the bottom of the tank, but more complex protocols can be carried out in which the container is lowered in several steps. These conventional methods of cooling with liquid nitrogen vapor are widely used and easy to apply in fish farms; however, the freezing volumes that can be managed are smaller than those of programmable automated systems. Nonetheless, these simple protocols are necessary to standardize techniques that will allow requirements to be projected at the industry level (Cabrita et al. 2009; Figueroa et al. 2020).

Non-conventional or automated freezing methods are devices capable of lowering or increasing the temperature of biological samples at a predefined rate. These freezing systems often have programming capabilities, so the required freezing protocols can be carried out with relative ease. As in the case of simple methods, there are several types of methods for controlling the freezing temperature through LN_2 or alcohol. Most freezers use LN_2 from a pressurized tank to chill samples.

The samples are placed in an insulated chamber containing several temperature sensors. LN_2 is injected via a solenoid control valve, and the system also includes a heater to increase the temperature when necessary. Liquid nitrogen vapor is allowed to enter the chamber at a rate set by the program to reduce the temperature of the sperm samples according to cooling requirements. The temperature variations are detected by the sensors, and the program software corrects deviations by increasing or suppressing the input of LN_2 vapor or activating the heating system. Depending on the model, the final temperature reached in the chamber ranges between –100°C and –190°C. The different freezing curves are programmed using software installed on a computer. Information on the freezing curves is integrated with the required freezing

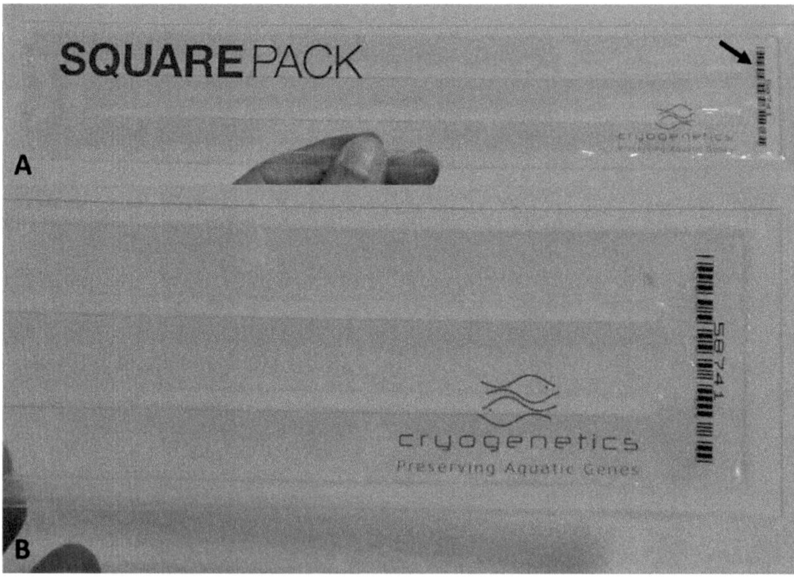

Figure 12. Macro milt container (Squarepack) of 12 mL from the company Cryogenetics used for the cryopreservation of salmonid milt on a large scale. A: 12 mL squarepack and barcode for traceability (Arrow). B: Detail of the squarepack. Further information can be obtained from www.cryogenetics.com.

speed, allowing real-time monitoring of the freezing process (Cabrita et al. 2009; Merino et al. 2011; Magnotti et al. 2018).

Although this type of equipment is more expensive to purchase, operate, and maintain, it offers advantages over conventional methods due to the possibility of controlling freezing rates and standardizing the protocols for larger volumes of fish sperm samples. Currently, this technology is being developed to provide freezing services for salmon farms in Chile at an industrial level by the Norwegian company Cryogenetics AS and the French IMV Technologies SA.

6. Results

Sperm cryopreservation has been gradually adopted in aquaculture, particularly in Chilean salmon farming, with the objective of maintaining genetics with special characteristics such as resistance to any or all pancreas disease (PD), bacterial kidney disease (BKD), infectious pancreatic necrosis virus (IPN), piscine reovirus (PRV), *piscirickettsia salmonis* (SRS), cardiomyopathy syndrome (CMS), and caligus (sea lice). Furthermore, to improve the economic efficiency and the effectiveness of the production processes of companies dedicated to embryo production through industrial-scale cryopreservation, one of the greatest challenges is to reduce the number of males used in each production cycle. However, salmonid sperm cryopreservation at commercial scale is poorly reported in the literature since this knowledge has been developed by large private companies, specifically Cryogenetics (Norway) and IMV Technologies (France).

Both Cryogenetics and IMV Technologies provide training to salmon farms as well as equipment and technology for the cryopreservation of salmonid milt. Cryogenetics also offers a frozen milt storage service for facilities that do not have on-site storage space. Cryogenetics has a dilution buffer (AquaBoots® Dilutor) that allows fresh milt to be diluted for same-day fertilization but not for preserving samples. According to the protocols established by Cryogenetics, a 1:2 (v/v) dilution of fresh milt and AquaBoots® Dilutor will result in a concentration of 2×10^9 spermatozoa/mL for Atlantic salmon, although this may vary between males, between strippings, and throughout the spawning season. In the case of IMV Technologies, milt dilution is normally performed in StorFish® solution, which allows milt to be stored for up to 4 days at 4°C. Their cryopreservation medium (FreezeFish) has been reported to allow fertilization rates above 90% after thawing. This value is higher than that reported experimentally in Chile for Atlantic salmon (80.3–85%) and for coho salmon (12–31%) by Figueroa et al. (2016, 2018 and 2019) and Sandoval-Vargas et al. (2021), respectively.

After the sperm are thawed and during fertilization, different sperm activators are regularly used to improve the fertilizing capacity of the sperm, such as Billard's solution (Sandoval-Vargas et al. 2020a,b), ActiFish (IMV Technologies), AquaBoost Activator or PowerMilt (Trademark of the Catholic University of Temuco). The sperm activator can be applied just before fertilization or immediately after mixing the thawed sperm with the oocytes (from Cryogenetics).

7. Conclusions

Cryopreservation is a biotechnological tool widely used in fish Chilean farming for storing the sperm of salmonid species, principally Atlantic salmon: either fresh milt (obtained by abdominal massage) or testicular macerate, since this species regularly releases very little milt under the application of abdominal massage. The results of fertility and embryo survival with cryopreserved sperm of this species do not present significant differences from fresh sperm, using cryopreservation protocols commercialized and developed by foreign companies such as IMV Technologies and Cryogenetics.

Acknowledgments

This work was supported by The National Fund for Scientific and Technological Development FONDECYT REGULAR N° 1211246, FONDECYT INICIACIÓN N° 11230690, FONDECYT POSTDOCTORAL N° 3180765 and the company IMV Technologies (Chile) which provided general information of its biotechnological processes.

References

Aas, Ø., A.E.S. Klemetsen, S. Einum and J. Skurdal. 2010. Atlantic Salmon Ecology. Willey-Blackwell, London.

Alderson, R. and A.J. Macneil. 1984. Preliminary investigations of cryopreservation of milt of Atlantic salmon (*Salmo salar*) and its application to commercial farming. Aquaculture 43: 351–354. https://doi.org/10.1016/0044-8486(84)90036-X.

Almendras, F. 1993. Ensayo de un sistema de criopreservación de semen de salmonídeos aplicado a la trucha arcoíris (*Oncorhynchus mykiss*). Ph.D. Thesis. Universidad de Chile, Santiago, Chile.

Babiak, I., J. Glogowski, K. Goryczko, S. Dobosz, H. Kuzminski, J. Strzezek et al. 2001. Effect of extender composition and equilibration time on fertilization ability and enzymatic activity of rainbow trout cryopreserved spermatozoa. Theriogenology 56: 177–192. https://doi.org/10.1016/S0093-691X(01)00553-2.

Babiak, I. 2002. Semen from rainbow trout produced using cryopreserved spermatozoa is more suitable for cryopreservation. J. Fish Biol. 60: 561–570.

Babin, P.J., J. Cerdà and E. Lubzens. 2007. The Fish Oocyte: From Basic Studies to Biotechnological Applications. Springer, Netherlands.

Balamurugan, R. and N. Munuswamy. 2017. Cryopreservation of sperm in Grey mullet *Mugil cephalus* (Linnaeus, 1758). Anim. Reprod. Sci. 185: 205–213. https://doi.org/10.1016/j.anireprosci.2017.08.022.

Betsy, J. and S. Kumar. 2020. Cryopreservation of fish gametes, Cryopreservation of Fish Gametes. Springer Nature Singapore Pte Ltd, Singapore. https://doi.org/10.1007/978-981-15-4025-7.

Bromage, N. and R. Roberts. 1995. Broodstock Management and Egg and Larval Quality. Wiley-Blackwell, London.

Cabrita, E., L. Anel and M.P. Herraéz. 2001. Effect of external cryoprotectants as membrane stabilizers on cryopreserved rainbow trout sperm. Theriogenology 56: 623–635. https://doi.org/https://doi.org/10.1016/S0093-691X(01)00594-5.

Cabrita, E. 2002. Identificación de los daños producidos por la congelación seminal en la membrana plasmática de espermatozoides de trucha arcoiris (*Oncorhynchus mykiss*) y optimización de protocolos de criopreservación. Ph.D. Thesis. Universidad de León. León, Spain.

Cabrita, E., V. Robles, L. Rebordinos, C. Sarasquete and M.P. Herráez. 2005. Evaluation of DNA damage in rainbow trout (*Oncorhynchus mykiss*) and gilthead sea bream (*Sparus aurata*) cryopreserved sperm. Cryobiology 50: 144–153. https://doi.org/10.1016/j.cryobiol.2004.12.003.

Cabrita, E., S. Engrola, L. Conceição, P. Pousão-Ferreira and M.T. Dinis. 2009. Successful cryopreservation of sperm from sex-reversed dusky grouper, *Epinephelus marginatus*. Aquaculture 287: 152–157.

Cervera, R. and F. Garcia-Ximénez. 2003. Vitrification of zona-free rabbit expanded or hatching blastocysts: a possible model for human blastocysts. Human. Reprod. 18: 2151–2156.

Ciereszko, A. and K. Dabrowski. 1993. Estimation of sperm concentration of rainbow trout, whitefish and yellow perch using a spectrophotometric technique. Aquaculture 109: 367–373.

Ciereszko, A., K. Dabrowski, F. Lin and L. Liu. 1999. Protective role of ascorbic acid against damage to male germ cells in rainbow trout (*Oncorhynchus mykiss*). Can. J. Fish. Aquat. Sci. 56: 178–183. https://doi.org/10.1139/f98-161.

Ciereszko, A., G.J. Dietrich, J. Nynca, S. Dobosz and T. Zalewski. 2014. Cryopreservation of rainbow trout semen using a glucose-methanol extender. Aquaculture 420–421: 275–281. https://doi.org/10.1016/j.aquaculture.2013.11.014.

Critser, J.K., Y. Agca and K.T. Gunasena. 1997. The cryobiology of mammalian oocytes. pp. 332–358. *In*: Karrow, A.M. and J.K. Critser [eds.]. Reproductive Tissue Banking. Academic Press, San Diego, USA.

Devismita, D. and A. Kumar. 2015. Effect of cryoprotectant on optimal cooling rate during cryopreservation. Cryobiology 70: 53–59. https://doi.org/10.1016/j.cryobiol.2014.12.002.

Díaz, R., M. Lee-Estevez, J. Quiñones, K. Dumorné, S. Short, P. Ulloa-Rodríguez, I. Valdebenito, N. Sepúlveda and J.G. Farías. 2019. Changes in Atlantic salmon (*Salmo salar*) sperm morphology and membrane lipid composition related to cold storage and cryopreservation. Anim. Reprod. Sci. 204: 50–59. https://doi.org/10.1016/j.anireprosci.2019.03.004.

Dietrich, G.J., J. Nynca, M. Szczepkowski, S. Dobosz, B. Szczepkowska and A. Ciereszko. 2016. The effect of cryopreservation of semen from whitefish (*Coregonus lavaretus*) and northern pike (*Esox lucius*) using a glucose-methanol extender on sperm motility parameters and fertilizing ability. Aquaculture 464: 60–64. https://doi.org/10.1016/j.aquaculture.2016.06.015.

Dong, Q.X., S.E. Rodenburg, D. Hill and C.A. Vandevoort. 2011. The role of low-density lipoprotein (LDL) and high-density lipoprotein (HDL) in comparison with whole egg yolk for sperm cryopreservation in rhesus monkeys. Asian J. Androl. 13: 459–464. https://doi.org/10.1038/aja.2010.145.

Ekici, A., G. Yamaner and M.D. Demircan. 2022. Cryopreservation studies in aquaculture from past to present: Scientific techniques and quality controls for commercial applications. pp. 1–33. *In*: Quain, M. [ed.]. Cryopreservation - Applications and Challenges. IntechOpen, Ghana, https://doi.org/10.5772/intechopen.108566.

Erraud, A., V. Cornet, S. Baekelandt, Y. Neus, S. Antipine, J. Lambert, S.N.M. Mandiki and P. Kestemont. 2022. Optimal sperm–egg ratios for successful fertilization using fresh and cryopreserved sperm in wild anadromous Atlantic salmon (*Salmo salar* L. 1758). Aquaculture 549. https://doi.org/10.1016/j.aquaculture.2021.737758.

Estay, F. 1988. Análisis de algunos factores determinantes de alteraciones en la maduración gonadal y fertilidad en *Salmo gairdneri* (trucha arco iris) cultivada. Tesis, Magister Cs. Biológicas,Mención Biologíaa de la Reproducción. Facultad de Medicina, Universidad de Chile. 185.

Estay, F., H. Cerisola and V. Téllez. 1994. Biología del desarrollo y reproducción artificial en la trucha arco iris. Conicyt-Fondef, Santiago, Chile.

Estay, F., N. Díaz, R. Neira and X. García. 1995. Reproductive performance of cultured female Coho Salmon in Chile. Prog Fish C. 59: 36–40.

Estay, F., C. Vergara and N. Díaz. 1999. Reproductive performance of cultures Atlantic Salmon *Salmo salar* L. 1758 in Chile. Aquac. Res. 10: 759–764.

Fabbrocini, A., L. Lavadera, S. Rispoli and G. Sansone. 2000. Cryopreservation of sea bream (*Sparus aurata*) spermatozoa. Cryobiology 40: 46–53.

Figueroa, E., J. Risopatrón, R. Sánchez, E. Isachenko, O. Merino, V. Isachenko et al. 2013. Spermatozoa vitrification of sex-reversed rainbow trout (*Oncorhynchus mykiss*): Effect of seminal plasma on physiological parameters. Aquaculture 372-375: 119–126.

Figueroa, E., I. Valdebenito, O. Merino, A. Ubilla, J. Risopatrón and J.G. Farias. 2016. Cryopreservation of Atlantic salmon *Salmo salar* sperm: effects on sperm physiology. Fish Biol. 89: 1537–50. https://doi.org/10.1111/jfb.13052.

Figueroa, E., J.G. Farias, M. Lee-Estevez, I. Valdebenito, J. Risopatrón, C. Magnotti, J. Romero, I. Watanabe and R.P.S. Oliveira. 2018. Sperm cryopreservation with supplementation of α-tocopherol and ascorbic acid in freezing media increase sperm function and fertility rate in Atlantic salmon (*Salmo salar*). Aquaculture 493: 1–8. https://doi.org/10.1016/j.aquaculture.2018.04.046.

Figueroa, E., M. Lee-Estevez, I. Valdebenito, I. Watanabe, R.P.S. Oliveira, J. Romero, R.L. Castillo and J.G. Farías. 2019. Effects of cryopreservation on mitochondrial function and sperm quality in fish. Aquaculture 511: 634190. https://doi.org/10.1016/j.aquaculture.2019.06.004.

Figueroa, E., M. Lee-Estévez, I. Valdebenito, J.G. Farías and J. Romero. 2020. Potential biomarkers of DNA quality in cryopreserved fish sperm: impact on gene expression and embryonic development. Rev. Aquac. 12: 382–391.

Gao, D., P. Mazur and J.K. Critser. 1997. Fundamental cryobiology of mammalian spermatozoa. pp. 263–328. *In*: Karrow A.M. and J.K. Critser [eds.]. Reproductive Tissue Banking. Academic Press, San Diego, USA.

Golshahi, K., N. Shabani, M.S. Aramli and E. Noori. 2015. Motility and fertilizing ability of cryopreserved Caspian brown trout (*Salmo trutta caspius*) sperm: Effect of post-thaw storage time and different sperm-to-egg ratios. Cryobiology 71: 360–363. https://doi.org/10.1016/j.cryobiol.2015.08.005.

Groot, C. and L. Margolis. 1991. Pacific Salmon, Life Histories. UBC Press, Vancouver, Canada.

Horváth, Á., D. Jesenšek, B. Csorbai, Z. Bokor, É. Raboczki, D. Kaczkó, G. Bernáth, G. Hoitsy, B. Urbányi, S.S. Bajec and A. Snoj. 2012. Application of sperm cryopreservation to hatchery practice and species conservation: A case of the Adriatic grayling (*Thymallus thymallus*). Aquaculture 358–359: 213–215. https://doi.org/10.1016/j.aquaculture.2012.07.012.

Horváth, Á., Z. Bokor, G. Bernáth, Z. Csenki, A. Gorjan, M.P. Herráez, B. Urbányi and D. Jesenšek. 2015. Very low sperm-egg ratios result in successful fertilization using cryopreserved sperm in the Adriatic grayling (*Thymallus thymallus*). Aquaculture 435: 75–77. https://doi.org/10.1016/j.aquaculture.2014.09.032.

Isachenko, E., V. Isachenko, J.M. Weiss, R. Kreienberg, I.I. Katkov, M. Schulz et al. 2008. Acrosomal status and mitochondrial activity of human spermatozoa vitrified with sucrose. Reproduction 136: 167–173.

Judycka, S., J. Nynca, E. Liszewska, S. Dobosz, T. Zalewski and A. Ciereszko. 2016. Potassium ions in extender differentially influence the post-thaw sperm motility of salmonid fish. Cryobiology 73: 248–256. https://doi.org/10.1016/j.cryobiol.2016.07.002.

Judycka, S., J. Nynca, E. Liszewska, S. Dobosz, J. Grudniewska and A. Ciereszko. 2018. Optimal sperm concentration in straws and final glucose concentration in extender are crucial for improving the cryopreservation protocol of salmonid spermatozoa. Aquaculture 486: 90–97. https://doi.org/10.1016/j.aquaculture.2017.12.019.

Lahnsteiner, F., T. Weismann and R.A. Patzner. 1997. Methanol as cryoprotectant and the suitability of 1.2 ml and 5 ml straws for cryopreservation of semen from salmonid fishes. Aquac. Res. 28: 471–479. https://doi.org/10.1111/j.1365-2109.1997.tb01065.x.

Lahnsteiner, F. 2000. Semen cryopreservation in the Salmonidae and in the Northern pike. Aquac. Res. 31: 245–258. https://doi.org/10.1046/j.1365-2109.2000.00452.x.

Lee, Y.H., J.Y. Park, I.Y. Lee, I. Zidni and H.K. Lim. 2021. Effects of cryoprotective agents and treatment methods on sperm cryopreservation of stone flounder, Kareius bicoloratus. Aquaculture 531: 735969. https://doi.org/10.1016/j.aquaculture.2020.735969.

Linhart, O., M. Rodina and J. Cosson. 2000. Cryopreservation of sperm in common carp *Cyprinus carpio*: Sperm motility and hatching success of embryos. Cryobiology 41: 241–250. https://doi.org/10.1006/cryo.2000.2284. Magnotti, C., V. Cerqueira, M. Lee-Estevez, J.G. Farias, I. Valdebenito and E. Figueroa. 2018. Cryopreservation and vitrification of fish semen: a review with special emphasis on marine species. Rev. Aquac. 10: 15–2.

Magnotti, C., V. Cerqueira, A. Villasante, J. Romero, I. Watanabe, R.P.S Oliveira et al. 2022. Spermatological characteristics and effects of cryopreservation in Lebranche mullet spermatozoa (*Mugil liza* Valenciennes, 1836): First report of ultra-rapid freezing. Anim. Reprod. Sci. 241: 106986.

Martino, A., N. Songsasen and S.P. Leibo. 1996. Development into blastocysts of bovine oocytes cryopreserved by ultra-rapid cooling. Biol. Reprod. 4: 1059–1069.

Mazur, P. 1963. Studies on rapidly frozen suspensions of yeast cell by differential thermal analysis and conductometry. Biophys. J. 3: 323–353.

Mazur, P., K.W. Cole, J.W. Hall, P.D. Schreuders and A.P. Mahowald. 1992. Cryobiological preservation of Drosophila embryos. Science 258: 1896–1897.

Merino, O., J. Risopatrón, R. Sánchez, E. Isachenko, E. Figueroa, I. Valdebenito et al. 2011. Fish (*Oncorhynchus mykiss*) spermatozoa cryoprotectant-fee vitrification: Stability of mitochondrion as criterion of effectiveness. Anim. Reprod. Sci. 124: 125–131.

Meryman, H.T. 1971. Cryoprotective agents. Cryobiology 8: 173–183. https://doi.org/10.1016/0011-2240(71)90024-1.

Mounib, M.S. 1978. Cryogenic preservation of fish and mammalian spermatozoa. J. Reprod. Fertil. 53: 13–18.

Muchlisin, Z. 2004. Preliminary study on the cryopreservation of tropical bagrid catfish (*Mystus nemurus*) spermatozoa; the effect of extender and cryoprotectant on the motility after short-term storage. Theriogenology 62: 25–34. https://doi.org/10.1016/s0093-691x(03)00422-9.

Muchlisin, Z. 2005. Review: Current status of extenders and cryoprotectants on fish spermatozoa cryopreservation. Biodiversitas J. Biol. Divers. 6: 66–69. https://doi.org/10.13057/biodiv/d060114.

Nynca, J., G.J. Dietrich, J. Grudniewska, S. Dobosz, E. Liszewska, M. Krzyś, R. Rózyński and A. Ciereszko. 2015. Efficient method for cryopreservation of European huchen (*Hucho hucho* L.) and grayling (Thymallus thymallus L.) semen. Aquaculture 435: 146–151. https://doi.org/10.1016/j.aquaculture.2014.09.031.

Nynca, J., S. Judycka, E. Liszewska, S. Dobosz, J. Grudniewska, K. Arai, T. Fujimoto and A. Ciereszko. 2016. Utility of different sugar extenders for cryopreservation and post-thaw storage of sperm from Salmonidae species. Aquaculture 464: 340–348. https://doi.org/10.1016/j.aquaculture.2016.07.014.

Parks, J.E. and N.A. Ruffing. 1992. Factors affecting low temperature survival of mammalian oocytes. Theriogenology 37: 59–73. https://doi.org/https://doi.org/10.1016/0093-691X(92)90247-O.

Parks, J.E., W.Y. Son, K. Lee, J.J. Ko and K. Cha. 1997. Chromosome and spindle configurations of human oocytes matured *in vitro* alter cryopreservation at the germinal vesicle stage. Fertil. Steril. 68: 920–926.

Paynter, S.J., A. Cooper, B.J. Fuller and R.W. Shaw. 1999. Cryopreservation of bovine ovarian tissue: structural normality of follicles after thawing and culture *in vitro*. Cryobiology 38: 301–309.

Pennell, W. and B. Barton. 1996. Principles of Salmonid Culture. Elsevier, Amsterdam, Netherland.

Pérez-Cerezales, S., S. Martínez-Páramo, J. Beirão and M.P. Herráez. 2010. Evaluation of DNA damage as a quality marker for rainbow trout sperm cryopreservation and use of LDL as cryoprotectant. Theriogenology 74: 282–289. https://doi.org/10.1016/j.theriogenology.2010.02.012.

Piironen, J. 1993. Cryopreservation of sperm from brown trout (*Salmo trutta* m. *lacustris* L.) and Arctic charr (*Salvelinus alpinus* L.). Aquaculture 116: 275–285.

Polge, C., A. Smith and A. Parkes. 1949. Revival of spermatozoa after vitrification and dehydration at low temperatures. Nature 666.

Sandoval-Vargas, L., K. Dumorné, P. Contreras, J. Farías, E. Figueroa, J. Risopatrón et al. 2020a. Cryopreservation of coho salmon sperm (*Oncorhynchus kisutch*): effect on sperm function, oxidative stress and fertilizing capacity. Aquaculture 533: 736151.

Sandoval-Vargas, L., E.V. Figueroa, F. Estay and I. Valdebenito. 2020b. Effect of different calcium concentration on sperm motility and fertilization capacity of rainbow trout (*Oncorhynchus mykiss*). J. Appl. Ichthyol. 37: 106–112.

Sandoval-Vargas, L., K. Dumorné, P. Contreras, J.G. Farías, E. Figueroa, J. Risopatrón and I. Valdebenito. 2021. Cryopreservation of coho salmon sperm (*Oncorhynchus kisutch*): Effect on sperm function, oxidative stress and fertilizing capacity. Aquaculture 533: 736151. https://doi.org/10.1016/j.aquaculture.2020.736151.

Salte, R., A. Galli, U. Falaschi, K.T. Fjalestad and R. Aleandri. 2004. A protocol for the on-site use of frozen milt from rainbow trout (*Oncorhynchus mykiss* Walbaum) applied to the production of progeny groups: Comparing males from different populations. Aquaculture 231: 337–345. https://doi.org/10.1016/j.aquaculture.2003.10.037.

Sarvi, K., H. Niksirat, B. Mojazi Amiri, S.M. Mirtorabi, G.R. Rafiee and M. Bakhtiyari. 2006. Cryopreservation of semen from the endangered Caspian brown trout (*Salmo trutta caspius*). Aquaculture 256: 564–569. https://doi.org/10.1016/j.aquaculture.2006.02.012.

Schneider, U. and P. Mazur. 1984. Osmotic consequences of cryoprotectant permeability and its relation to the survival of frozen-thawed embryos. Theriogenology 21: 68–79.

Sernapesca. Available online: http://www.sernapesca.cl/informacion-utilidad/anuarios-estadisticos-de-pesca-y-acuicultura (accessed on 05 May 2023).

Shaw, J.M., L.L. Kuleshova, D.R. MacFarlane and A.O. Trounson. 1997. Vitrification properties of solutions of ethylene glycol in saline containing PVP, Ficoll, or dextran. Cryobiology 35: 219–229.

Shaw, J.M., A. Oranratnachai and A.O. Trounson. 2000. Fundamental cryobiology of mammalian oocytes and ovarian tissue. Theriogenology 53: 59–72.

Stoss, J. and W. Holtz. 1983. Cryopreservation of rainbow trout (*Salmo gairdneri*) sperm: III. Effect of proteins in the diluent, sperm from different males and interval between sperm collection and freezing. Aquaculture 31: 275–282.

Suquet, M., C. Dreanno, C. Fauvel, J. Cosson and R. Billard. 2000. Cryopreservation of sperm in marine fish. Aquac. Res. 31: 231–243. https://doi.org/10.1046/j.1365-2109.2000.00445.x.

Tiersch, T.R. and P.M. Mazik. 2000. Cryopreservation in Aquatic Species. World Aquaculture Society, Baton Rouge.

Tiersch, T.R. and C.C. Green. 2011. Cryopreservation in Aquatic Species 2nd edition. A Comprehensive Overview of Current Practices, Programmatic Development and Future Directions for Cryopreservation of Gametes, Embryos and Larvae of Aquatic Species, 2nd ed. World Aquaculture Society, Louisiana, USA.

Torres, L., E. Hu and T.R. Tiersch. 2016. Cryopreservation in fish: Current status and pathways to quality assurance and quality control in repository development. Reprod. Fertil. Dev. 28: 1105–1115. https://doi.org/10.1071/RD15388.

Truscott, B. and D.R. Idler. 1969. An improved extender for freezing Atlantic salmon spermatozoa. J. Fish. Res. Board Canadá 26.

Valdebenito, I., C. Fletcher, V. Vera and J. Fernández. 2009. Factores físico-químicos que regulan la motilidad espermática en peces: aspectos básicos y aplicados. Una revisión. Arch. Med. Vet. 41: 97–106.

Valdebenito, I., C. Moreno, C. Lozano and A. Ubilla. 2010. Effect of l-glutamate and glycine incorporated in activation media, on sperm motility and fertilization rate of rainbow trout (*Oncorhynchus mykiss*) spermatozoa. J. Appl. Ichthyol. 26: 702–706.

Vanderzwalmen, P., G. Bertin, C.H. Debauche, V. Standaert, N. Bollen, E. Van Roosendaal et al. 2003. Vitrification of human blastocysts with the Hemi-Straw carrier: application of assisted hatching after thawing. Human. Reprod. 18: 1504–1511.

Weng, L., W. Li and J. Zuo. 2011. Two applications of the thermogram of the alcohol/water binary system with compositions of cryobiological interests. Cryobiology 62: 210–217. https://doi.org/10.1016/j.cryobiol.2011.03.005.

Yang, H., E. Hu, J.T. Buchanan and T.R. Tiersch. 2018. A strategy for sperm cryopreservation of Atlantic salmon, *Salmo salar*, for remote commercial-scale high-throughput processing. J. World Aquac. Soc. 49: 96–112. https://doi.org/10.1111/JWAS.12431.

Yildiz, C., Y. Bozkurt and I. Yavas. 2013. An evaluation of soybean lecithin as an alternative to avian egg yolk in the cryopreservation of fish sperm. Cryobiology 67: 91–94. https://doi.org/10.1016/j.cryobiol.2013.05.008.

CHAPTER 3

Biobanks

Application in Environmental Conservation

Borys Dzyuba[1,*] and *William V. Holt*[2]

INTRODUCTION

As the conservation of the environment is an extremely wide topic, it is necessary to explain that, in this chapter, we will consider just one of the elements of this problem—fish conservation. That point, in turn, is directly related to the global concept of biodiversity, which, according to the Convention on Biological Diversity of the UN, should be considered at three levels; "within species, between species and of ecosystems" (Feest et al. 2010). Considering the enormous evolutionary diversity of fishes (Nelson et al. 2016), allowing them to occupy numerous ecosystems, and which has resulted in the huge diversity of reproductive systems (Reid and Hall 2003), fishes are a source of complication for conservation measures as they exemplify all three levels of diversity.

In the context of fish conservation, cryopreservation of biomaterials (i.e., storage at low temperatures) is regarded as a reliable method for preserving genetic material and preventing degradation. Many organizations worldwide have now established banks of cryopreserved genetic materials (often known as "biobanks") from a variety of sources and for different purposes. Some specialize in preserving seeds and plants, while many others are focused on preserving agriculturally important or

[1] University of South Bohemia in České Budějovice, Faculty of Fisheries and Protection of Waters, Research Institute of Fish Culture and Hydrobiology, Zátiší 728/II, 38901, Vodňany, Czech Republic.

[2] Department of Oncology and Metabolism, The Medical School, Beech Hill Road, Sheffield S10 2RX, UK.

Emails: bdzyubat@frov.jcu.cz, bill2holt@gmail.com

* Corresponding author: bdzyubat@frov.jcu.cz

conservation-sensitive animal breeds and species (Bennett 2001; Watson and Holt 2001; Rawson et al. 2011; Holt and Comizzoli 2021).

At the start of the 21st century, it became clear that the application of biobanking is complicated because species conservation programs can only be implemented with input from many varied institutions involved in complex cross-disciplinary collaborations. These require significant financial and intellectual investment to make the outcomes practical and achievable (Pickard and Holt 2004). Collaborations have to involve geneticists, population biologists, reproductive biologists, and cryobiologists playing their respective roles in the orchestra named the conservation program. This chapter is focused on the critical points to be considered when establishing conservation biobanking programs for fish species. It sequentially combines the answers to the main questions related to biobanking, namely, "Why?"—addressing the reasoning behind the prioritizations of species to be involved in biobanking. This should be based on an adequate estimation of the risks affecting species and include factors such as phylogenetics and geographical location, population size, degree of endangerment, and possibly the extent of inbreeding. These, in turn, lead to precise determinations of "What?"—i.e., the type of genetic material to be cryopreserved (sperm, eggs, primordial germ cells, and somatic cells) and a clear plan for future application. Finally, these two questions are related to "How?"—i.e., which biotechnological approaches are available for implementing successful biobank activities.

1. Reasons for Prioritizing Which Species Are to be Involved in Environmental Conservation via Biobanking (Why?)

1.1 General Considerations

The only way to achieve effective biodiversity conservation in the face of opposing anthropogenic influences is to understand the fundamental ecological processes, ecosystem functioning, and arrangement of international activities (Cardinale et al. 2012). Additionally, it is clear that the reasoning for conservation measures should be considered at different organizational levels, from international to local and governmental. That is why the first step in arranging any conservation activity is evaluating the level of importance of the action (global/regional/local).

The risk of species diversity deterioration is closely related to human activities of high economic value while also causing many types of aggressive environmental damage. Deforestation, damming (Winemiller et al. 2016), and fisheries fall into this group of damaging processes. Unfortunately, this problem cannot be solved solely by a biological scientific approach to conservation measures. That is because economic priorities for society are usually so demanding that there is almost no realistic chance to improve the ecological situation quickly and stop species diversity degradation. It is proposed that evaluating risks of new and already running human activities leading to biodiversity degradation should be explored as priority aspects in long-term conservation actions (Reis et al. 2016; Hilsdorf and Hallerman 2017). Historically, this point was clearly formulated internationally in the first global convention on biological diversity (CBD 1992). However, reliable approaches for developing conservation strategies via monitoring natural areas still require future elaboration (Miranda and Miqueleiz 2021) to attract attention at the top levels of international management (Brondízio et al. 2019).

One of the essential concepts related to species conservation based on the geographical zoning of species richness is "biodiversity hotspots." This concept has already been described in detail in a separate book (Zachos and Habel 2011), and there is no need for a comprehensive description here. At the same time, the main provisions of the concept related to the selection of criteria for conservation priorities should be mentioned. For example, according to Brooks et al. (2006), ecological criteria of "vulnerability" and "replaceability" are core elements to be considered when planning conservation measures. Vulnerability can be simplified as a temporal characteristic of risk to biodiversity deterioration due to threatening processes. Irreplaceability describes the importance of an area in terms of species richness and the presence of endangered, critically endangered, and endemic species. At the same time, these two criteria are just "starting points" that should be followed by additional evaluations as a biodiversity hotspot paradigm cannot be considered universal. The biodiversity hotspots approach currently represents a shortcut for a more complicated concept that includes consideration of ecosystem services, policies, costs, social preferences, and climate change (Marchese 2015).

The idea of applying particular sets of criteria for evaluating biodiversity status and conservation action needs has a long history of elaboration (Usher 1986; Schmitt 2011). Nowadays, a well-developed system of these criteria involves using numerous indicators describing the current status of biodiversity and the development of appropriate conservation actions (Strassburg et al. 2020). Elaboration and the correct application of these criteria have become integral to modern ecology, conservation biology, and integral biology, indicating this area's complexity and importance for human beings. As a result, the comprehensive review of biodiversity status and its significance for future human civilization has been elaborated in the global assessment report on biodiversity and ecosystem services from the Intergovernmental Science-Policy Platform on Biodiversity and Ecosystem Services (IPBES) (Brondízio et al. 2019). This source of information, among other aspects, can be considered as a modern, extremely clear message on the necessity for developing conservation measures as a net of activities among science, business, and politics.

Moreover, the IPBES review clearly shows that this activity should be global (international) if it is to be effective. Additionally, the number of species requiring urgent care to avoid extinction is taxon-specific. Protecting natural landscapes can provide the most effective way to save species diversity. That is why the first question in the biodiversity conservation program is to understand why some species are selected for attention and others are not. Is that influenced by the ease and/or practicality of protecting their natural habitats?

The goal of the conservation program becomes critically important in the current chapter of the book as precisely the value of biobanking for whatever conservation program should be considered in advance. Evaluating the importance of biobanking as a part of *ex-situ* conservation activity was intensively discussed at the end of the last century when it was argued that biobanks fulfill a limited but potentially unique role in species preservation (Holt et al. 2003; Pickard and Holt 2004). Simultaneously, it was evident that conservation approaches cannot be effectively performed without a strong taxonomic background (Godfray et al. 2004). Nowadays, biobanking is considered a part of a complicated network of activities required to achieve success in conservation action (Hughes et al. 2020). That is why we will consider some aspects of biobanking related to fish conservation below.

1.2 The Case of Fishes

Description of the state of the global diversity of fishes and the conservation status of particular species have become essential components of international nature conservation. State of the art can be found in several reviews from which it is clear that, within the last two decades, the number of recognized fish species has been increasing each year [e.g., from 25,157 in 2003 (Reid and Hall 2003) to 35,558 in 2019 (Keppeler and Silvano 2020)]. This growing knowledge of fish diversity is associated with difficulties in evaluating priorities for conservation measures; as more species are recognized, more efforts should be made to assess their conservation status for developing conservation programs. For example, Neotropical fishes consist of some 5,500 previously described species, but it is anticipated that the total number is 8,000–9,000 (Reis et al. 2016) and that a variety of different conservation approaches will be needed. For such a considerable species number (even if they belong "just" to 17 orders), prioritizing the primary candidates involved in conservation actions is challenging.

Moreover, the Neotropical region can be taken as an example for consideration of the pros and cons of biobanking activity, as the considerable variability of fishes is accompanied by general difficulties arising from human activity destroying natural landscapes. These difficulties are based on several aspects that should be considered in decision-making policies for any conservation program. For example, the possibility to quantitatively evaluate diversity has already been demonstrated for the Amazon Basin (Jézéquel et al. 2020). In this work, using a fish-occurrence database and ecological criteria (i.e., degree of deforestation and habitat fragmentation by dams) allowed the development of planning solutions for freshwater fish protection and prediction of outcomes of conservation actions. This is just an example of modern quantitative approaches to exploring natural habitats in terms of species richness and risks of their deterioration by human activity.

Help in estimating species priority for conservation programs can be found in reports from several nature conservation organizations, such as the famous international IUCN (The IUCN Red List of Threatened Species), and from local and regional Red Lists, which are currently (2022) being elaborated in 44 countries. However, the correct reasoning can only be reached by taking into account the whole spectrum of associated aspects, such as habitat deterioration due to climate change, damming, reservoir building, or other human activity; pollution; the presence of invasive species; sanitary reasons (infections and parasites), overfishing, etc. (Tyus 2011). This assessment should be made with the current taxonomic consideration and evaluation of population-specific genetic descriptors. The latter becomes vital as intraspecific diversity should be considered one of the primary conservation targets (Paz-Vinas et al. 2018). All these considerations become a potent background for conserving fishes as a part of global nature biodiversity (Marchese 2015) and understanding why using biobanking for conservation should be considered as a tool.

Generally, fish conservation has a long history and, nowadays, acts via numerous managing international and regional organizations (Helfman 2007; Tyus 2011); however, biobanking is involved in the activity of just a few of them. In order to understand the reasons for this situation, two main goals of cryobanking can be distinguished. First, and maybe the most easily appreciated, is cryobanking, which is needed for scientific purposes. This kind of cryobank allows the storage of genetic material for taxonomic and comparative studies as well as biomedical research. For these purposes, cryobanking is oriented toward storing genetic information (mainly DNA and RNA). That makes them practical from a cryobiological perspective, as no cell-type specific cryopreservation methods are needed for this approach. The actual state of the art of this kind of biobank and its applied importance is reviewed by Pérez-Espona (2021). The most representative of this kind of cryobank is projects such as the Frozen Ark, in association with the Natural History Museum of London in the United Kingdom (Clarke 2009), Frozen Zoo at the San Diego Zoo, and Ambrose Monell Cryo Collection at the American Museum of Natural History in the USA (Amato 2009; Hanner et al. 2009; Chemnick et al. 2009). However, a priori, these cryobanks (ARK-type) cannot be used directly for biodiversity restoration as they are focused biomaterial collections for use in modern genetic, genomic, and taxonomic studies rather than on the restoration of biological individuals needed for biodiversity conservation measures. Funding the upkeep of these cryobanks is a major problem that has to be solved for effective and fruitful activity of this kind of cryobanks.

The second kind of biobank is oriented toward the possibility of restoring individuals from cryopreserved biological materials, thus leading to the restoration or protection of wild populations. Precisely, this kind of cryobank can be considered an approach to species conservation. One of these is the Norwegian Gene Bank (NGB) for Atlantic salmon (Bøe et al. 2021). This bank is focused on sperm cryopreservation from a total of 174 Atlantic salmon populations in Norway. An important aspect of this cryobank activity is establishing strong relationships between sperm cryobanking and its application of cryopreserved sperm in breeding programs. The potential of participation of sperm cryo-banks in the conservation of aquatic species has been reviewed in several works (Tiersch 2011; Asturiano et al. 2017; Martínez-Páramo et al. 2017). Still, the description of the robust and active involvement of gene banking in fish conservation (in the sense of protecting population genetic diversity) is not so easy to find. That is why the Norwegian Gene Bank can be considered a very rare example of an institution performing wild species conservation by applying biobanking. Below we will refer to this kind of institution as "Norway salmon-type" biobanks. Generally, we propose accepting the point that differences in biobank type (its purpose of activity) dictate the selection of biological material type for biobanking and methods of its storage.

2. Selection of the Type of Genetic Material to be Preserved, and Biotechnological Approaches That Will Help to Establish Working Biobanks (What and How?)

2.1 The Main Difference in Materials to be Included in Biobanking

For "ARK-type biobanks," for any type of tissue, no cryobiologically-specific methods of cryopreservation are required. The main request/challenge here is the correct cryofixation of biological material (tissue) to avoid *in vitro* DNA damage, compromising further analysis and database growth. Ideally, methods for the collection of biological materials should be unified, but this task is not technically difficult. Direct freezing of samples as soon as possible after extraction and avoiding temperature fluctuation during storage are enough for effective future analysis. Simultaneously, the methods of biomaterial preservation oriented toward further DNA/RNA analysis are diverse and do not necessarily apply at low temperatures when the aim is short-term storage. These methods have been comprehensively described and are readily available for biobanking activity (Nagy 2010). Further, the availability of appropriate methods of sampling has made progress in modern DNA analysis very applicable for use alongside conservation measures. Examples include the application of DNA barcoding databases to prevent the illegal trade of endangered species (Amato 2009; Pérez-Espona 2021), enhancing the taxonomy of endangered species (Amato 2009), and aquatic biodiversity monitoring via eDNA utilization (Schenekar 2022).

Other methods are required for biobanks of "Norway salmon-type," oriented on applying stored biological materials for massive progeny production in breeding programs. Nowadays, this type of cryobank is mainly oriented toward preserving viable cells, which is possible via cryopreservation. Some success in methods of prolonged storage without cryopreservation has been achieved (e.g., freeze-drying) (Gil et al. 2014). However, the application of freeze-drying as a method for biobanking still requires future development if it is to become a favorable alternative to cryopreservation in animal conservation programs (Wakayama et al. 2022).

2.2 Fish Germplasm Feasible for Cryobanking

Sperm

Sperm, in the case of fishes, are the most "popular" and easy-to-use germplasm for cryobanking. That is because robust sperm cryopreservation methods have been developed for many fish orders. The diversity of methods is probably explained because of the variety of different problems and purposes, such as ensuring there are enough sperm in a sample and that enough samples are preserved. Based on general cryobiological assumptions, such parameters as sample volume, sperm concentration, and ability of cells to withstand the changes of environmental osmolality determine the parameters of the freezing procedure (Holt and Penfold 2014). The parameters of the freezing procedure are the chemical composition of the cryoprotective medium (including cryoprotectant) and freezing/thawing rates. As these parameters are interrelated, it is not surprising that methods are developed for applying a particular type of container, as the container volume and dimensions may dictate freezing conditions. Among them are methods for large [10 ml, example of salmon, a large volume is required for hatchery practice (Bøe et al. 2021)] and small [1–2 μl, example of zebrafish, small fish model in biomedical research (Carmichael et al. 2009)] volume samples; freezing rate controlling and no-freezing rate controlling methods (Tiersch 2011); whole sample vitrification methods (Xin et al. 2017). Generally, after the first successful fish sperm cryopreservation in 1953 (Blaxter 1953), protocols have been developed for the representatives of 22 fish orders (around 30% of extant fish orders), accounting for more than 220 species (Stechkina 2019). Additionally, we remark that sperm cryopreservation protocols have also been developed for fishes possessing an internal mode of fertilization (Liu et al. 2019; García-Salinas et al. 2021). So, generally, fish sperm cryopreservation for the diversity of fish taxa is a technically well-developed and progressing area (Zhang 2004; Tiersch 2011).

There is a risk that the elaboration of new taxon-specific methods of sperm cryopreservation would be required for the biobanking of rare species. Exactly this situation is known for mammals (Prieto et al. 2014), and the diversity of sperm cryopreservation methods in fishes is explained by the taxon-specific properties of spermatozoa (Kopeika and Kopeika 2008). At the same time, it seems likely that elaborating new sperm cryopreservation methods should not be a major challenge as fish spermatological knowledge is progressing quite well. At least examples of failure to perform sperm cryopreservation in fishes are quite rare, and for the authors of this chapter, they are known as "personal communications." Details of fish sperm cryopreservation will be exemplified in detail for Neotropical fishes in Chapter 10; here, we would like to mention two general aspects of fish sperm cryopreservation that are to be considered in biobanking.

First, applying androgenesis in fish presents an additional important prospect in fish sperm cryobanking for species conservation. This point was already evident at the start of the development of the general concept of cryobanking for conservation measures (Veprintsev and Rott 1979). The importance of androgenesis is related to the fact that the cryopreservation of eggs and embryos in fish is a challenging task (Zhang 2004). These complications require challenging reproductive technologies to conserve maternal genomes in fishes. Elaborate methods for obtaining progeny solely from the sperm genome (androgenesis) can partially solve the problem. In species with an XY sex-determination system, female individuals can be produced by androgenesis from cryopreserved sperm for further breeding programs (Komen and Thorgaard 2007). The main challenges here are the low survival rate of androgenetic progeny and the low number of fish taxa in which the development of procedures (protocols) has been performed (Franěk et al. 2020). Additionally, this option is a priori not possible for fishes with the WZ sex-determination system.

Second, whatever plan for cryopreserving sperm is used, one of the challenges of cryobanking for species conservation is the risk of genetic material cryoalteration. Cryopreservation is potentially hazardous for genetic material; DNA cryodamage occurs at different organizational levels of the genetic material (single- and double-strand DNA chain breaks, chromatin deterioration (Kopeika et al. 2015) and, additionally, cryopreservation can lead to epigenetic modifications) (Labbé et al. 2017). While these aspects are reviewed in the specialized literature, for this chapter, it is needed only to remark that even though fish sperm cryopreservation is considered a "safe" procedure (Martínez-Páramo et al. 2017), genetic material cryoalteration has been demonstrated in several studies [see review (Riesco et al. 2022)]. These alterations can be repaired by egg DNA reparases, at least partially (Kopeika et al. 2004; Pérez-Cerezales et al. 2010a), securing this way progeny from the adverse effects of sperm cryopreservation. However, abnormalities in embryo development of progenies obtained with cryopreserved sperm have been documented in numerous studies, indicating that egg reparase activity cannot always be good enough to avoid an expression of cryodamaged sperm genetic material. These kinds of results were obtained in several taxonomically distant fishes such as African catfish (Miskolczi et al. 2005), Levantine scraper (Zadmajid et al. 2019), or rainbow trout (Pérez-Cerezales et al. 2010b).

In contrast, no differences in the morphology of developing embryos and hatched larvae were found in Tiete tetra (Viveiros et al. 2012) and rainbow trout (Young et al. 2009). The results mentioned above suggest that the possible impact of sperm cryopreservation on the performance of resultant progeny should be considered during the planning of sperm cryobanking. This point was also highlighted by Nusbaumer et al. (2019), who demonstrated on brown trout that fertilization by cryopreserved sperm led to significantly reduced larval growth after hatching. We consider this message "alarming" and request a further understanding of the risks of genetic alterations in populations created from cryopreserved sperm.

Recently, several studies focused on avoiding the risks of introducing sperm genetic material cryo-deterioration have been performed in fishes. These studies are based on two assumptions: (1) after cryopreservation, sperm fractions of differentially cryo-damaged spermatozoa appear in the sample, and (2) separation of spermatozoa possessing minimally damaged genetic material could improve fertilization outcomes. This idea has been checked in just a few studies on fish. In these studies, the reduction of embryo development abnormalities after applying sperm fractionation techniques was reported (Pérez-Atehortúa et al. 2022; Sotnikov et al. 2022a; Sotnikov et al. 2022b). However, this direction of studies in fish is not well developed and requires future elaboration. Here, it is important to remark on the concept of sperm heterogeneity in the ejaculate, which initially was elaborated in mammals (Holt and Van Look 2004; Martínez-Pastor 2022). According to this concept, only a small fraction of spermatozoa from the whole ejaculate possesses biologically relevant properties, allowing them to be involved in selecting the best spermatozoa for fertilization by the female reproductive tract. To get this fraction for *in vitro* fertilization, several techniques, which are based on different sperm cell properties, have been elaborated (Katigbak et al. 2019) and currently these methods could become the goal for the creation of additional useful approaches for the avoidance of harmful cryo-damage when producing progeny from fishes (Li et al. 2010; Valcarce et al. 2016; Horokhovatskyi et al. 2018; Sotnikov et al. 2022c).

At the same time, even with the risk of introducing "cryo-mutations," sperm cryobanking is a strong candidate for participation in conservation programs because of its easy, realizable techniques, while the history of application and achievements in agriculture and medicine indeed leaves space for optimism. The above-mentioned is true, even as it is evident that both good (Maitland and Lyle 2013; Bøe et al. 2021) and suboptimal results (Petersson et al. 2022) of reintroduction to the wild progeny obtained with cryopreserved sperm are known.

Egg and Embryo

In these, cryopreservation would be helpful in the cryobanking of genetic resources for restoring the maternal genomes. However, this is challenging in fish because of the particular cryobiological properties of the object (Zhang 2004). While being of evident importance, studies in this area sometimes led to scientific anecdotes on post-thaw embryo survival (Chen and Tian 2005), which were not readily accepted (Edashige et al. 2006). Currently, fish embryo cryopreservation can be implemented by applying pre-incubation of an embryo in cryo-protective media followed by quick cooling (by immersing the embryo directly in liquid nitrogen). At the same time, it is also proven that only a small fraction of embryos survive for a short post-thawing time. The list of 11 fish species in which this kind of result has been achieved so far can be found in the recent publication by Li et al. (2022).

While the scientific enthusiasm of adherents of the above method is respected, a long-time history of failures in this field has pushed the development of technically more complicated but potentially more efficient cryo-biological methods. Nowadays, the only reliable method that provides successive and repeatable embryo cryopreservation in fishes is the one described by Khosla et al. (2017). This method applies the microinjection of nanogold particles and cryo-protectants into the embryos, fast cooling, and hit with a laser, speeding up the warming process to avoid ice crystal formation. While this method has only been tested in zebrafish embryos, its physical background allows us to look forward to progress in the application for other species (Neff 2017). Moreover, elaborated technological steps allowing quick injection of numerous embryos by robotic devices provide additional technical possibilities for future involvement of fish egg cryopreservation in biobanking oriented toward species preservation programs. However, this technology is at an initial stage of development in terms of species specificity, and a lot of effort will be required to make it a vital participant in biobanking for species conservation.

Germ-Line Precursors of Gametes

As candidates for cryobanking of genetic resources, they have become an essential alternative to the problems of embryo and egg cryopreservation in fishes. Embryonic primordial germ cells (PGS) or spermatogonial/oogonial stem cells are the primary cells that can be applied for biobanking aimed at further breeding and restocking programs. However, this application is now possible only via surrogacy when the surrogate host parent produces gametes of donor germ-line cells from cryopreserved stock cells. That makes applying this genetic material for species restoration technically tricky. At the same time, due to significant efforts in the field performed over the last 50 years, it has become clear that germ-line precursors in cryobanking should be considered as a specific biotechnological approach. The principles of this approach can be found in recent reviews (Franěk and Pšenička 2020; Pšenička and Saito 2020; Jin et al. 2021), and the main principles can be simplified as follows. In fish, surrogate production means that a surrogate host parent produces the gametes of a donor. The host individual carrying the germ line of a donor is called a germ-line chimera. To obtain germ-line chimeras, germ stem cells (GSC) must be isolated from a donor and transplanted into a host. The host becomes a germ-line chimera if the transplanted GSCs become correctly localized to the genital ridge. Donor genotypes can be restored in the next generation if the transplanted germ cells differentiate into functional gametes (Pšenička and Saito 2020). This short description determines the main key elements of germ-cell transplantation technology, such as germ-stem cell isolation, host sterilization, isolated cell transplantation into the sterile host, germ-line chimera growth, and finally, collection of gametes for the next massive production of donor-genotype individuals. Incorporation into technology donor cell cryopreservation made it applicable for cryobanking (Franěk and Pšenička 2020). Each of these elements is crucial for technological development, and they have been studied in tens of species (Goto and Saito 2019; Franěk and Pšenička 2020), allowing us to conclude that this technology becomes a good "partner" to sperm cryopreservation to be considered as an element of the conservation program. However, studies related to the evaluation of risks of genome cryo-damage are required for germ-line precursors of gametes to the same extent as for sperm.

3. Summary

The considerable diversity of fishes—in the sense of their evolution and species richness in higher taxa and habitat diversity and species population size—determines the case-sensitive conservation approaches that are needed. Two of them, considered in this chapter, are compared in Figure 1.

In the case of biobanking aimed at creating databases for future phylogenetic studies, the number of individuals—the source of genetic materials—is not crucial. This is because the collection of genetic material from as many species as possible is a priority. Moreover, a simple cryo-fixation method is sufficient for this kind of cryobanking, which simplifies the biobanking activity. In contrast, when cryobanking is focused on species conservation, it should be considered part of an entire conservation program, in which the prioritization of species to be involved and precise strategy of cryopreserved material use should be determined in advance. For this kind of cryobanking, the number of individuals becomes crucial for answering the question, "Can sufficient genetically diverse samples be preserved for species restoration?" In the case of fish, this question can be solved on different levels of cooperation (local or international) and, as a result, a realistic plan for the conservation of certain species in particular localities must be created.

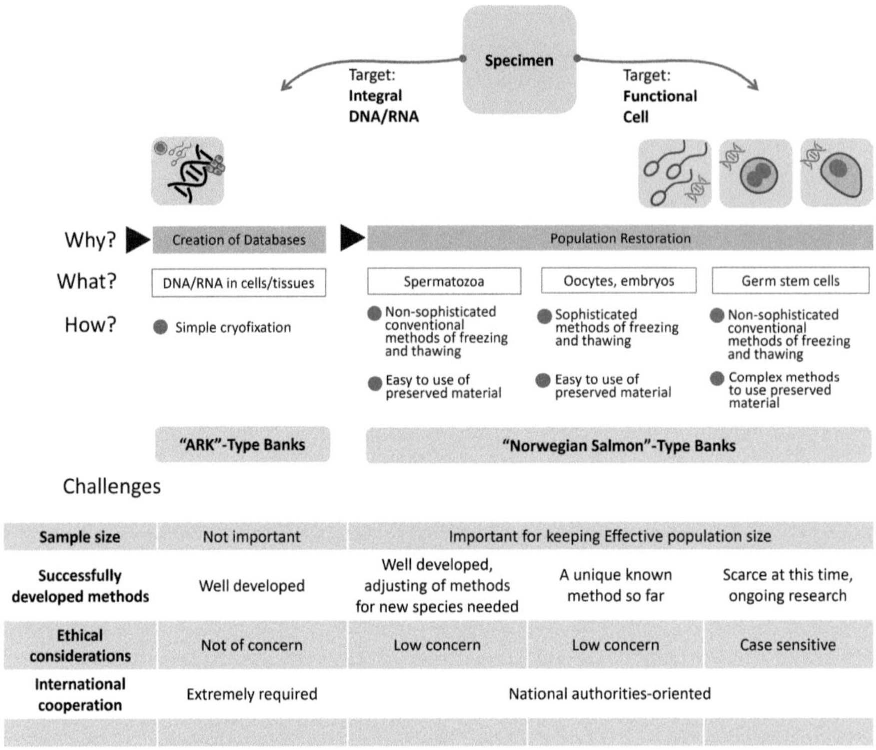

Figure 1. Two types of biobanks for wild species conservation. Cryobanking in fish biodiversity conservation: priorities ("Why?"), types of material ("What?"), approaches applied ("How?"), and challenges associated with or for supporting the activities.

Acknowledgments

The authors acknowledge the help of Dr. Vitaliy Kholodnyy in reviewing the manuscript and preparing the figure. BD was supported by projects from the Czech Science Foundation (No. 22-14069S), the Ministry of Agriculture of the Czech Republic (NAZV QK21010141), and Biodiversity (CZ.02.1.01./0.0/0.0/16025/0007370).

References

Amato, G. 2009. Conservation Genetics in the Age of Genomics. Columbia University Press, New York.

Asturiano, J.F., E. Cabrita and Á. Horváth. 2017. Progress, challenges and perspectives on fish gamete cryopreservation: A mini-review. Gen. Comp. Endocrinol. 245: 69–76.

Bennett, P.M. 2001. Establishing animal germplasm resource banks for wildlife conservation: genetic, population and evolutionary aspects. pp. 48–67. *In*: Watson, P.F. and W.V. Holt [eds.]. Cryobanking the Genetic Resource. Wildlife Conservation for the Future? Taylor and Francis, London, UK.

Blaxter, J.H.S. 1953. Sperm storage and cross-fertilization of spring and autumn spawning herring. Nature 172: 1189–1190.

Bøe, K., B. Bjøru, M. Tangvold Bårdsen, A. Nordtug Wist, S. Wolla and A. Sivertsen. 2021. Opportunities and challenges related to sperm cryopreservation in Atlantic salmon gene banks. Conserv. Sci. Pract. 3: e552.

Brondízio, E.S., J. Settele, S. Diaz and H.T. Ngo. 2019. Global Assessment Report on Biodiversity and Ecosystem Services of the Intergovernmental Science-Policy Platform on Biodiversity and Ecosystem Services. IPBES secretariat, Bonn.

Brooks, T.M., R.A. Mittermeier, G.A.B. da Fonseca, J. Gerlach, M. Hoffmann, J.F. Lamoreux et al. 2006. Global biodiversity conservation priorities. Science 313: 58–61.

Cardinale, B.J., J.E. Duffy, A. Gonzalez, D.U. Hooper, C. Perrings, P. Venail et al. 2012. Biodiversity loss and its impact on humanity. Nature 486: 59–67.

Carmichael, C., M. Westerfield and Z.M. Varga. 2009. Cryopreservation and *in vitro* fertilization at the zebrafish international resource center. Methods Mol. Biol. 546: 45–65.

CBD. 1992. The Convention on Biological Diversity of 1992. United Nations, New York, USA.

Chemnick, L.G., M.L. Houck and O.A. Ryder. 2009. Banking of genetic resources: the frozen zoo at the San Diego Zoo. pp. 124–130. *In*: Amato, G., O. Ryder, H. Rosenbaum and R. DeSalle [eds.]. Conservation Genetics in the Age of Genomics. Columbia University Press, New York, USA.

Chen, S.L. and Y.S. Tian. 2005. Cryopreservation of flounder (*Paralichthys olivaceus*) embryos by vitrification. Theriogenology 63: 1207–1219.

Clarke, A.G. 2009. The Frozen Ark Project: the role of zoos and aquariums in preserving the genetic material of threatened animals. Inter. Zoo Yearb. 43: 222–230.

Edashige, K., D.M. Valdez, T. Hara, N. Saida, S. Seki and M. Kasai. 2006. Japanese flounder (*Paralichthys olivaceus*) embryos are difficult to cryopreserve by vitrification. Cryobiology 53: 96–106.

Feest, A., T. Aldred and K. Jedamzik. 2010. Biodiversity quality: a paradigm for biodiversity. Ecol. Indic. 10: 1077–1082.

Franěk, R., A.R. Baloch, V. Kašpar, T. Saito, T. Fujimoto, K. Arai et al. 2020. Isogenic lines in fish—a critical review. Rev. Aquac. 12: 1412–1434.

Franěk, R. and M. Pšenička. 2020. Cryopreservation of germ stem cells in fish. pp. 285–312. *In*: Betsy, J. and S. Kumar [eds.]. Cryopreservation of Fish Gametes. Springer Nature Singapore Pte Ltd., Singapore, Singapore.

García-Salinas, P., V. Gallego and J.F. Asturiano. 2021. Development of sperm cryopreservation protocols for sharks and rays: new tools for elasmobranch conservation. Front. Mar. Sci. 8: 689089.

Gil, L., M. Olaciregui, V. Luño, C. Malo, N. González and F. Martínez. 2014. Current status of freeze-drying technology to preserve domestic animals sperm. Reprod. Domest Anim. 49: 72–81.

Godfray, H.C.J., S. Knapp and G.M. Mace. 2004. The role of taxonomy in species conservation. Philos. Trans. R. Soc. Lond., B, Biol Sci. 359: 711–719.

Goto, R. and T. Saito. 2019. A state-of-the-art review of surrogate propagation in fish. Theriogenology 133: 216–227.

Hanner, R., A. Corthals and R. DeSalle. 2009. Biodiversity, conservation, and genetic resources in modern museum and herbarium collections. pp. 115–123. *In*: DeSalle, R., G. Amato, O. Ryder and H. Rosenbaum [eds.]. Conservation Genetics in the Age of Genomics. Columbia University Press, New York, USA.

Helfman, G. 2007. Fish Conservation: A Guide to Understanding and Restoring Global Aquatic Biodiversity and Fishery Resources, 1 ed. Island Press, Washington.

Hilsdorf, A.W.S. and E.M. Hallerman. 2017. Prospective views and recommendations. pp. 211–238. *In*: Hilsdorf, A.W.S. and E.M. Hallerman [eds.]. Genetic Resources of Neotropical Fishes. Springer International Publishing AG, Cham, Switzerland.

Holt, W.V., T. Abaigar, P.F. Watson and D.E. Wildt. 2003. Genetic resource banks for species conservation. pp. 267–280. *In*: Holt,W.V., A.R. Pickard, J.C. Rodger and D.E. Wildt [eds.]. Reproductive Science and Integrated Conservation. Cambridge University Press, New York, USA.

Holt, W.V. and K.J. Van Look. 2004. Concepts in sperm heterogeneity, sperm selection and sperm competition as biological foundations for laboratory tests of semen quality. Reproduction 127: 527–535.

Holt, W.V. and L.M. Penfold. 2014. Fundamental and practical aspects of semen cryopreservation. pp. 76–99. *In*: Chenoweth, P.J. and S.P. Lorton [eds.]. Animal Andrology: Theories and Applications. CAB International, Wallingford, UK.

Holt, W.V. and P. Comizzoli. 2021. Genome resource banking for wildlife conservation: promises and caveats. Cryo Lett. 42: 309–320.

Horokhovatskyi, Y., M.A. Dietrich, I. Lebeda, P. Fedorov, M. Rodina and B. Dzyuba. 2018. Cryopreservation effects on a viable sperm sterlet (*Acipenser ruthenus*) subpopulation obtained by a Percoll density gradient method. PLoS One 13: e0202514.

Hughes, A.C., H. Qiao and M.C. Orr. 2020. Extinction Targets Are Not SMART (Specific, Measurable, Ambitious, Realistic, and Time Bound). Bioscience 71: 115–118.

Jézéquel, C., P.A. Tedesco, W. Darwall, M.S. Dias, R.G. Frederico, M. Hidalgo et al. 2020. Freshwater fish diversity hotspots for conservation priorities in the Amazon Basin. Conserv. Biol. 34: 956–965.

Jin, Y.H., D. Robledo, J.M. Hickey, M.J. McGrew and R.D. Houston. 2021. Surrogate broodstock to enhance biotechnology research and applications in aquaculture. Biotechnol. Adv. 49: 107756.

Katigbak, R.D., G.M. Turchini, S.P. de Graaf, L. Kong and L.F. Dumée. 2019. Review on sperm sorting technologies and sperm properties toward new separation methods via the interface of biochemistry and material science. Adv Biosyst. 3: 1900079.

Keppeler, F.W. and R.A.M. Silvano. 2020. Introduction: Fishes of the Brazilian Amazon. pp. 1–11. *In*: Silvano, R.A.M. [ed.]. Fish and Fisheries in the Brazilian Amazon: People, Ecology and Conservation in Black and Clear Water Rivers. Springer International Publishing AG, Cham, Switzerland.

Khosla, K., Y. Wang, M. Hagedorn, Z. Qin and J. Bischof. 2017. Gold nanorod induced warming of embryos from the cryogenic state enhances viability. ACS Nano. 11: 7869–7878.

Komen, H. and G.H. Thorgaard. 2007. Androgenesis, gynogenesis and the production of clones in fishes: A review. Aquaculture 269: 150–173.

Kopeika, J., E. Kopeika, T. Zhang, D.M. Rawson and W.V. Holt. 2004. Effect of DNA repair inhibitor (3-aminobenzamide) on genetic stability of loach (*Misgurnus fossilis*) embryos derived from cryopreserved sperm. Theriogenology 61: 1661–1673.

Kopeika, E. and J. Kopeika. 2008. Variability of sperm quality after cryopreservation in fish. pp. 347–396. *In*: Alavi, S.M.H., J.J. Cosson, K. Coward and G. Rafiee [eds.]. Fish Spermatology. Alpha Science International Ltd., Oxford, UK.

Kopeika, J., A. Thornhill and Y. Khalaf. 2015. The effect of cryopreservation on the genome of gametes and embryos: principles of cryobiology and critical appraisal of the evidence. Hum. Reprod. Update 21: 209–227.

Labbé, C., V. Robles and M.P. Herraez. 2017. Epigenetics in fish gametes and early embryo. Aquaculture 472: 93–106.

Li, L., Y. Tian, Z. Li, Z. Li, P. Duan, X. Wang et al. 2022. Cryopreservation of embryos of humpback grouper (*Cromileptes altivelis*) using combinations of non-permeating cryoprotectants. Aquaculture 548: 737524.

Li, P., B. Dzyuba, M. Hulak, M. Rodina, S. Boryshpolets, Z.H. Li et al. 2010. Percoll gradient separation of cryopreserved common carp spermatozoa to obtain a fraction with higher motility, velocity and membrane integrity. Theriogenology 74: 1356–1361.

Liu, Y., H. Blackburn, S.S. Taylor and T.R. Tiersch. 2019. Development of germplasm repositories to assist conservation of endangered fishes: Examples from small-bodied livebearing fishes. Theriogenology 135: 138–151.

Maitland, P.S. and A.A. Lyle. 2013. *Ex situ* and *in situ* approaches, including assisted reproduction, for the conservation of native species of charr (Salmonidae) and whitefish (Coregonidae) in Scotland. Int. Zoo Yearb. 47: 129–139.

Marchese, C. 2015. Biodiversity hotspots: A shortcut for a more complicated concept. Glob. Ecol. Conserv. 3: 297–309.

Martínez-Páramo, S., Á. Horváth, C. Labbé, T. Zhang, V. Robles, P. Herráez et al. 2017. Cryobanking of aquatic species. Aquaculture 472: 156–177.

Martínez-Pastor, F. 2022. What is the importance of sperm subpopulations? Anim. Reprod. Sci. 246: 106844.

Miranda, R. and I. Miqueleiz. 2021. Ecology and conservation of freshwater fishes biodiversity: we need more knowledge to develop conservation strategies. Water 13: 1929.

Miskolczi, E., S. Mihalffy, E.P. Varkonyi, B. Urbanyi and A. Horvath. 2005. Examination of larval malformations in African catfish *Clarias gariepinus* following fertilization with cryopreserved sperm. Aquaculture 247: 119–125.

Nagy, Z.T. 2010. A hands-on overview of tissue preservation methods for molecular genetic analyses. Org. Divers. Evol. 10: 91–105.

Neff, E.P. 2017. All that glitters is gold-frozen fish. Lab. Anim. 46: 355–355.

Nelson, J.S., T.C. Grande and M.V.H. Wilson. 2016. Fishes of the World. John Wiley & Sons, Inc., New Jersey.

Nusbaumer, D., L. Marques da Cunha and C. Wedekind. 2019. Sperm cryopreservation reduces offspring growth. Proc. Royal Soci. B. 286: 20191644.

Paz-Vinas, I., G. Loot, V. Hermoso, C. Veyssière, N. Poulet, G. Grenouillet et al. 2018. Systematic conservation planning for intraspecific genetic diversity. Proc. Royal Soci. B. 285: 20172746.

Pérez-Atehortúa, M., A. Giannotti Galuppo, R. Batista Rodrigues, T. de Souza França, N. dos Santos Teixeira, T. Rodrigues de Freitas et al. 2022. The use of differential separation and density gradient with AllGrad 90% after thawing improves the sperm quality of South American catfish (*Rhamdia quelen*). Aquaculture 553: 738072.

Pérez-Cerezales, S., S. Martínez-Páramo, J. Beirão and M.P. Herráez. 2010a. Fertilization capacity with rainbow trout DNA-damaged sperm and embryo developmental success. Reproduction 139: 989–997.

Pérez-Cerezales, S., S. Martínez-Páramo, J. Beirão and M.P. Herráez. 2010b. Evaluation of DNA damage as a quality marker for rainbow trout sperm cryopreservation and use of LDL as cryoprotectant. Theriogenology 74: 282–289.

Pérez-Espona, S. 2021. Conservation-focused biobanks: A valuable resource for wildlife DNA forensics. Forensic Sc. Int.: Anim. Environ. 1: 100017.

Petersson, E., J. Rask and P. Dębowski. 2022. Failed protective effort of *ex situ* conservation of River Vistula trout (*Salmo trutta*) in Sweden. Environ. Biol. Fishes. 105: 1099–1109.

Pickard, A.R. and W.V. Holt. 2004. Cryopreservation as a supporting measure in species conservation: "Not the Frozen Zoo!". pp. 393–413. *In*: Fuller, B.J., N. Lane and E.E. Benson [eds.]. Life in the Frozen State. CRC Press, Boca Raton, LA, USA.

Prieto, M.T., M.J. Sanchez-Calabuig, T.B. Hildebrandt, J. Santiago-Moreno and J. Saragusty. 2014. Sperm cryopreservation in wild animals. Eur. J. Wildl. Res. 60: 851–864.

Pšenička, M. and T. Saito. 2020. Specificity of germ cell technologies in sturgeons. pp. 335–356. *In*: Yoshida, M. and J.F. Asturiano [eds.]. Reproduction in Aquatic Animals: From Basic Biology to Aquaculture Technology. Springer Nature Singapore, Singapore, Singapore.

Rawson, D.M., G.M. Reid and R.E. Lloyd. 2011. Conservation rationale, research applications and techniques in the cryopreservation of lower vertebrate biodiversity from marine and freshwater environments. Int. Zoo Yearb. 45: 108–123.

Reid, G.M. and H. Hall. 2003. Reproduction in fishes in relation to conservation. pp. 375–393. *In*: Holt, W.V., A.R. Pickard, J.C. Rodger and D.E. Wildt [eds.]. Reproductive Science and Integrated Conservation. Cambridge University Press, New York, NY, USA.

Reis, R.E., J.S. Albert, F. Di Dario, M.M. Mincarone, P. Petry and L.A. Rocha. 2016. Fish biodiversity and conservation in South America. J. Fish Biol. 89: 12–47.

Riesco, M.F., D.G. Valcarce and V. Robles. 2022. Molecular approaches on DNA damage evaluation after primordial germ cell cryopreservation in zebrafish. pp. 49–68. *In*: Fernández, I. and J. Fernandes [eds.]. Cellular and Molecular Approaches in Fish Biology. Elsevier Science, Amsterdam, Netherland.

Schenekar, T. 2023. The current state of eDNA research in freshwater ecosystems: are we shifting from the developmental phase to standard application in biomonitoring? Hydrobiologia 850: 1263–1282.

Schmitt, C.B. 2011. A tough choice: approaches towards the setting of global conservation priorities. pp. 23–42. *In*: Zachos, F.E. and J.C. Habel [eds.]. Biodiversity Hotspots: Distribution and Protection of Conservation Priority Areas. Springer, Berlin, Germany.

Sotnikov, A., O. Linhart, S. Boryshpolets, M. Rodina, T. Policar, V. Dzyuba et al. 2022a. Motile sperm fraction separation from cryopreserved samples in taxonomically distant fishes, Joint Meeting of the Society for Cryobiology and the Society for Low Temperature Biology (In person and virtual meeting CRYO2022). Dublin, Ireland: pp. 70.

Sotnikov, A., J. Sterba, S. Boryshpolets, M. Rodina and B. Dzyuba. 2022b. Separation of highly motile spermatozoa from cryopreserved sterlet sperm is a promising method for improving offspring quality. 8th International Workshop on the Biology of Fish Gametes. Gdansk, Poland: pp. 102.

Sotnikov, A., A. Urbansky, J. Alsved, S. Boryshpolets, M. Rodina and B. Dzyuba. 2022c. Sperm fractionation in fishes: density gradient centrifugation and microfluidic separation approaches for cryopreserved samples, 8th International Workshop on the Biology of Fish Gametes. Gdansk, Poland: pp. 52.

Stechkina, T. 2019. Advances in Fish Sperm Cryopreservation: Taxonomical Consideration. B.Sc. Thesis, University of South Bohemia in České Budějovice, České Budějovice, Czech Republic.

Strassburg, B.B.N., A. Iribarrem, H.L. Beyer, C.L. Cordeiro, R. Crouzeilles, C.C. Jakovac et al. 2020. Global priority areas for ecosystem restoration. Nature 586: 724–729.

Tiersch, T.R. 2011. Cryopreservation in Aquatic Species, 2 ed. World Aquaculture Society, Baton Rouge.

Tyus, H.M. 2011. Ecology and Conservation of Fishes 1ed. CRC Press, Boca Raton.

Usher, M.B. 1986. Wildlife Conservation Evaluation. Chapman and Hall, London.

Valcarce, D.G., M.P. Herráez, O. Chereguini, C. Rodríguez and V. Robles. 2016. Selection of nonapoptotic sperm by magnetic-activated cell sorting in Senegalese sole (*Solea senegalensis*). Theriogenology 86: 1195–1202.

Veprintsev, B.N. and N.N. Rott. 1979. Conserving genetic resources of animal species. Nature 280: 633–634.

Viveiros, A., Z. Isau, D. Caneppele and M. Leal. 2012. Sperm cryopreservation affects postthaw motility, but not embryogenesis or larval growth in the Brazilian fish *Brycon insignis* (Characiformes). Theriogenology 78: 803–810.

Wakayama, S., D. Ito, E. Hayashi, T. Ishiuchi and T. Wakayama. 2022. Healthy cloned offspring derived from freeze-dried somatic cells. Nat. Commun. 13: 3666.

Watson, P.F. and W.V. Holt. 2001. Cryobanking the Genetic Resource; Wildlife Conservation for the Future? Taylor and Francis, London.

Winemiller, K.O., P.B. McIntyre, L. Castello, E. Fluet-Chouinard, T. Giarrizzo, S. Nam et al. 2016. Balancing hydropower and biodiversity in the Amazon, Congo, and Mekong. Science 351: 128–129.

Xin, M., M.A.M. Siddique, B. Dzyuba, R. Cuevas-Uribe, A. Shaliutina-Kolešová and O. Linhart. 2017. Progress and challenges of fish sperm vitrification: A mini review. Theriogenology 98: 16–22.

Young, W.P., K. Frenyea, P.A. Wheeler and G.H. Thorgaard. 2009. No increase in developmental deformities or fluctuating asymmetry in rainbow trout (*Oncorhynchus mykiss*) produced with cryopreserved sperm. Aquaculture 289: 13–18.

Zadmajid, V., E. Falahipour, E. Ghaderi, S.R. Sørensen and I.A.E. Butts. 2019. Outcomes of *in vitro* fertilization with frozen-thawed sperm: An analysis of post-thaw recovery of sperm, embryogenesis, offspring morphology, and skeletogenesis for a cyprinid fish. Dev. Dyn. 248: 449–464.

Zachos, F.E. and J.C. Habel. 2011. Biodiversity Hotspots: Distribution and Protection of Conservation Priority Areas. Springer, Berlin.

Zhang, T. 2004. Cryopreservation of gametes and embryos of aquatic species. pp. 415–435. *In*: Fuller, B.J., N. Lane and E.E. Benson [eds.]. Life in the Frozen State CRC Press, Boca Raton, USA.

Chapter 4

Biobanks of Biological and Aquarium Model Species

Patrícia Diogo[1,*] and *Rômulo B. Rodrigues*[2,*]

INTRODUCTION

Animal models are essential for research since they are used to investigate multiple areas, such as physiology, metabolism, molecular mechanisms, and different biological functions. The use of animal models started in Greece more than 2400 years ago in comparative medicine. The basis of the utilization of model species is the use of interspecies extrapolation when there is significant conservation between cells, genes, or biological functions between two species (Ericsson et al. 2013). In the 20th century, the use of animal models increased dramatically, not only to support research in biomedicine but also to model functions of other species such as aquaculture species. The selection of the most informative model species is highly relevant since there are several necessary characteristics to improve the reliability of the comparison between the model and the species of interest. Many species were used throughout time, such as goats, cattle, chicken, sheep, pigs, flies, and fish (e.g., zebrafish). Although mouse models were abundantly used in the 20th century, the use of small-bodied fish has been increasing due to ethical, zootechnical, and economic reasons (Ericsson et al. 2013; Choi et al. 2021).

Several fish species were used as models, such as trout (Bailey et al. 1996), puffer fish (*Fugu rubripes*) (Brenner et al. 1993), and seabream (*Sparus aurata*) (Houston et al. 2022). However, small-bodied fish species are more advantageous since they require less space, water, and food for maintenance, which is suitable for research facilities conditions. Some small-bodied fish that are used as model

[1] Companhia Portuguesa de Culturas Marinhas S.A, Belamandil s/n, 8700-152 Olhão, Portugal.

[2] Departamento de Zootecnia e Ciências Biológicas, Universidade Federal de Santa Maria - Campus de Palmeira das Missões, RS, Brazil.

* Corresponding authors: patricia.diogo@necton.pt, romulo.batista@ufsm.br

species are gold fish (*Carassius auratus*) (Popesku et al. 2008), Japanese medaka (*Oryzias latipes*) (Nor Azra et al. 2022), flathead minnow (*Pimephales promelas*) (Miller and Ankley 2004), killifish (*Fundulus auroguttatus*) (Planchart et al. 2016), guppy (*Poecilia reticulata*), Siamese fighting fish (*Betta splendens*) (Wang et al. 2022), and zebrafish (*Danio rerio*) (Adhish and Manjubala 2023). However, zebrafish started to be considered the second most used model species in the first two decades of the 21st century. Its increasing usage and popularity mostly occurred because its genome is fully sequenced; around 70% of human genes have at least one zebrafish ortholog, and there is a well-developed target gene editing technology for this species (Howe et al. 2013; Liu et al. 2017; Choi et al. 2021). Due to these facts, thousands of new mutants and transgenic (reporter) strains were developed. The exponential generation of newly established zebrafish lines for research brought space and management constraints for maintaining these valuable resources (Silva et al. 2019). Therefore, it is essential to preserve these valuable genetic resources through biobanking methodologies.

Zebrafish (*Danio rerio*) have become a widely used model species established in research centers worldwide to study vertebrate mechanisms in areas such as development, regeneration, toxicology, and pathologies (Driever et al. 1996; Haffter et al. 1996; Gemberling et al. 2013; Meyers 2018; Patton and Tobin 2019; Sieber et al. 2019; Tanguay 2018). Zebrafish is a robust, small-bodied vertebrate with high fecundity and year-round reproduction, external fertilization, rapid development, transparency of embryos, and ease of experimental manipulation (Lieschke and Currie 2007; Ribas and Piferrer 2013). Altogether, these characteristics make zebrafish a particularly useful model species. The methodological tools to generate zebrafish lines are continuously expanding. After zebrafish genome sequencing, reverse genetics techniques have been developed successfully. The reverse genetics methods use specific knock-out or knock-down of genes of interest, allowing a rapid understanding of the functions of the predicted/known genes. Although there are several reverse genetics methodologies, the most used currently in zebrafish is CRISPR/Cas9 system and transcription activator-like effector nucleases (TALENs) (Dahm and Geisler 2006; Housden et al. 2016). The research performed with zebrafish contributes to the understanding of basic vertebrate biology and development as well as to the understanding of factors controlling the specification of cell types, organ systems, and body axis (Talbot et al. 1995; Resende et al. 2014; Lleras Forero et al. 2018). Consequently, zebrafish are now considered a valuable model organism to investigate human diseases (Lieschke and Currie 2007) and aging (Gerhard 2003). Zebrafish is relevant in research areas, such as aquatic toxicology (Hill et al. 2005; Laizé et al. 2014), drug screenings (Vaz et al. 2018), regenerative medicine (Azevedo et al. 2012; Goessling and North 2014; Cardeira et al. 2016), and neuroscience and behavior (Oliveira et al. 2016). Recent advancements in next-generation sequencing (NGS) supported the development of new approaches such as personalized medicine, and zebrafish are being used to identify relationships between the genotype and phenotype of human diseases (Choi et al. 2021).

The cryopreservation of germ cells is a valuable methodology for adequately preserving these valuable genotypes. Germ cells are responsible for transmitting genetic information from one generation to another. Mature germ cells, such as mature spermatozoa, are uniquely specialized to overcome the challenges associated with the fertilization process and provide a single chromosomal complement to offspring (Murphy et al. 2014). Several laboratories and research centers are working in this direction to preserve important lineages of the species, including the Zebrafish International Resource Center (ZIRC), the European Zebrafish Resource Center (EZRC), and the Zebrafish Information Network (ZFIN). There have been considerable efforts of the scientific community on fish embryo cryopreservation. Although partial improvements have been made, until today, no cryopreservation protocol for teleost's mature oocytes and embryos has successfully obtained viable post-thaw oocytes or embryos able to generate offspring. Therefore, the cryopreservation methodologies target primordial germ cells (PGCs) and spermatozoa. Male germ cells include spermatozoa and their previous stages of maturation, such as primordial germ cells (PGCs) and spermatogonia. Several cryopreservation protocols are developed for testis, PGCs (Marinović et al. 2018a,b), or spermatozoa (Draper and Moens 2009; Matthews et al. 2018; Yang et al. 2007). Spermatozoa have a small size and a relatively high resistance to chilling, therefore being more advantageous compared to other cell types. For this reason, the cryopreservation of spermatozoa is the most widely used technique in aquatic species (Martínez-Páramo et al. 2017). Spermatozoa are the most practical cells routinely cryopreserved in zebrafish facilities to reconstitute their different lines. The main advantage of sperm over PGCs cryopreservation is that sperm collection can be performed through a non-lethal method (Jing et al. 2009) and does not require laborious cell transplantation methodologies.

The development of cryopreservation techniques for aquarium fishes is hindered by their small body sizes and sperm volumes, but given their usefulness in research and the existence of stock centers, the development of these genotype preservation techniques is essential (Tiersch 2001). Long-established breeding programs and detailed biological databases exist for lines of zebrafish, swordtails, and Japanese medaka. However, most of the development of cryopreservation methodologies uses mostly wild-type strains, which are generally more resilient than transgenic strains. Considering the fact that cryopreservation can promote damage from cold exposure, the application of cryopreservation methodologies should also be validated in valuable zebrafish strains with reproductive constraints for the full comprehension of the effectiveness of the cryopreservation protocol (Diogo et al. 2019; Diogo et al. 2023). Here, we present a compilation of data from scientific articles on the cryopreservation of reproductive tissues or cells of aquarium fish species. In this chapter, we will present and discuss the status of research on cryopreservation with these aquarium model species and how the findings in these studies can help and serve as a basis for studies with cryopreservation of Neotropical fish. In addition, we provide details on cryopreservation protocols currently in use, allowing for a critical assessment of their limitations and possible trends and paths to be followed in this research area.

1. Zebrafish Model Species

Zebrafish is a member of teleost infraclasses, a monophyletic group estimated to have diverged from a common ancestor approximately 340 million years ago (Amores et al. 2011). This ancestor underwent an additional round of whole genome duplication denominated teleost-specific genome duplication (Meyer and Schartl 1999). Zebrafish was first described by Hamilton as a small freshwater species found naturally in rivers, small streams, channels, and paddy fields from Myanmar, Pakistan, India, Bangladesh, and Nepal (Arunachalam et al. 2013).

Zebrafish is a species tolerant to a wide range of environmental conditions and food resources. Its natural habitat has a wide range of temperatures from 12.3–28.4°C, pH 6.2–9.8 (Arunachalam et al. 2013), and water conductivity 10–280 µS/cm (Engeszer et al. 2007). The species is found in the wild in schools of varying sizes, such as small groups of 6–7 fish in slow-moving waters, groups of 22 fish in still waters, and up to 300 fish in waters with high flow (Graham et al. 2018). This species feeds naturally on allochthonous materials, such as ants and other insects falling into streams, secondary channels, and pools (Arunachalam et al. 2013). Zebrafish colonies in research facilities are maintained in recirculation systems with controlled environmental parameters such as photoperiod, water temperature, pH, and conductivity. Although this species is established in research centers worldwide, the rearing procedures still lack methodological standardization (Lawrence 2016). Consequently, high variability is observed in biological and reproductive performances (Robles et al. 2009; Lawrence 2016; Tsang et al. 2017; Torres et al. 2017; Torres and Tiersch 2018). Due to its high tolerance and robustness, different protocols and methodologies have been employed between facilities, according to each laboratory's specific needs, influenced by personal biases and traditions of specific laboratories (Tsang et al. 2017).

The domestication of wild species requires the ability to control the organism's husbandry and comprehend its specific nutritional and reproductive requirements (T Gjedrem 2005; Duarte et al. 2007). The domestication of a species can improve the application of the cryopreservation methodologies since the physiological conditions of the donors affect the initial cells/tissues quality and for cryopreservation purposes it is recommended the highest gametes or cell quality possible to resist the stress to cold exposure.

Zebrafish are able to reproduce from 4 months after fertilization and up to 3 years of age; however, reproduction capacity decreases with advancing age, and the recommendation of several laboratories is that reproduction be carried out with animals between 6 to 8 months of age in the case of sperm collection at least 14 days of rest between sperm collection events is necessary to obtain high-quality sperm for cryopreservation (Nasiadka and Clark 2012; Diogo et al. 2018). Aquacultured species have high commercial value and, therefore, require increased efficiency of fish production to improve the profit (Duarte et al. 2007). Since most aquaculture species breed at specific times of the year and are much more vulnerable compared to zebrafish, the efforts in the optimization and standardization of procedures by the scientific community were very efficient (Martínez-Páramo et al. 2017). There are several zebrafish strains considered domesticated, such as AB, TU, SJA, or

TM1, that reveal genomic variations in relation to wild populations, which are typically observed in domesticated species (Whiteley et al. 2011). According to a report by Kinth et al. (2013), zebrafish is used in research laboratories in more than 100 countries in research. Therefore, the zebrafish scientific community would benefit greatly by investing in optimizing and standardizing procedures for this species to obtain higher biological efficiency and scientific replicability.

2. Cryopreservation Methodologies

Cryopreservation uses low temperatures to preserve structurally intact living cells and tissues (Pegg 2007). Most cryopreservation applications are related to the ability to stop the biological clock at very low temperatures (Mazur 1984; Benson et al. 2013). However, the cell should be able to survive the extreme environmental shifts that cryopreservation produces (Mazur 1984). Cryobiology is a multidisciplinary science that studies cells' and tissues' biological and physical behavior at low temperatures (Gao and Critser 2000). Cabrita et al. (2010) summarized the benefits of cryopreservation in teleost species as follows:

1) synchronization of gametes availability of both sexes;
2) sperm economy;
3) simplification of broodstock management;
4) simpler transport of gametes;
5) simplification of broodstock management;
6) germplasm storage for genetic selection programs or species conservation.

The cellular biophysics response during freezing is essential to developing successful cryopreservation protocols. The cryopreservation methodology is specific for each species cell type, depending on cell properties such as morphometry, density, organelles, cell compartmentalization, plasma membrane structure, and components (Mazur 1963, 1984). Because of these biophysical characteristics, the water permeability of a type of cell is species-specific; therefore, the cell requires different cryopreservation methodological improvements, such as different cryoprotectants and cooling/thawing rates (Mazur 1984; Mazur et al. 2008). Gao and Cristser (2000) considered cryopreservation key points, i.e., the addition of cryoprotective agents to cells/tissues before cooling, the cooling rate toward the temperature at which the biological material will be stored and the warming of the biological material. Sperm is a particularly challenging cell type to cryopreserve under optimal conditions due to its small size and non-spherical shape (Hagiwara et al. 2009).

Attempts in cryopreservation of teleost's mature oocyte and embryo failed (Asturiano et al. 2017). Two types of zebrafish male germ cells are possible to cryopreserve, namely spermatogonia and mature spermatozoa (Harvey et al. 1982; Wang et al. 2015; Marinović et al. 2018a, b; Matthews et al. 2018). Developing new technological methodologies for spermatogonia cryopreservation constitutes a great advance in research (Asturiano et al. 2017). After thawing the cells, they are transplanted onto a surrogate host that will spawn part of the progeny with the desired genotype; if a triploid surrogate host can be used, all the offspring will carry the desired genotype (Yoshizaki et al. 2016; Yoshizaki and Lee 2018). This technique is particularly useful for complex double mutant lines. Nowadays, these technologies are being successfully developed in zebrafish (Marinović et al. 2018a, b; Franěk et al. 2019). The main disadvantage is that it requires an experienced practitioner, particularly for cell transplantation and therefore is mainly used experimentally. Alternatively, zebrafish sperm cryopreservation is easier and has, therefore, higher practical applicability in zebrafish facilities worldwide.

In research, the reporter zebrafish lines are generally heterozygotic; however, mutant lines frequently require homozygosity to express the desired phenotype. *In vitro* fertilization must be performed with freshly collected oocytes. Consequently, if homozygous or heterozygous females of the same zebrafish line are available in the facility, an *in vitro* fertilization can be performed with cryopreserved sperm to obtain heterozygous and homozygous offspring, according to Mendelian principles of inheritance. In this scenario, freshly collected oocytes from wild-type females can be used to perform *in vitro* fertilization with the cryopreserved sperm from the target zebrafish line. This method allows to obtain heterozygous offspring in a Mendelian proportion. The ultimate objective of a cryopreservation methodology is to obtain an integrated approach that promotes an adequate balance of the biophysical events during cryopreservation, successfully protecting the cell throughout the cryopreservation process. Therefore, this methodology should ensure spermatozoa structural and genomic integrity as well as its fertilization competence and offspring health.

3. Zebrafish (*Danio rerio*) Sperm Cryopreservation

Sperm was the most used research material in zebrafish cryopreservation methodologies. This was expected, given that sperm is the most explored material for cryopreservation in fish (Tiersch et al. 2007; De Carvalho et al. 2014). However, despite the structural, morphological, and biophysical similarities observed between zebrafish sperm and other cyprinid species (Zhang et al. 2014), it is important to recognize that cryopreservation of zebrafish sperm is difficult because of the low volume of sperm, low quality after thawing, and high variation in the success of *in vitro* fertilization for both fresh and cryopreserved sperm (Diogo et al. 2019). The studies demonstrate a great variation in the results obtained with the tested protocols. An effective method to perform zebrafish sperm cryopreservation has been one of the most important and difficult objectives to achieve in the past years by the zebrafish scientific community. Zebrafish sperm cryopreservation methodologies lack standardization, and therefore there is high variability in post-thaw sperm quality and *in vitro* fertilization success. The sources of variability reported in the literature are not only related to key factors affecting cryopreservation success but also to zebrafish management features, such as nutrition and dietary protocols, sperm collection method, sperm quality analysis method, and *in vitro* fertilization procedures (Paull et al. 2008; Robles et al. 2009; Torres et al. 2017; Varga et al. 2018; Martins et al. 2018).

Since the development of the first zebrafish sperm cryopreservation protocol (Harvey et al. 1982) more than 30 years ago, there have been many attempts to improve and standardize the methodology (Harvey et al. 1982; Ransom and Zon 1999; Morris et al. 2003; Yang et al. 2007, 2016, 2020, 2022; Carmichael et al. 2009; Draper and Moens 2009; Hagedorn et al. 2009, 2012; Bai et al. 2013; Wang et al. 2015; Matthews et al. 2018; Diogo et al. 2018, 2019; Depincé et al. 2019; Silva et al. 2019; da Costa et al. 2020; Rodrigues et al. 2020, 2021).

Several extenders and cryoprotectants have been tested in zebrafish sperm with similar results, but it has been difficult to identify the specific effects of cryoprotectants on cryopreservation success (Robles et al. 2009). When reviewing the literature, it is remarkably difficult to compare studies of zebrafish sperm cryopreservation since all steps of the methodologies used vary considerably. This fact is more evident in studies where application of different solutions and methods are used without studying their individual effect on freezing success. Because of these facts, the beneficial or detrimental effects of the solution and cryopreservation steps are not evident due to biases associated with the interactions between them. Therefore, a consistent strategy to develop protocols according to cryobiological principles is necessary. The first step in sperm cryopreservation protocols is sperm collection, which is an important source of variability in the cryopreservation protocol. There is a non-lethal technique to collect sperm by abdominal massage and a lethal method through testis dissection. Jing et al. (2009) observed that sperm collection through abdominal massage yielded lower sperm volumes when compared to testis dissection.

Although testis dissection improves one of the major bottlenecks of sperm cryopreservation, which is the low sperm volumes yielded per male, it implies the sacrifice of the fish. Additionally, using this sperm collection method, a heterogeneous mixture of cells in different maturation stages will be obtained, therefore declining the overall sperm quality of the sample. In many cases, sacrificing highly valuable transgenic and mutant fish is not a viable option due to the scarcity of highly valuable individuals of these specific strains. After sperm collection, the sample is immediately diluted in a solution with composition and osmolarity similar to seminal plasma, thus preventing motility activation and maintaining normal cell functions. In zebrafish, several solutions were tested, such as Ginsburg Ringer solution and Hank's Balanced Salt Solution (HBSS) (Jing et al. 2009). Previous studies considered HBSS the most suitable diluent for zebrafish sperm (Jing et al. 2009); in more than half of the studies, the authors chose HBSS as an extender. Since the development of the first zebrafish sperm cryopreservation protocol, a variety of extender media and permeating cryoprotectants were tested, such as methanol, dimethyl sulfoxide (DMSO), N,N-dimethylformamide (DMF), dimethylacetamide (DMA) glycerol, and ethylene glycol at concentrations ranging 4% to 20%. When analyzing the data from the cryopreserved sperm samples, we observed that methanol was used in almost half of the protocols, followed by DMSO and DMF. The use of non-permeating cryoprotectants is potentially a powerful tool if used in a comprehensive form.

One of the most relevant and practical applications of cryopreservation in zebrafish facilities is the germplasm conservation of valuable zebrafish mutant and transgenic strains. There are numerous situations where the facility manager has very few individuals of a valuable strain, such as buying or receiving a few adults to establish a cohort (but requiring the proper safeguard of the germplasm) or in the case of a sudden pathological outbreak. Since one zebrafish male produces only 0.8 µl to 3 µl of sperm, the packaging must meet this species' sperm volume to be useful in the challenging situations mentioned above. Considering zebrafish average sperm volume per male, using French straws (250 µl or 500 µl) is a high investment since it requires high sperm volume to be used in a single *in vitro* fertilization. The first cryopreservation protocols used capillary tubes (Harvey et al. 1982; Ransom and Zon 1999; Morris et al. 2003). There were several interesting approaches with French straws (Yang et al. 2007, 2016, 2020, 2022; Hagedorn et al. 2012; Bai et al. 2013; Wang et al. 2015; Silva et al. 2019; Depincé et al. 2019) which are useful for large facilities and central repositories; however, they are not useful for small facilities and challenging situations. Nevertheless, several protocols were proposed using cryovials (Carmichael et al. 2009; Draper and Moens 2009; Diogo et al. 2018, 2019, 2023; Matthews et al. 2018; Rodrigues et al. 2019, 2020; da Costa et al. 2020) for smaller sperm volumes and lower number of males. This is a useful packaging method with high applicability to zebrafish facilities. Therefore, developing a successful cryopreservation protocol must meet both cryobiological principles and practical and objective applications in zebrafish facilities.

The cooling rate is one of the major factors affecting sperm cryopreservation, as discussed previously. In zebrafish, both the dry ice method (Harvey et al. 1982; Morris et al. 2003; Carmichael et al. 2009; Draper and Moens 2009; Rodrigues et al. 2019, 2020; da Costa et al. 2020) and controlled cooling through biofreezers have been performed. The cooling rates tested with programable biofreezers ranged from –5 to –35°C/min in one step (Yang et al. 2007; Hagedorn et al. 2009; Matthews et al. 2018; Diogo et al. 2023) or two-step protocols (Bai et al. 2013; Wang et al. 2015). Interestingly, throughout all the highly variable cooling rates and solutions used, the post-thaw sperm quality seemed similar, with average sperm total motility values around 20%. One of the most relevant bottlenecks in germ cell cryopreservation is the necessity for liquid nitrogen. Liquid nitrogen is a refrigerator medium that can store samples at very low temperatures and is traditionally used for cryopreservation and germ cell storage. However, it is hazardous with a high evaporation rate, requires appropriate equipment and management, and requires a continuous supply, which is one of the main costs of the methodology. Although tissues and bacteria can be stored without significant loss of cell viability at higher temperatures (–80°C), germ cells are more vulnerable and require temperatures below –153°C (Mazur 1984). Nowadays, ultrafreezer storage systems refrigerated at –150°C are commonly found in research facilities dedicated to zebrafish research. Some attempts in mammalian sperm cryopreservation in ultrafreezers (Medrano et al. 2002; Álamo et al. 2005; Batista et al. 2006, 2009; Yavaş and Daskin 2012) have been noticed with encouraging results, which was successfully applied in zebrafish (Diogo et al. 2018, 2019). Since this temperature is below the threshold necessary for germ cell storage stability, it can potentially be an alternative storage method for zebrafish sperm.

The number of larvae obtained from cryopreserved sperm is still poorly reported. However, in terms of viability after thawing, studies that showed positive results used dry ice (Harvey et al. 1982; Morris et al. 2003; Carmichael et al. 2009), crushed dry ice (Ransom and Son 1999; Draper and Moens 2009; Matthews et al. 2018; Silva et al. 2019) a controlled-rate freezer (Yang et al. 2007, 2016; Hagedorn et al. 2012; Bai et al. 2013; Wang et al. 2015; Diogo et al. 2018), and more recently an –150°C ultrafreezer (Diogo et al. 2018, 2019) as a cryopreservation method. The absence of vitrification tests for zebrafish sperm opens an opportunity for studies to be carried out in this area. Studies on sperm vitrification of other fish species are being used with interesting results, as reviewed by Xin et al. (2017).

In Table 1, we can observe the main results of the zebrafish sperm cryopreservation protocols. The results obtained with the cryopreservation of

Table 1. Main results observed in articles published with cryopreservation of zebrafish sperm.

Reference	**Cryoprotective Solution Chosen**	**Motility (%)**	**Viability**	**Fertilization**	**Hatching**
Harvey et al. (1982)	GFR + MET (10%) + PM (15%)	43 ± 12[1]	-	51 ± 36	-
Morris et al. (2003)	HBSS + DMA(10%)	11.6 ± 5.8[1]	-	14 ± 10.1	-
Yang et al. (2007)	HBSS + MET (8%)	78 ± 10[1]	-	33 ± 20	-
Draper and Moens (2009)	GFR + MET (10%) + PM (15%)	-	-	25	-
Hagedorn et al. (2012)	HBSS + MET (8%) + CAT (200 U/mL)	~ 34[1]	~ 66[3]	-	-
Bai et al. (2013)	HBSS+ DMSO (8%)	46.8 ± 6.40[2]	-	-	-
Wang et al. (2015)	HBSS+ DMSO (8%)	28 ± 4[2]	-	62 ± 6	~ 60
Yang et al. (2016)	HBSS + MET (8%)	28 ± 15[1]	45 ± 4[4]	73 ± 21	-
Diogo et al. (2018)	HBSS + DMF (10%)	~ 22[1]	~ 25[4]	~ 90	~5
Matthews et al. (2018)	E400 + RMMB	20 ± 13[1]	-	68 ± 16	-
Silva et al. (2019)	Grayling + MET (8%)	-	-	16 ± 9	-
Diogo et al. (2019)	HBSS + DMF (10%) + BIC (50 mM)	~ 16[2]	~ 8[4]	~ 18	~ 8
da Costa et al. (2020)	GFR + MET (8%) + PM (15%)	22.40 ± 6.03[1]	55.00 ± 8.25[5]	-	-
Rodrigues et al. (2020)	HBSS + MET (8%) + PM (15%)	24.38 ± 9.04[1]	46.4 ± 9.6[6]; 60 ± 10.6[5]	-	-
Yang et al. (2020)	HBSS + MET (5%)	14 ± 4[2]	45 ± 11[4]	-	-
Yang et al. (2022)	HBSS + DMF (10%)	~ 10[1]	~ 50[4]	~ 60	-

Ginsburg Fish Ringer (GFR); Methanol (MET); Powered Milk (PM); Hank's Balanced Salt Solution (HBSS); Catalase (CAT); Dimethylsulfoxide (DMSO); Dimethylformamide (DMF); Bicine (BIC); [1]Subjective Motility Analysis; [2]Objective Motility Analysis (Software CASA); [3]Flow Cytometry (BODIPY/PI); [4]Flow Cytometry (SYBR14/PI); [5]Eosine-Nigrosine Stained; [6]Fluorescence Microscope (Sybr 14/PI). Approximately (~).

zebrafish sperm showed the highest observed sperm motility was 78% (Yang et al. 2007). The highest recorded viability was approximately 66% (Hagedorn et al. 2012), the highest fertilization rate was approximately 90% (Diogo et al. 2018), and the highest hatching rate was approximately 60% (Wang et al. 2015). In general, it is difficult to directly compare the results presented by the different published works, as the techniques used for evaluation, as well as the period and form of observation of fertilization between the studies, were different.

4. Sperm Cryopreservation of Other Aquarium Model Fish Species

When it comes to the cryopreservation of sperm from other species of aquarium model fish, we found a total of five published articles. For goldfish, three articles were published focusing on developing seminal cryopreservation protocols (Kutluyer et al. 2015, 2016; Marinović et al. 2017). As for medaka, three articles were published with the development of cryopreservation protocols (Aoki et al. 1997; Krone and Wittbrodt 1997; Yang et al. 2010).

In a study by Aoki et al. (1997), the researchers tested different cooling rates through several cooling methods (nitrogen vapor, liquid nitrogen, and dry ice) using dimethyl sulfoxide (DMSO) and dimethylformamide (DMF) as cryoprotectants. The best results were obtained using the following methodology: Medaka's sperm was collected in a plastic tube containing 50 µl of fetal bovine serum supplemented with 10% DMF by compressing an isolated testis. The sperm suspension was frozen by holding the tube for 10 or 20 minutes in the vapor phase of the LN at a depth of 9 or 10 cm from the top of a container. The frozen sample was immersed and stored in LN. The sample was thawed at 30°C for one minute. The fertilization rate with this protocol ranged from 96–100% and the hatching rate from 84–100%. In the same year, Krone and Wittbrodt (1997) an alternative medaka sperm cryopreservation protocol was published. In this protocol, sperm is mixed with freezing medium (0.6 M sucrose and 10% DMSO) in a glass capillary. The capillaries containing sperm are successively brought to lower temperatures inside a cryovial on dry ice, and subsequently stored in liquid nitrogen. The authors claim to obtain a fertilization rate above 85% with this protocol. More than a decade later, Yang et al. (2010) tested the cooling rate and cryoprotectant in medaka seminal cryopreservation protocol. First, the authors tested the toxicity of several cryoprotectants at different concentrations, and based on the results, they defined that methanol and 2-methoxyethanol would be tested at a concentration of 10% at cooling rates from 5 to 25°C/minute, compared with 10% DMSO and DMF, based on published previous studies. The highest post-thawing motility (50 ± 10%) was observed at a cooling rate of 10°C/minute with methanol as cryoprotectant. The fertility test of sperm thawed and cryopreserved with 10% methanol at a rate of 10°C/minute showed mean hatching of 70 ± 30%, comparable to that of fresh sperm (86 ± 15%). Thus, this research was able to improve the medaka sperm cryopreservation protocol, reaching levels of effectiveness comparable to the quality of fresh semen.

The research by Kutluyer et al. (2015) aimed to analyze the effect of diluent supplementation with L-methionine in the cryoprotectant solution and the cryopreservation device on post-thawing sperm motility, duration, and DNA damage in goldfish sperm. Briefly, the authors used the diluent Kokura as an extender and the cryoprotective solution composed of DMSO (12.5%) and egg yolk (10%). The cryoprotectant solutions were supplemented with different levels of L-methionine (1 mM, 1.5 mM, 3 mM, and 6 mM). After diluting the sperm sample to the extender, the sample was packaged in 0.25 mL pallets or frozen as pellets in nitrogen vapor (3 cm above liquid nitrogen). The results indicate that the increase in L-methionine concentration caused a significant increase in motility rate and duration. Adding the diluent with L-methionine reduced DNA damage compared to the control group.

Moreover, freezing in the form of pellets, without using straws, also provided better results. Just one year later, Kutluyer et al. (2016) published new research with the same protocol proposed in the previous publication with cryoprotectant solution modification by introducing taurine supplementation (1 mM, 2 mM, or 4 mM). The percentage and duration of sperm motility were significantly increased by taurine. Additionally, sperm motility and the motility period in pellets were higher than in straws. The best concentration of taurine was 4 mM, and the highest post-thaw motility rate (72.50 ± 3.54%) and duration (17.50 ± 0.71 s) were obtained from the extender with 4 mM in pellets. DNA damage was decreased after taurine supplementation. In both studies, the authors verified that using an amino acid with antioxidant action in the cryoprotective solution improves sperm motility and reduces DNA damage, being an interesting alternatives for improving the goldfish sperm cryopreservation protocol. In 2017, Marinović and collaborators tested cryopreservation protocols for the entire testicular tissue and testicular cells isolated from the species. In this study, the authors concluded that cryopreservation of the entire testis is an interesting alternative for preserving stem spermatogonia but not mature spermatozoa.

5. Zebrafish (*Danio rerio*) Oocytes Cryopreservation

Several sperm cryopreservation protocols have been widely studied for many fish species, mainly freshwater salmonids and cyprinids, with efficient and replicable results for several species, making the preservation of male gametes feasible. However, sperm cryopreservation only preserves the paternal genome, which is insufficient to maintain genetic diversity and depends on the genome of maternal origin (Zhang et al. 2008). Thus, improving the cryopreservation of female gametes or embryos is important for preserving the genome of both sexes for the species.

Cryopreservation protocols for mature oocytes and fish embryos are inefficient, as these samples are large and compartmented cells with high intracytoplasmic lipid content and low membrane permeability (Hagedorn et al. 1998). Thus, using ovarian tissue fragments containing immature oocytes or early stages of maturation is the best way to preserve the maternal genome of fish.

For zebrafish, twelve studies were performed evaluating the cryopreservation of ovarian follicles (dissociated oocytes) (Guan et al. 2008, 2010; Zhang et al. 2008; Tsai et al. 2009, 2010; Zampolla et al. 2012) or ovarian tissue fragments (Zampolla et al. 2011; Godoy et al. 2013; Marques et al. 2015; Anil et al. 2018; Marques et al. 2019; Freitas et al. 2023). As we can see, the first studies were focused on the cryopreservation of ovarian follicles, while more recent studies use fragments of ovarian tissue containing oocytes in different stages of maturation in the protocol.

The viability of oocytes after freezing/thawing or vitrification/heating in the cryopreservation processes of ovarian follicles or ovarian tissue is shown in the table below. Most studies evaluated viability by staining with trypan blue. The results presented by the studies show a variation in viability from approximately 10% (Anil et al. 2018) to 90% (Zampolla et al. 2011).

In the cryopreservation methodologies used in zebrafish ovarian follicles, most studies (60%) applied methanol as a cryoprotectant. The most used extender was the KCl buffer, accounting for 44.5% of the total. Regarding the storage devices used, we observed that only straws were tested (0.25 mL and 0.5 mL), and a controlled-rate freezer was the most used cryopreservation method, according to our survey.

By overviewing the available studies focused on the cryopreservation of zebrafish ovarian tissue fragments, it was possible to observe the right range of compounds that were used as cryoprotectants. According to our survey, the most used permeable cryoprotectant was methanol (27, 27%), and sucrose was the non-permeable cryoprotectant (18, 18%) used most often. When we analyzed the data referring to cryopreservation of ovarian tissue, we observed a different approach from studies with already dissociated ovarian follicles, with the L-15 medium being the most used. A variety of devices were also tested for the cryopreservation of ovarian tissue, with straws once again being the most used, followed by the cryovials. We observed that both freezing and vitrification techniques were also tested, with predominance for the vitrification technique in four of the six studies. For slow freezing, a controlled-rate freezer was preferentially used as the cryopreservation method (Table 2).

Regarding the cryopreservation of reproductive tissues and cells of female zebrafish, studies focused on the cryopreservation of immature oocytes (stages I and II) either in stage III of development or fragments of ovarian tissue without dissociating the cells and selecting follicles prior to cryopreservation. The techniques of slow freezing and vitrification have also been applied in this species (Marques et al. 2019). Cryopreservation techniques have been improved, and positive results regarding cell viability after cryopreservation have been obtained. The highest rates of cell viability observed after the cryopreservation process were published by Guan et al. (2008), who reported 88% oocyte viability after freezing of ovarian follicles. The study carried out by Zampolla et al. (2011) obtained approximately 90% oocyte viability after freezing of ovarian tissue.

The results are still limited to a baseline knowledge of the effect of cryopreservation on cells, as the practical use of these cells for fertilization is not yet possible or is unlikely, given that the cryopreservation of mature oocytes has not been successful for those that are ready to be fertilized. Therefore, studies are being carried out focusing on the maturation of oocytes *in vitro* and have shown promising results (Seki et al. 2008; Welch et al. 2017; Silva et al. 2018). This may soon lead to viable immature oocytes after cryopreservation, which can then be used for fertilization after being matured *in vitro*.

Table 2. Viability of oocytes after cryopreservation of ovarian follicles and ovarian tissue of zebrafish.

Reference	Method	Cryoprotective Solution Chosen	Viability (%)	
			TB	**FDA/PI**
Ovarian Follicle				
Guan et al. (2008)	Frz.	KCl Buffer + MET (4 M)	88 ± 1.7	–
Zhang et al. (2008)	Frz.	HBSS + DMSO (2 M)	14.9 ± 2.6	–
Tsai et al. (2009)	Frz.	KCl Buffer + MET (4 M)	69.4 ± 2.2	54.3 ± 3.7
Guan et al. (2010)	Vtr.	KCl Buffer + DMSO (3 M) + MET (2 M) + PG (2 M)	~ 80	–
Tsai et al. (2010)	Frz.	KCl Buffer + MET (4 M)	20.6 ± 3.1	–
Zampolla et al. (2012)	Frz.	FBS (10%) + MET (2 M)	70.2 ± 4.2	–
Fragment of ovarian tissue				
Zampolla et al. (2011)	Frz.	HBSS+ MET (1 M)	~ 90	~ 90
Godoy et al. (2013)	Vtr.	L-15 + MET (1.5 M) + PG (4.5 M)	59.9 ± 18.4	–
Marques et al. (2015)	Vtr.	L-15 + MET (1.5 M) + PG (4.5 M)	–	76.8 ± 2.3
Anil et al. (2018)	Frz.	L-15 + FBS (20%) + MET (2 M)	~ 10	–
Marques et al. (2019)	Vtr.	L-15 + MET (1.5 M) + DMSO (5.5 M) + SUC (0.5 M) + EY (10%)	56.3 ± 19.1	–
Freitas et al. (2023)	Vtr.	L-15 + L-15 + MET (1.5 M) + DMSO (5.5 M) + SUC (0.5 M) - encapsulated in alginate	37.7 ± 3.7	33.0 ± 5.5

Freezing (Frz); Vitrification (Vtr); Methanol (MET); Hank's Balanced Salt Solution (HBSS); Dimethylsulfoxide (DMSO); Propylene glycol (PG); Fetal Bovine Serum (FBS); Leibovitz medium (L-15); Sucrose (SUC); Egg yolk (EY); Tripan Blue (TB); Fluorescein diacetate (FDA); Propidium iodide (PI). Approximately (~).

The most recent article (Freitas et al. 2023) sought to reduce the concentration of permeable cryoprotectant by encapsulating the ovarian tissue fragment with alginate hydrogel. In this study, it was impossible to observe an improvement in the cell's viability with the use of encapsulation. An important observation was made by Freitas et al. (2023), where the use of the alginate hydrogel encapsulation technique reduced the loss of oocytes during the manipulation of ovarian tissue fragments by promoting greater support. This is important, as zebrafish ovaries have a similar cluster of grapes, in which the oocytes easily detach from the tissue. An interesting alternative for encapsulating fragments of zebrafish ovarian tissue in alginate hydrogel would be *in vitro* oocyte maturation, promoting oocyte support.

As we can see, new studies seeking an efficient cryopreservation protocol for mature oocytes or those that can carry out the maturation of immature oocytes after cryopreservation should be the focus of this area of research. Thus, improving existing protocols that present good cell viability after thawing may allow the preservation of the maternal genome of the species.

6. Oocyte Cryopreservation in Other Aquarium Model Species

When we consider other model species of aquarium fish, only one study has been published on the cryopreservation of oocytes with medaka (*Oryzias latipes*). However, two other articles were published by the same research group in Japan to evaluate the permeability of cryoprotectants in oocytes of the species and define some standards for developing the cryopreservation protocol.

In 2005 it was investigated in medaka the immature oocytes' permeability (water and to glycerol) and cryopreservation (ethylene glycol, propylene glycol, and DMSO) in a solution containing sucrose (Valdez Jr et al. 2005). The authors found that the permeability to solutes of mature oocytes is extremely low, making the cryopreservation of this cell type unfeasible; however, immature oocytes already present good permeability in the cryoprotectants ethylene glycol, DMSO, and propylene glycol, which was the cryoprotectant that presented the highest permeability and, therefore, a greater potential to be used in a cryopreservation protocol. The second article published in 2006 (Valdez Jr et al. 2006) found that the permeability of medaka oocytes to water and cryoprotectants was improved by the artificial expression of aquaporin-3 (AQP3), and the oocytes maintained the ability to develop. The third work published in 2013 (Valdez Jr et al. 2013) used the knowledge obtained in previous studies to carry out a vitrification protocol for immature medaka oocytes. However, even using the information previously obtained, such as the artificial use of aquaporin 3 and a cryoprotectant solution with propylene glycol (15%), which was the most promising cryoprotectant, there was no viability of the oocytes after thawing. Immature oocytes with AQP3 expression, which were placed in 0.5 mL straws and vitrified, did not show ice crystal formation during cooling, but during heating, they did and lost viability.

These studies demonstrate that despite finding more permeable cryoprotectants and methods that further increase permeability to the plasma membrane, which is one of the challenges in cryopreservation of fish oocytes and embryos. Moreover, the use of vitrification to reduce the chances of intracellular ice generation was still unsuccessful. These facts demonstrate that cryopreservation of female genetic backgrounds, such as oocytes, is necessary not only in zebrafish but also in other aquarium fish models.

7. Cryopreservation of Embryos, Cells, and Tissues in Zebrafish

Seminal cryopreservation is already a reality for several species of fish, and despite the difficulties encountered in zebrafish semen cryopreservation, advances in this area have been made. On the other hand, the cryopreservation of mature fish oocytes is not a reality, being possible only through the successful cryopreservation of immature oocytes, which requires further maturation techniques (Seki et al. 2008). Therefore, *in vitro* maturation protocols for these oocytes must be improved so that the use of these samples for subsequent fertilization becomes an applicable reality and the preservation of maternal genetic material possible.

Embryo cryopreservation would be the most direct way to preserve diploid cells (paternal and maternal DNA); however, in fish, there are difficulties in obtaining protocols that are efficient and replicable. In view of these limitations in embryo cryopreservation, primordial germ cells (PGCs) and spermatogonia stem (SSC) are an alternative for cryopreservation, in which the complete genetic material of an individual is preserved and when transplanted into another animal, due to their genetic plasticity, these cells can differentiate into spermatozoa or oocytes, depending on the somatic environment in which they develop (Okutsu et al. 2006; Lee et al. 2013). However, cryopreservation of spermatogonia may be a more convenient approach than PGCs to preserve fish genetic resources, as mass isolation of spermatogonia from donor testes is much easier than isolation of PGCs from larval gonads in an initial stage (Yoshizaki et al. 2011). In this context, the constitution of a genetic bank through the preservation of spermatogonia stem, and subsequent reconstitution through mediated propagation (xenogenesis) would be a solid strategy to conserve and reconstitute endangered fish, including preserving important genetic components, such as germ plasma and mitochondrial DNA, making it possible to transfer genetic information to future generations (Reunov et al. 2000; Cinalli et al. 2008).

There are ten studies on zebrafish embryo cryopreservation (Harvey 1983; Zhang et al. 1993; Zhang and Rawson 1996; Janik et al. 2000; Robles et al. 2004; Martinez et al. 2005; Martínez-Páramo et al. 2009; Higaki et al. 2013; Khosla et al. 2020; Castro et al. 2021). We found eight studies focusing on blastomere preservation (Harvey 1983; Kopeika et al. 2005; Cardona-Costa and García-Ximénez 2007; Lin et al. 2009a, b, c; Martínez-Páramo et al. 2009; Cardona-Costa et al. 2009). The cryopreservation of primordial germ cells (PGCs) was applied in seven studies with the species (Higaki et al. 2009, 2010, 2013; Riesco et al. 2012; Riesco and Robles 2013; Marinović et al. 2018, 2019). As we can observe, the beginning of the studies with cryopreservation of cells other than gametes occurred through embryos and blastomeres' cryopreservation attempts. Due to failures and lack of efficacy in the protocols, studies began to be carried out with the cryopreservation of primordial germ cells (PGCs). This strategy aimed to preserve the genetic material to perform surrogate production, transferring and implanting the PGCs in recipient animals in the future.

Through the analysis of the available investigation in zebrafish embryo cryopreservation, we initially observed a wide variety of methodologies using different extenders, cryoprotectants, devices, and techniques (Table 3). DMSO and methanol was the most utilized cryoprotectant (21, 21%), while sucrose was the most used non-permeable cryoprotectant. Among the extenders, filtered water and Ringer's solution were utilized in more than 60% of the studies. A 0.5 mL straw was used in half of the tests, followed by a 0.25 mL straw. As for the cryopreservation method, we observed that most studies used liquid nitrogen in vitrification tests, while slow freezing was preferably performed in a controlled-rate freezer, followed by refrigerated alcohol, liquid nitrogen vapor, and ice.

Several cryoprotectants have already been tested for cryopreservation of blastomeres, with DMSO being the most used in general (38%). Sucrose was the most commonly used non-permeable cryoprotectant. Unlike what was observed in other tissues and cells analyzed, fetal bovine serum solution was the most widely used extender for cryopreservation of blastomeres. As for cryopreservation

Table 3. Cell viability after cryopreservation of primordial germ cells (PGCs), blastomeres, and zebrafish embryos.

References	Method	Cryoprotective Solution Chosen	Viability (%)
Embryo			
Zhang and Rawson (1996)	Vtr	FW + BD (2 M) + PD (3 M) + PEG (0.15 M)	~ 32[1]
Martínez-Páramo et al. (2009)	Vtr	DMSO (5 M) + MET (2 M) + EG (1 M) + SUC (10%)	~ 51[1]
Khosla et al. (2020)	Vtr	PG (10.6 M) + MET (4.8 M)	~ 40[1]
Castro et al. (2021)	Vtr	L15 + PG (3 M) + MET (1.5 M) + PEN (0.09 mM) + DIH (0.06 M) +EY (15 mL) + MEL (1 μM)	0[1]
Blastomeres			
Kopeika et al. (2005)	Frz	PBS + FBS (10%) + DMSO (1.4 M)	25.2 ± 4.3[2]
Cardona-Costa and García-Ximénez (2007)	Vtr	HBSS + FBS (20%) + DMSO (5 M)	~ 93.4[2]
Lin et al. (2009a)	Frz	PBS + FBS (10%) + DMSO (1.5 M) + SUC (0.1 M)	70.2 ± 3.2[2]
Cardona-Costa et al. (2009)	Vtr	HBSS + FBS (10%) + DMSO (5 M)	~ 90%[2]
Martínez-Páramo et al. (2009)	Frz	DMSO (2 M) + AFP (10 mg/mL)	54.2 ± 0.97[3]
Primordial Germ Cells - PGCs			
Higaki et al. (2009)	Vtr	BSA (0.7%) + EG (3 M) + PG (3 M)	~ 25[2]
Higaki et al. (2010)	Vtr	BSA (0.7%) + EG (5 M) + SUC (0.5 M)	~ 40[2]
Riesco et al. (2012)	Frz	DMSO (5 M) + EG (1 M) + PVP (4%)	92.3 ± 13.9[2]
Higaki et al. (2013)	Vtr	FBS (0.7%) + EG (3 M) + DMSO (2 M) + SUC (0.5 M)	~ 81[2]
Spermatogonia			
Marinović et al. (2018)	Vtr	DMSO (3 M) + PG (3 M)	58 ± 9[2]
Marinović et al. (2019)	Frz	DMSO (1.3 M)	~ 60[2]

Freezing (Frz.); Vitrification (Vtr.); Filtered Water (FW); Methanol (MET); 1,4-Butanediol (BD); 1,3-Propanediol (PD); Polyethylene Glycol (PEG); Dimethylsulfoxide (DMSO); Ethylene Glycol (EG); Sucrose (SUC); Melatonine (MEL); Phosphate Buffered Saline (PBS); Fetal Bovine Serum (FBS); Hank's Balanced Salt Solution (HBSS); Antifreeze Protein (AFP); Bovine Serum Albumin (BSA); Propylene glycol (PG); Polyvinylpyrrolidone (PVP); Penicillin (PEN); Dihydrostreptomycin (DIH); Egg Yolk (EY); Melatonin (MEL). Viability (%) = [1]Morphological survival; [2]Trypan blue; [3]Diacetato fluorescein - FDA/Propidium iodide - PI. Approximately (~).

devices and maintenance of the samples, we observed that 0.25 mL straws were the most used. However, it is interesting that the micro-droplet technique was the second most used. This was the first time this technique appeared in the zebrafish cryopreservation protocols, and we did not observe its use in any other sample types besides blastomeres. Regarding the cryopreservation method, we observed greater use of liquid nitrogen in vitrification tests and controlled-rate freezers for slow freezing, with most tests being carried out by slow freezing.

The analysis of cryopreservation protocols for PGCs revealed a variety of cryoprotectants, with DMSO being the most used. We observed that most protocols used protein-based extenders. Many studies used bovine serum albumin; the remainder utilized fetal bovine serum or HBSS. Among the devices used to cryopreserve PGCs, a nylon mesh was present in most studies. Regarding the cryopreservation method, we observed a greater use of liquid nitrogen, indicating that vitrification was the preferred technique in most studies.

The viability of embryos, blastomere, PGCs, and spermatogonia after the cryopreservation processes found in the available research is highly variable (Table 3).

Although the results obtained in these studies with different types of cells and tissues were quite variable and, in many cases, had few positive responses, cryopreservation of the zebrafish embryo was the only one that was unsuccessful. Unlike cryopreservation of mammalian embryos, cryopreservation of fish embryos has yielded a few positive results. For the protocols applied to date in the cryopreservation of zebrafish embryos, only Zhang and Rawson (1996) and Martínez-Páramo et al. (2009) observed the morphological viability of embryos after the cryopreservation process; however, they were unable to obtain embryo development and hatching. This shows that cryopreservation of embryos for this species is not yet effective. This is reflected in the reduction of studies evaluating embryo cryopreservation over time since the protocols had not been successful due to the lack of cell viability. Studies in this area have become increasingly scarce, and there is a need for attempts at cryopreservation of other sample types, such as gametes and primordial cells. Cryopreservation protocols that have been tested so far have already shown efficiency in blastomeres and primordial germ cells. The studies by Cardona-Costa and García-Ximénez (2007) reported approximately 93% viability of blastomeres after vitrification. A study by Riesco et al. (2012) reported approximately 92% viability of PGCs post-thawing in zebrafish. These are the highest levels of cell viability reported after cryopreservation. Primordial cells have limitations regarding their use after cryopreservation. The maturation and implantation techniques for these cells in recipient materials are still very new and have many application difficulties. This is a field of research that must still evolve over the upcoming years, and must be explored more frequently, considering that the studies carried out so far are already efficient in terms of the cell viability of these materials after cryopreservation. Recently, two studies published by Marinović et al. (2018, 2019) demonstrated the possibility of cryopreservation in male zebrafish by developing protocols for the preservation of spermatogonia. The results showed > 50% rate of viability of spermatogonia after the samples were heated. The protocol was based on the cryopreservation of the zebrafish testis with the objective of transplanting these types of cells into recipient animals. This technique would make it easier to preserve the female and diploid genomes of the species.

Fish embryo, PGCs, and spermatogonia cryopreservation protocols are far from being effective and replicable, and for zebrafish, this is no different. Although studies have shown some PGCs post-thawing viability, there is still much to be improved until this method can be used for transplantation in recipient animals. The preservation of

PGCs and spermatogonia requires much investigation by the scientific community so that, in the future, it can be commonly applied in fish facilities. Preserving embryos is so challenging that if no new technology emerges immediately, it will be on standby until our technology evolves and provides adequate conditions for this research. Undeniably, the current focus of research is on PGCs cryopreservation and transplantation of these cells.

8. Cryopreservation of Embryos, Cells, and Tissues in Other Aquarium Model Species

When we deal with the cryopreservation of embryos, PGCs, and blastomeres in other aquatic aquarium species, we observed few publications with few advances regarding the determination of efficient and replicable cryopreservation protocols. When we searched for publications with alternative tissue samples for medaka, we found one publication testing embryo cryopreservation protocols (Valdez Jr et al. 2005), one publication testing cryopreservation of PGCs (Seki et al. 2017), and one with blastomeres (Strüssmann et al. 1999). For goldfish (*Carassius auratus*), we found a publication with the cryopreservation of testicular tissue for the preservation of spermatogonia and spermatozoa (Marinović et al. 2017) and a publication with the cryopreservation of blastomeres (Kusuda et al. 2004). For Siamese fighting fish (*Betta splendens*), only one study testing embryo cryopreservation has been published (Macoretta and Miranda 2021). In medaka, it is possible to find research in the cryopreservation of blastomeres, embryos and testicular tissues fragments. In 1999, Strüssmann and collaborators tested the survival of blastomeres isolated from three fish species, including the medaka, in relation to DMSO toxicity and cryopreservation damage. The authors observed that older blastomeres, cryopreserved with a slower cooling rate, with DMSO (9–18%), showed greater viability after thawing, reaching a percentage above 30% for medaka.

In a study by Valdez Jr et al. (2005), the sensitivity to the cold of medaka embryos at different stages of development was observed to enable the determination of the most suitable developmental stage for the application of a cryopreservation protocol. Embryos at stages 2–4 cells, 8–16 cells, morula, blastula, and early gastrula were suspended in Hanks' solution and cooled at different temperatures. In conclusion, the authors highlighted that chilling injury is less severe in early-stage embryos, and this stage can be potentially used in cryopreservation studies in the future. In the research by Seki et al. (2017), they developed a protocol for the cryopreservation of medaka testicular tissue fragments, aiming to preserve spermatogonia stem cells such as PGCs. The authors achieved success with viable PGCs post-thawing when a vitrification protocol was applied using ethylene glycol (30%, v/v) or propylene glycol (30%, v/v), with Ficoll PM-70 (21%, w/v) and sucrose (0.35 M). This study was a great advance for the preservation of different medaka lineages since, until this research was performed, no other cryopreservation protocol was efficient and practical.

In studies that focused on goldfish blastomeres, PGCs and testicular tissue cryopreservation were observed. Kusuda et al. (2004) developed a blastomere cryopreservation protocol using the Eagle's minimum essential medium (MEM) extender, containing DMSO (10%) and fetal bovine serum (10%). The authors used 0.25 mL French straws and carried out the freezing with a programmable freezer. The methodology applied by these authors promoted 44% of blastomeres' survival post-thawing. Cryopreserved blastomeres were transplanted into the blastula of triploid carp (*C. a. longsdorfii*), reproducing gynogenetically in nature. Transplanted blastomeres were identified histologically in various organs derived from all three germ layers. A primordial germ cell differentiated from a cryopreserved blastomere was detected in one of the 32 chimeric fish examined. These results suggest that blastomeres that survive after cryopreservation retain their pluripotency and can differentiate into somatic and germ cell lines. In 2017, Marinović and collaborators tested cryopreservation protocols for the entire testicular tissue and testicular cells isolated from the species, aiming to preserve spermatogonia steam cells. Additionally, the effects of three cryoprotectants (DMSO, methanol, and ethylene glycol) at three concentrations (1 M, 2 M, and 3 M) on post-thawing cell viability were evaluated. The authors observed greater survival of testicular germ cells than spermatozoa and greater survival when DMSO and ethylene glycol were used as cryoprotectants instead of methanol. The authors concluded that the cryopreservation of the entire testicular tissue is indeed a valid alternative method to the cryopreservation of testicular cells, although further studies on testicular cell function are needed.

In Siamese fighting fish (*Betta splendens*), only one study evaluating the cooling of embryos at low temperatures was found. The research by Macoretta and Miranda (2021) aimed at developing a cooling protocol for Siamese fighting fish embryos, as a first step toward a future cryopreservation protocol. The authors observed the best results when the cryoprotectant solution was composed of methanol (0.75 M) and sucrose (0.25 M) and a cooling of 5°C was applied for 9 hours. Under these conditions these authors observed 77% of hatching rate, 52% of larvae with a normal appearance phenotype and inflated swim bladder. In the same publication the use of methanol (0.375 M) and sucrose (0.125) as cryoprotectant solution with cooling at 14°C for 24 hours obtaining 88% hatching and 81% larvae with mild malformation phenotype. This work allowed to establish base conditions for further studies on the development of cryopreservation protocols for Siamese fighting fish embryos.

9. Final Considerations

Several cryopreservation protocols for reproductive cells and tissues of aquarium fish species considered biological models for science, especially zebrafish, have already been developed. Due to the advances obtained, particularly in zebrafish sperm cryopreservation, these preservation methodologies were applied in the biggest research centers but also medium and small-size facilities. Zebrafish sperm cryopreservation presents challenges that lead to high post-thaw variability. Sperm cryopreservation is already used to maintain germplasm banks of these species wild type, mutant, and transgenic (reporter) lines since sperm cryopreservation is the most reliable method established for this species. It is possible to observe the efforts of researchers to optimize cryopreservation methods to improve post-thawing cell quality, and this trend should continue in the future. Therefore, the scientific community started to focus on optimizing this species' husbandry and assisted reproduction practices toward the standardization of practices. These efforts aim to improve experimental reliability and replicability in this model species. Nevertheless, there are still poor reporting on zebrafish hatching rate, larvae survival and the quality of the progeny sired by cryopreserved sperm. The stability of the zebrafish mutant and reporter lines, as well as their polymorphisms is still to be investigated in animals sired by cryopreserved sperm which requires further investigation. This matter is particularly relevant for the end users of this cryopreservation technology which are the zebrafish facilities managers.

The development of cryopreservation methods that allow the preservation of diploid genomes is still one of the most significant challenges in this research field since the cryopreservation of fish embryos continues to be unsuccessful. Cryopreservation of primordial cells shows promising results and can be potentially used as an alternative to embryo cryopreservation. Therefore, more research efforts are expected in the upcoming years in PGC cryopreservation and transplantation for surrogate production. The cryopreservation of the female genome already presents some positive results; however, the successes obtained until this moment are only focused on immature oocyte cryopreservation, which makes its use impossible after thawing the samples for fertilization. Therefore, studies focusing on developing *in vitro* oocyte maturation or establishing new cryopreservation protocols to obtain viable mature oocytes are highly needed in future studies.

Beyond zebrafish, the use of small-bodied fishes as animal models, such as killifish and Siamese fighting fish, has been increasingly used in research areas, such as the genes associated to specific behaviors, pigmentation patterns, and hereditability. However, their use in research is far lower when compared to zebrafish, therefore it is expected that the maturity of the cryopreservation methods development is considerably lower when compared to zebrafish. In fact, several studies focus on investigating the chilling sensitivity of cells and tissues to support future studies in cryopreservation. Nevertheless, it was possible to identify some advances in cryopreservation of other aquarium fish species in different types of cells and tissues, particularly in sperm, PGCs, and testicular tissue cryopreservation.

Due to the importance of zebrafish as a model species in science, we are confident in the efforts of researchers to improve protocols and develop efficient methods that can be commonly applied in research centers of different dimensions worldwide. With the development of new technologies and new information being published constantly, it will only be a matter of time before the cryopreservation protocols reach the highest possible efficiency and guarantee the preservation of lineages and strains of the different and valuable genotypes, which are fundamental for all academic communities.

References

Adhish, M. and I. Manjubala. 2023. Effectiveness of zebrafish models in understanding human diseases—A review of models. Heliyon 9: e14557.

Álamo, D., M. Batista, F. González, N. Rodríguez, G. Cruz, F. Cabrera et al. 2005. Cryopreservation of semen in the dog: Use of ultra-freezers of –152°C as a viable alternative to liquid nitrogen. Theriogenology 63: 72–82.

Amores, A., J. Catchen, A. Ferrara, Q. Fontenot and J.H. Postlethwait. 2011. Genome evolution and meiotic maps by massively parallel DNA sequencing: spotted gar, an outgroup for the teleost genome duplication. Genetics 188: 799 LP–808.

Anil, S., D. Rawson and T. Zhang. 2018. Development of molecular markers for zebrafish (*Danio rerio*) ovarian follicle growth assessment following *in-vitro* culture in cryopreservation studies. Cryobiology 83: 75–83.

Aoki, K., M. Okamoto, K. Tatsumi and Y. Ishikawa. 1997. Cryopreservation of medaka spermatozoa. Zool. Sci. 14: 641–644.

Arunachalam, M., M. Raja, C. Vijayakumar, P. Malaiammal and R.L. Mayden. 2013. Natural history of zebrafish (*Danio rerio*) in India. Zebrafish 10: 1–14.

Asturiano, J.F., E. Cabrita and Á. Horváth. 2017. Progress, challenges and perspectives on fish gamete cryopreservation: a mini-review. Gen. Comp. Endocrinol. 245: 69–76.

Azevedo, A.S., S. Sousa, A. Jacinto and L. Saúde. 2012. An amputation resets positional information to a proximal identity in the regenerating zebrafish caudal fin. BMC Dev. Biol. 12: 24.

Bai, C., X. Wang, G. Lu, L. Wei, K. Liu, H. Gao et al. 2013. Cooling rate optimization for zebrafish sperm cryopreservation using a cryomicroscope coupled with SYBR14/PI dual staining. Cryobiology 67: 117–123.

Bailey, G.S., D.E. Williams and J.D. Hendricks. 1996. Fish models for environmental carcinogenesis: the rainbow trout. Environ. Health Perspect. 104: 5–21.

Batista, M., D. Alamo, F. González, M.G. Cruz and A. Gracia. 2006. Influence of the freezing technique (nitrogen liquid vs ultrafreezer of –152°C) and male-to-male variation over the semen quality in Canarian Mastiff breed dogs. Reprod. Domest. Anim. 41: 423–428.

Batista, M., T. Niño, D. Alamo, N. Castro, M. Santana, F. González et al. 2009. Successful artificial insemination using semen frozen and stored by an ultrafreezer in the Majorera goat breed. Theriogenology 71: 1307–1315.

Benson, E.E., F. Betsou, B.J. Fuller, K. Harding and O. Kofanova. 2013. Translating cryobiology principles into trans-disciplinary storage guidelines for biorepositories and biobanks: A concept paper. Cryo-Lett. 34: 277–312.

Brenner, S., G. Elgar, R. Sanford, A. Macrae, B. Venkatesh and S. Aparicio. 1993. Characterization of the pufferfish (*Fugu*) genome as a compact model vertebrate genome. Nature 366: 265–268.

Cabrita, E., C. Sarasquete, S. Martínez-Páramo, V. Robles, J.,Beirão, S. Pérez-Cerezales et al. 2010. Cryopreservation of fish sperm: applications and perspectives. J. Appl. Ichthyol. 26: 623–635.

Cardeira, J., P.J. Gavaia, I. Fernández, I.F. Cengiz, J. Moreira-Silva, J.M. Oliveira et al. 2016. Quantitative assessment of the regenerative and mineralogenic performances of the zebrafish caudal fin. Sci. Rep. 6: 39191.

Cardona-Costa, J. and F. García-Ximénez. 2007. Vitrification of zebrafish embryo blastomeres in microvolumes. Cryo-Lett. 28: 303–309.

Cardona-Costa, J., M. Francisco-Simão and F. García-Ximénez. 2009. Can vitrified zebrafish blastomeres be used to obtain germ-line chimaeras? Cryo-Lett. 30: 422–428.

Carmichael, C., M. Westerfield and Z.M. Varga. 2009. Cryopreservation and in vitro fertilization at the zebrafish international resource center. Methods Mol. Biol. 546: 45–65.

Castro, P.L., A.L.J. Ferraz, J.G. Patil and R.P. Ribeiro. 2021. Use of melatonin as an inhibitor of apoptotic process for cryopreservation of zebrafish (*Danio rerio*) embryos. Braz. J. Biol. 82: e241081.

Cinalli, R.M., P. Rangan and R. Lehmann. 2008. Germ cells are forever. Cell 132: 559–562.

Choi, T.Y., T.I. Choi, Y.R. Lee, S.K. Choe and C.H. Kim. 2021. Zebrafish as an animal model for biomedical research. Exp. Mol. Med. 53: 310–317.

da Costa, B.B., D.L. de Oliveira, R.B. Rodrigues, I.C. Gomes and D.P. Streit Jr. 2020. Morphological abnormalities in zebrafish cryopreserved sperm. Cryobiology 97: 235–237.

Dahm, R. and R. Geisler. 2006. Learning from small fry: The zebrafish as a genetic model organism for aquaculture fish species. Mar. Biotechnol. 8: 329–345.

de Carvalho, A.F.S., S.E. Ramos, T.S.G. de Carvalho, Y.C.P. de Souza, M.G. Zangeronimo, R. Pereira et al. 2014. Efficacy of fish embryo vitrification protocols in terms of embryo morphology—A systematic review. Cryo-Lett. 35: 361–370.

de Freitas, T.R., A.G. Galuppo, L.S. Marques, R.B. Rodrigues, M.P. Atehortúa, T.S. França et al. 2023. Hydrogel encapsulation as a handling and vitrification tool for zebrafish ovarian tissue. Theriogenology 198: 153–163.

Depincé, A., A. Gabory, K. Dziewulska, P.Y. Le Bail, H. Jammes and C. Labbé. 2020. DNA methylation stability in fish spermatozoa upon external constraint: Impact of fish hormonal stimulation and sperm cryopreservation. Mol. Reprod. Dev. 87: 124–134.

Diogo, P., G. Martins, I. Quinzico, R. Nogueira, P.J. Gavaia and E. Cabrita. 2018. Electric ultrafreezer (–150°C) as an alternative for zebrafish sperm cryopreservation and storage. Fish Physiol. Biochem. 44: 1443–1455.

Diogo, P., G. Martins, R. Nogueira, A. Marreiros, P.J. Gavaia and E. Cabrita. 2019. Cryoprotectants synergy improve zebrafish sperm cryopreservation and offspring skeletogenesis. Cryobiology 91: 115–127.

Diogo, P., G. Martins, M. Simão, A. Marreiros, A.C. Eufrásio, E. Cabrita et al. 2023. Type I Diabetes in zebrafish reduces sperm quality and increases insulin and glucose transporter transcripts. Int. J. Mol. Sci. 24: 7035.

Draper, B.W. and C.B. Moens. 2009. A high-throughput method for zebrafish sperm cryopreservation and *in vitro* fertilization. J. Vis. Exp. 29: 1–5.

Driever, W., L. Solnica-Krezel, A.F. Schier, S.C. Neuhauss, J. Malicki, D. Stemple et al. 1996. A genetic screen for mutations affecting embryogenesis in zebrafish. Development 123: 37–46.

Duarte, C.M., N. Marbá and M. Holmer. 2007. Rapid domestication of marine species. Science 316: 382–383.

Engeszer, R.E., L.B. Patterson, A.A. Rao and D.M. Parichy. 2007. Zebrafish in the wild: a review of natural history and new notes from the field. Zebrafish 4: 21–40.

Ericsson, A.C., M.J. Crim and C.L. Franklin. 2013. A brief history of animal modeling. Mo. Med. 110: 201.

Franěk, R., T. Tichopád, M. Fučíková, C. Steinbach and M. Pšenička. 2019. Production and use of triploid zebrafish for surrogate reproduction. bioRxiv 608687.

Gao, D. and J.K. Critser. 2000. Mechanisms of cryoinjury in living cells. ILAR Journal 41: 187–196.

Gemberling, M., T.J. Bailey, D.R. Hyde and K.D. Poss. 2013. The zebrafish as a model for complex tissue regeneration. Trends Genet. 29: 611–620.

Gerhard, G.S. 2003. Comparative aspects of zebrafish (*Danio rerio*) as a model for aging research. Exp. Gerontol. 38: 1333–1341.

Gjedrem. T. 2005. Selection and Breeding Programs in Aquaculture. Selection and Breeding Programs in Aquaculture. Springer, Netherlands.

Godoy, L.C., D.P. Streit Jr., T. Zampolla, A. Bos-Mikich and T. Zhang. 2013. A study on the vitrification of stage III zebrafish (*Danio rerio*) ovarian follicles. Cryobiology 67: 347–354.

Goessling, W. and T.E. North. 2014. Repairing quite swimmingly: advances in regenerative medicine using zebrafish. Dis. Model. Mech. 7: 769–776.

Graham, C., M.A. Von Keyserlingk and B. Franks. 2018. Zebrafish welfare: Natural history, social motivation and behaviour. Appl. Anim. Behav. Sci. 200: 13–22.

Guan, M., D.M. Rawson and T. Zhang. 2008. Cryopreservation of zebrafish (*Danio rerio*) oocytes using improved controlled slow cooling protocols. Cryobiology 56: 204–208.

Guan, M., D.M. Rawson and T. Zhang. 2010. Cryopreservation of zebrafish (*Danio rerio*) oocytes by vitrification. Cryo-Lett. 31: 230–238.

Haffter, P., M. Granato, M. Brand, M.C. Mullins, M. Hammerschmidt, D.A. Kane et al. 1996. The identification of genes with unique and essential functions in the development of the zebrafish, *Danio rerio*. Development 123: 1–36.

Hagedorn, M., F.W. Kleinhans, D. Artemov and U. Pilatus. 1998. Characterization of a major permeability barrier in the zebrafish embryo. Biol. Reprod. 59: 1240–1250.

Hagedorn, M., J. Ricker, M. McCarthy, S.A. Meyers, T.R. Tiersch, Z.M. Varga et al. 2009. Biophysics of zebrafish (*Danio rerio*) sperm. Cryobiology 58: 12–19.

Hagedorn, M., M. McCarthy, V.L. Carter and S.A. Meyers. 2012. Oxidative stress in Zebrafish (*Danio rerio*) sperm. PLoS One 7: e39397.

Hagiwara, M., J.H. Choi, R.V. Devireddy, K.P. Roberts, W.F. Wolkers, A. Makhlouf et al. 2009. Cellular biophysics during freezing of rat and mouse sperm predicts post-thaw motility. Biol. Reprod. 81: 700–706.

Harvey, B., R.N. Kelley and M.J. Ashwood-Smith. 1982. Cryopreservation of zebra fish spermatozoa using methanol. Can. J. Zool. 60: 1867–1870.

Harvey, B. 1983. Cooling of embryonic cells, isolated blastoderms, and intact embryos of the zebra fish *Brachydanio rerio* to –196°C. Cryobiology 20: 440–447.

Higaki, S., K. Mochizuki, H. Baba, Y. Akashi, E. Yamaha, S. Katagiri et al. 2009. Feasibility of cryopreservation of zebrafish (*Danio rerio*) primordial germ cells by whole embryo freezing. Jpn. J. Vet. Res. 57: 119–128.

Higaki, S., K. Mochizuki, Y. Akashi, E. Yamaha, S. Katagiri and Y. Takahashi. 2010. Cryopreservation of primordial germ cells by rapid cooling of whole zebrafish (*Danio rerio*) embryos. J. Reprod. Dev. 56: 212–218.

Higaki, S., Y. Kawakami, Y. Eto, E. Yamaha, M. Nagano, S. Katagiri et al. 2013. Cryopreservation of zebrafish (*Danio rerio*) primordial germ cells by vitrification of yolk-intact and yolk-depleted embryos using various cryoprotectant solutions. Cryobiology 67: 374–382.

Hill, A.J., H. Teraoka, W. Heideman and R.E. Peterson. 2005. Zebrafish as a model vertebrate for investigating chemical toxicity. Toxicol. Sci. 86: 6–19.

Housden, B.E., M. Muhar, M. Gemberling, C.A. Gersbach, D.Y.R. Stainier, G. Seydoux et al. 2016. Loss-of-function genetic tools for animal models: cross-species and cross-platform differences. Nat. Rev. Genet. 18: 24.

Houston, S.J., V. Karalazos, J. Tinsley, D.R. Tocher, B.D. Glencross and Ó. Monroig. 2022. A comparison of regression models for defining EPA+ DHA requirements using the gilthead seabream (*Sparus aurata*) as a model species. Aquaculture 556: 738308.

Howe, K., M.D. Clark, C.F. Torroja, J. Torrance, C. Berthelot, M. Muffato et al. 2013. The zebrafish reference genome sequence and its relationship to the human genome. Nature 496: 498–503.

Janik, M., F.W. Kleinhans and M. Hagedorn. 2000. Overcoming a permeability barrier by microinjecting cryoprotectants into zebrafish embryos (*Brachydanio rerio*). Cryobiology 41: 25–34.

Jing, R., C. Huang, C. Bai, R. Tanguay and Q. Dong. 2009. Optimization of activation. Collection, dilution, and storage methods for zebrafish sperm. Aquaculture 290: 165–171.

Khosla, K., J. Kangas, Y. Liu, L. Zhan, J. Daly, M. Hagedorn et al. 2020. Cryopreservation and laser nanowarming of zebrafish embryos followed by hatching and spawning. Adv. Biosyst. 4: 2000138.

Kinth, P., G. Mahesh and Y. Panwar. 2013. Mapping of zebrafish research: a global outlook. Zebrafish 10: 510–517.

Kopeika, J., T. Zhang, D.M. Rawson and G. Elgar. 2005. Effect of cryopreservation on mitochondrial DNA of zebrafish (*Danio rerio*) blastomere cells. Mutat. Res.- Fundam. Mol. Mech. Mutagen. 570: 49–61.

Krone, A. and J. Wittbrodt. 1997. A simple and reliable protocol for cryopreservation of medaka *Oryzias latipes* spermatozoa. Fish Biol. J. Medaka 9: 47–48.

Kusuda, S., T. Teranishi, N. Koide, T. Nagai, K. Arai and E. Yamaha. 2004. Pluripotency of cryopreserved blastomeres of the goldfish. J. Exp. Zool. A Comp. Exp. Biol. 301: 131–238.

Kutluyer, F., F. Öğretmen and B.E. Inanan. 2015. Effects of semen extender supplemented with l-methionine and packaging methods (straws and pellets) on post-thaw goldfish (*Carassius auratus*) sperm quality and DNA damage. Cryo-Lett. 36: 336–343.

Kutluyer, F., F. Öğretmen and B.E. Inanan. 2016. Cryopreservation of goldfish (*Carassius auratus*) spermatozoa: effects of extender supplemented with taurine on sperm motility and DNA damage. Cryo-Lett. 37: 41–46.

Laizé, V., P.J. Gavaia and M.L. Cancela. 2014. Fish: a suitable system to model human bone disorders and discover drugs with osteogenic or osteotoxic activities. Drug Discov. Today Dis. Model. 13: 29–37.

Lawrence, C. 2016. New frontiers for zebrafish management. Methods Cell Biol. 135: 483–508.

Lee, S., Y. Iwasaki, S. Shikina and G. Yoshizaki. 2013. Generation of functional eggs and sperm from cryopreserved whole testes. PNAS 110: 1640–1645.

Lin, C., T. Zhang and D.M. Rawson. 2009a. Cryopreservation of zebrafish (*Danio rerio*) blastomeres by controlled slow cooling. Cryo-Lett. 30: 132–141.

Lin, C., E. Spikings, T. Zhang and D.M. Rawson. 2009b. Effect of chilling and cryopreservation on expression of Pax genes in zebrafish (*Danio rerio*) embryos and blastomeres. Cryobiology 59: 42–47.

Lin, C., E. Spikings, T. Zhang and D.M. Rawson. 2009c. Housekeeping genes for cryopreservation studies on zebrafish embryos and blastomeres. Theriogenology 71: 1147–1155.

Lieschke, G.J. and P.D. Currie. 2007. Animal models of human disease: Zebrafish swim into view. Nat. Rev. Genet. 8: 353–367.

Liu, J., Y. Zhou, X. Qi, J. Chen, W. Chen, G. Qiu et al. 2017. CRISPR/Cas9 in zebrafish: an efficient combination for human genetic diseases modeling. Hum. Genet. 136: 1–12.

Lleras Forero, L., R. Narayanan, L.F.A. Huitema, M. VanBergen, A. Apschner, J. Peterson-Maduro et al. 2018. Segmentation of the zebrafish axial skeleton relies on notochord sheath cells and not on the segmentation clock. Elife 7: e33843.

Macoretta, C.L. and L.A. Miranda. 2021. Cooling of Siamese fighting fish *Betta splendens* (Teleostei, Osphronemidae) embryos at low temperatures. Cryobiology 101: 78–86.

Marinović, Z., J. Lujić, E. Kása, G. Bernáth, B. Urbányi and Á. Horváth. 2017. Cryosurvival of isolated testicular cells and testicular tissue of tench *Tinca tinca* and goldfish *Carassius auratus* following slow-rate freezing. Gen. Comp. Endocrinol. 245: 77–83.

Marinović, Z., Q. Li, J. Lujić, Y. Iwasaki, Z. Csenki, B. Urbányi et al. 2018a. Criopreservação de testículos e transplante de espermatogônias como ferramenta para reconstituição da linha do peixe-zebra. Criobiologia 85: 145–146.

Marinović, Z., J. Lujić, E. Kása, Z. Csenki, B. Urbányi and Á. Horváth. 2018b. Criopreservação de espermatogônias de peixe-zebra por meio de resfriamento ultrarrápido imerso em agulha de testículos inteiros. J. Vis. Exp. 133: e56118.

Marinović, Z., J. Lujić, E. Kása, Z. Csenki, B. Urbányi and Á. Horváth. 2018. Cryopreservation of zebrafish spermatogonia by whole testes needle immersed ultra-rapid cooling. J. Vis. Exp. 4: 1–7.

Marinović, Z., Q. Li, J. Lujić, Y. Iwasaki, Z. Csenki, B. Urbányi et al. 2019. Preservation of zebrafish genetic resources through testis cryopreservation and spermatogonia transplantation. Sci. Rep. 9: 13861.

Marques, L.S., A. Bos-Mikich, L.C. Godoy, L.A. Silva, D. Maschio, T. Zhang et al. 2015. Viability of zebrafish (*Danio rerio*) ovarian follicles after vitrification in a metal container. Cryobiology 71: 367–373.

Marques, L.S., A.A.N. Fossati, R.B. Rodrigues, H.T. Da Rosa, A.P. Izaguirry, J.B. Ramalho et al. 2019. Slow freezing versus vitrification for the cryopreservation of zebrafish (*Danio rerio*) ovarian tissue. Sci. Rep. 9: 1–11.

Martinez, S., V. Robles, L. Anel, P. De Paz and M. Herráez. 2005. Cell viability after vitrification of zebrafish embryos. Reprod. Domest. Anim. 40: 402.

Martínez-Páramo, S., V. Barbosa, S. Pérez-Cerezales, V. Robles and M.P. Herráez. 2009. Cryoprotective effects of antifreeze proteins delivered into zebrafish embryos. Cryobiology 58: 128–133.

Martínez-Páramo, S., Á. Horváth, C. Labbé, T. Zhang, V. Robles, P. Herráez et al. 2017. Cryobanking of aquatic species. Aquaculture 472: 156–177.

Martins, G., P. Diogo, W. Pinto and P.J. Gavaia. 2018. Early transition to microdiets improves growth, reproductive performance and reduces skeletal anomalies in zebrafish (*Danio rerio*). Zebrafish 16: 300–307.

Matthews, J.L., J.M. Murphy, C. Carmichael, H. Yang, T. Tiersch, M. Westerfield et al. 2018. Changes to extender, cryoprotective medium, and *in vitro* fertilization improve zebrafish sperm cryopreservation. Zebrafish 15: 279–290.

Mazur, P. 1963. Kinetics of water loss from cells at subzero temperatures and the likelihood of intracellular freezing. J. Gen. Physiol. 47: 347 LP–369.

Mazur, P. 1984. Freezing of living cells: mechanisms and implications. Am. J. Physiol. Cell Physiol. 247: C125–C142.

Mazur, P., S.P. Leibo and G.E. Seidel. 2008. Cryopreservation of the germplasm of animals used in biological and medical research: importance, impact, status, and future directions. Biol. Reprod. 78: 2–12.

Medrano, A., F. Cabrera, F. González, M. Batista and A. Gracia. 2002. Is sperm cryopreservation at − degree a feasible alternative? Cryo-Lett. 23: 167–172.

Meyer, A. and M. Schartl. 1999. Gene and genome duplications in vertebrates: the one-to-four (to eight in fish) rule and the evolution of novel gene functions. Curr. Opin. Cell Biol. 11: 699–704.

Meyers, J.R. 2018. Zebrafish: development of a vertebrate model organism. Curr. Protoc. Essent. Lab. Tech. 16: e19.

Miller, D.H. and G.T. Ankley. 2004. Modeling impacts on populations: fathead minnow (*Pimephales promelas*) exposure to the endocrine disruptor 17β-trenbolone as a case study. Ecotoxicol. Environ. Saf. 59: 1–9.

Morris, J.P., S. Berghmans, D. Zahrieh, D.S. Neuberg, J.P. Kanki and A.T. Look. 2003. Zebrafish sperm cryopreservation with N,N-dimethylacetamide. Biotechniques 35: 956–968.

Murphy, P.J., K.E. Murphy and D.T. Carrell. 2014. Epigenetic changes in the paternal germline. pp. 43–55. *In*: Tollefsbol, T. [ed.]. Transgenerational Epigenetics. Academic Press, Oxford, UK.

Nasiadka, A. and M.D. Clark. 2012. Zebrafish breeding in the laboratory environment. ILAR Journal 53: 161–168.

Nor Azra, M., M.I. Mohd Noor, M. Pau Tan, M. Dawood, M. Amin, I. Zekker et al. 2022. Zebrafish and Medaka as model organisms for climate change research: Global literature scientometric analysis. Front. Earth Sci. 10: 988710.

Okutsu, T., A. Yano, K. Nagasawa, S. Shikina, T. Kobayashi, Y. Takeuchi, et al. 2006. Manipulation of fish germ cell: visualization, cryopreservation and transplantation. J. Reprod. Dev. 52: 685–693.

Oliveira, R.F., J.M. Simões, M.C. Teles, C.R. Oliveira, J.D. Becker and J.S. Lopes. 2016. Assessment of fight outcome is needed to activate socially driven transcriptional changes in the zebrafish brain. Proc. Natl. Acad. Sci. 113: E654–E661.

Patton, E.E. and D.M. Tobin. 2019. Spotlight on zebrafish: the next wave of translational research. Dis. Model. Mech. 12: dmm039370.

Paull, G.C., K.J.W. Van Look, E.M. Santos, A.L. Filby, D.M. Gray, J.P. Nash et al. 2008. Variability in measures of reproductive success in laboratory-kept colonies of zebrafish and implications for studies addressing population-level effects of environmental chemicals. Aquat. Toxicol. 87: 115–126.

Pegg, D.E. 2007. Principles of cryopreservation. pp. 39–57. *In*: Day, J., G. Stacey and N. Glyn [eds.]. Cryopreservation and Freeze-Drying Protocols. Humana Press Inc., Totowa, USA.

Planchart, A., C.J. Mattingly, D. Allen, P. Ceger, W. Casey, D. Hinton et al. 2016. Advancing toxicology research using *in vivo* high throughput toxicology with small fish models. Altex. 33: 435.

Popesku, J.T., C.J. Martyniuk, J. Mennigen, H. Xiong, D. Zhang, X. Xia et al. 2008. The goldfish (*Carassius auratus*) as a model for neuroendocrine signaling. Mol. Cell. Endocrinol. 293: 43–56.

Ransom, D.G. and Z. Li. 1999. Collection, storage, and use of zebrafish sperm. Methods Cell Biol. 60: 365–372.

Resende, T.P., R.P. Andrade and I. Palmeirim. 2014. Timing embryo segmentation: dynamics and regulatory mechanisms of the vertebrate segmentation clock. Biomed Res. Int. 2014: 718683.

Reunov, A., V. Isaeva, D. Au and R. Wu. 2000. Nuage constituents arising from mitochondria: is it possible? Dev. Growth Differ. 42: 139–143.

Ribas, L. and F. Piferrer. 2013. The zebrafish (*Danio rerio*) as a model organism, with emphasis on applications for finfish aquaculture research. Rev. Aquac. 6: 209–240.

Riesco, M.F., F. Martínez-Pastor, O. Chereguini and V. Robles. 2012. Evaluation of zebrafish (*Danio rerio*) PGCs viability and DNA damage using different cryopreservation protocols. Theriogenology 77: 122–130.

Riesco, M.F. and V. Robles. 2013. Cryopreservation causes genetic and epigenetic changes in zebrafish genital ridges. PloS One 8: e67614.

Robles, V., E. Cabrita, P. De Paz, S. Cuñado, L. Anel and M.P. Herráez. 2004. Effect of a vitrification protocol on the lactate dehydrogenase and glucose-6-phosphate dehydrogenase activities and the hatching rates of Zebrafish (*Danio rerio*) and Turbot (*Scophthalmus maximus*) embryos. Theriogenology 61: 1367–1379.

Robles, V., E. Cabrita and M.P. Herráez. 2009. Germplasm cryobanking in zebrafish and other aquarium model species. Zebrafish 6: 281–293.

Rodrigues, R.B., M. Uczay, V.B. Brito, A.A.N. Fossati, A.C. Godoy, D.J. Moura et al. 2020. Skim milk powder used as a non-permeable cryoprotectant reduces oxidative and DNA damage in cryopreserved zebrafish sperm. Cryobiology 97: 76–84.

Rodrigues, R.B., M. Uczay, V.B. Brito, A.C. Godoy, D.J. Moura, C. Vogel et al. 2021. Oxidative stress and DNA damage of zebrafish sperm at different stages of the cryopreservation process. Zebrafish 18: 97–109.

Seki, S., T. Kouya, R. Tsuchiya, D.M. Valdez Jr., B. Jin, T. Hara et al. 2008. Development of a reliable *in vitro* maturation system for zebrafish oocytes. Reproduction 135: 285–292.

Seki, S., K. Kusano, S. Lee, Y. Iwasaki, M. Yagisawa, M. Ishida et al. 2017. Production of the medaka derived from vitrified whole testes by germ cell transplantation. Sci. Rep. 7: 43185.

Sieber, S., P. Grossen, J. Bussmann, F. Campbell, A. Kros, D. Witzigmann et al. 2019. Zebrafish as a preclinical *in vivo* screening model for nanomedicines. Adv. Drug Deliv. Rev. 151: 152–168

Silva, L.A., L.S. Marques, T. Zhang, R.B. Rodrigues, L.R. Eloy and D.P. Streit Jr. 2018. Production of live larvae following *in vitro* maturation of zebrafish oocytes. Theriogenology 105: 115–119.

Silva, C., T. Kollár, Z. Csenki-Bakos, B.H.V. Fernandes, A. Horvath and L.R. Carvalho. 2019. An easy method for cryopreservation of zebrafish (*Danio rerio*) sperm. Zebrafish 16: 321–323.

Strüssmann, C.A., H. Nakatsugawa, F. Takashima, M. Hasobe, T. Suzuki and R. Takai. 1999. Cryopreservation of isolated fish blastomeres: effects of cell stage, cryoprotectant concentration, and cooling rate on post-thawing survival. Cryobiology 39: 252–261.

Talbot, W.S., B. Trevarrow, M.E. Halpern, A.E. Melby, G. Farr, J.H. Postlethwait et al. 1995. A homeobox gene essential for zebrafish notochord development. Nature 378: 150.

Tanguay, R.L. 2018. The rise of zebrafish as a model for toxicology. Toxicol. Sci. 163: 3–4.

Tiersch, T.R. 2001. Cryopreservation in aquarium fishes. Mar. Biotechnol. 3: S212–S223.

Tiersch, T.R., H. Yang, J.A. Jenkins and Q. Dong. 2007. Sperm cryopreservation in fish and shellfish. Soc. Reprod. Fertil. Suppl. 65: 493–508.

Torres, L., Y. Liu, A. Guitreau, H. Yang and T.R. Tiersch. 2017. Challenges in development of sperm repositories for biomedical fishes: quality control in small-bodied species. Zebrafish 14: 552–560.

Torres, L. and T.R. Tiersch. 2018. Addressing reproducibility in cryopreservation, and considerations necessary for commercialization and community development in support of genetic resources of aquatic species. J. World Aquac. Soc. 49: 644–663.

Tsai, S., D.M. Rawson and T. Zhang. 2009. Development of cryopreservation protocols for early-stage zebrafish (*Danio rerio*) ovarian follicles using controlled slow cooling. Theriogenology 71: 1226–1233.

Tsai, S., D.M. Rawson and T. Zhang. 2010. Development of *in vitro* culture method for early-stage zebrafish (*Danio rerio*) ovarian follicles for use in cryopreservation studies. Theriogenology 74: 290–303.

Tsang, B., H. Zahid, R. Ansari, R.C.Y. Lee, A. Partap and R. Gerlai. 2017. Breeding zebrafish: a review of different methods and a discussion on standardization. Zebrafish 14: 561–573.

Valdez, D.M. Jr., A. Miyamoto, T. Hara, S. Seki, M. Kasai and K. Edashige. 2005. Water- and cryoprotectant-permeability of mature and immature oocytes in the medaka (*Oryzias latipes*). Cryobiology 50: 93–102.

Valdez, D.M. Jr., A. Miyamoto, T. Hara, K. Edashige and M. Kasai. 2005. Sensitivity to chilling of medaka (*Oryzias latipes*) embryos at various developmental stages. Theriogenology 64: 112–122.

Valdez, D.M. Jr., T. Hara, A. Miyamoto, S. Seki, B. Jin, M. Kasai et al. 2006. Expression of aquaporin-3 improves the permeability to water and cryoprotectants of immature oocytes in the medaka (*Oryzias latipes*). Cryobiology 53: 160–168.

Valdez, D.M. Jr., R. Tsuchiya, S. Seki, N. Saida, S. Niimi, C. Koshimoto et al. 2013. A trial to cryopreserve immature medaka (*Oryzias latipes*) oocytes after enhancing their permeability by exogenous expression of aquaporin 3. J. Reprod. Dev. 59: 205–213.

Varga, Z.M., S.C. Ekker and C. Lawrence. 2018. Workshop report: zebrafish and other fish models—description of extrinsic environmental factors for rigorous experiments and reproducible results. Zebrafish 15: 533–535.

Vaz, R.L., T.F. Outeiro and J.J. Ferreira. 2018. Zebrafish as an animal model for drug discovery in Parkinson's disease and other movement disorders: a systematic review. Front. Neurol. 9: 347.

Wang, G., N. Kang, H. Gong, Y. Luo, C. Bai, Y. Chen et al. 2015. Upregulation of uncoupling protein Ucp2 through acute cold exposure increases post-thaw sperm quality in zebrafish. Cryobiology 71: 464–471.

Wang, L., F. Sun, M. Lee and G.H. Yue. 2022. Whole-genome resequencing infers genomic basis of giant phenotype in Siamese fighting fish (*Betta splendens*). Zool. Res. 43: 78.

Welch, E.L., C.C. Eno, S. Nair, R.E. Lindeman and F. Pelegri. 2017. Functional manipulation of maternal gene products using *in vitro* oocyte maturation in zebrafish. J. Vis. Exp. 122: e55213.

Whiteley, A.R., A. Bhat, E.P. Martins, R.L. Mayden, M. Arunachalam, S. Uusi-Heikkila et al. 2011. Population genomics of wild and laboratory zebrafish (*Danio rerio*). Mol. Ecol. 20: 4259–4276.

Xin, M., M.A.M. Siddique, B. Dzyuba, R. Cuevas-Uribe, A. Shaliutina-Kolešová and O. Linhart. 2017. Progress and challenges of fish sperm vitrification: A mini review. Theriogenology 98: 16–22.

Yang, H., C. Carmichael, Z.M. Varga and T. Tiersch. 2007. Development of a simplified and standardized protocol with potential for high throughput for sperm cryopreservation in zebrafish *Danio rerio*. Theriogenology 68: 128–136.

Yang, H. and T.R. Tiersch. 2009. Sperm motility initiation and duration in a euryhaline fish, medaka (*Oryzias latipes*). Theriogenology 72: 386–392.

Yang, H., M. Norris, R. Winn and T.R. Tiersch. 2010. Evaluation of cryoprotectant and cooling rate for sperm cryopreservation in the euryhaline fish medaka *Oryzias latipes*. Cryobiology 61: 211–219.

Yang, H., J. Daly, C. Carmichael, J. Matthews, Z.M. Varga e T. Tiersch. 2016. Um procedimento que abrange a análise da integridade da membrana plasmática para avaliação da viabilidade celular na criopreservação de espermatozoides de peixe-zebra *Danio rerio*. Peixe-zebra 13: 144–151.

Yang, H., E. Hu, T. Tiersch, C. Carmichael, J. Matthews and Z.M. Varga. 2020. Temporal and concentration effects of methanol on cryopreservation of zebrafish (*Danio rerio*) sperm. Zebrafish 17: 233–242.

Yang, H., E. Hu, J.L. Matthews, Z.M. Varga and T.R. Tiersch. 2022. Is catalase an effective additive to alleviate oxidative stress during cryopreservation of zebrafish sperm at the repository level? Cryobiology 104: 70–78.

Yavaş, T.K. and A. Daskin. 2012. Effect of alternative cryopreservation procedures on bull semen. Ankara Univ. Vet. Fak. Derg. 59: 231–234.

Yoshizaki, G., K. Fujinuma, Y. Iwasaki, T. Okutsu, S. Shikina, R. Yazawa et al. 2011. Spermatogonial transplantation in fish: a novel method for the preservation of genetic resources. Comp. Biochem. Physiol. Part D Genomics Proteomics 6: 55–61.

Yoshizaki, G., K. Takashiba, S. Shimamori, K. Fujinuma, S. Shikina, T. Okutsu et al. 2016. Production of germ cell-deficient salmonids by dead end gene knockdown, and their use as recipients for germ cell transplantation. Mol. Reprod. Dev. 83: 298–311.

Yoshizaki, G. and S. Lee. 2018. Production of live fish derived from frozen germ cells via germ cell transplantation. Stem Cell Res. 29: 103–110.

Zampolla, T., E. Spikings, S. Srirangarajah, D.M. Rawson and T. Zhang. 2011. Impact of cryoprotectants and cryopreservation on metabolic activity and cytoskeleton proteins of zebrafish (*Danio rerio*) ovarian fragments. Cryo-Lett. 32: 525–536.

Zampolla, T., D.M. Rawson and T. Zhang. 2012. The use of fetal bovine serum for cryopreservation of stage III zebrafish (*Danio rerio*) ovarian follicles. Cryo-Lett. 33: 435–42.

Zhang, T., D.M. Rawson and G.J. Morris. 1993. Cryopreservation of pre-hatch embryos of zebrafish (*Brachydanio rerio*). Aquat. Living Resour. 6: 145–153.

Zhang, T. and D.M. Rawson. 1996. Feasibility studies on vitrification of intact zebrafish (*Brachydanio rerio*) embryos. Cryobiology 33: 1–13.

Zhang, T., D.M. Rawson, L. Tosti and O. Carnevali. 2008. Cathepsin activities and membrane integrity of zebrafish (*Danio rerio*) oocytes after freezing to –196°C using controlled slow cooling. Cryobiology 56: 138–143.

Zhang, L., S. Wang, W. Chen, B. Hu, S. Ullah, Q. Zhang et al. 2014. Pakistan veterinary journal fine structure of zebrafish (*Danio rerio*) spermatozoa. Pak. Vet. J. 34: 518–521.

CHAPTER 5

Gonads and Gametes of South American Neotropical Species

Diógenes H. Siqueira-Silva,[1,*] *Maira da Silva Rodrigues,*[2] *Jeane Rodrigues,*[3] *Daniel Fernandes Da Costa,*[4] *Maria Clara Rosa-Silva*[5] and *Rafael Henrique Nóbrega*[6]

INTRODUCTION

The knowledge about fish reproductive biology is of paramount importance because its close association with other studies, such as the phylogenetic (Martins et al. 2012) and the taxonomic (Parenti and Grier 2004), helping the comprehension of the relations among the species of this group, which shows 36.367 described species, being the most numerous and biodiverse among the Vertebrata filum (Fricke et al. 2022). Moreover, reproductive biology has been an important and irreplaceable ally for biotechnological development, as the gametes cryopreservation aims to maintain and conserve fish species in nature.

By diving into the greater diversity of reproductive strategies among the vertebrates, a variety of reproductive tactics can be observed in the fish group

[1] Research Group of Studies on the Reproduction of Amazon fish (GERPA/LANEC), University of South and Southern of Pará, Marabá, 68507-590, Brazil.

[2] Biological Sciences Program (Applied and General Biology), São Paulo State University (UNESP), Botucatu, 18618-689, Brazil.

[3] PPG in Biodiversity and Biotechnology (BIONORTE), University of Pará, Belém, 66075-110, Brazil.

[4] Biological Sciences Program (Genetics), São Paulo State University (UNESP), Botucatu, 18618-689, Brazil.

[5] Research Group of Studies on the Reproduction of Amazon fish (GERPA/LANEC), University of South and Southern of Pará, Marabá, 68507-590, Brazil.

[6] Reproductive and Molecular Biology Group, Department of Structural and Functional Biology, Institute of Biosciences, São Paulo State University (UNESP), Botucatu, 18618-689, Brazil.

Emails: maira.rodrigues@unesp.br, jeane1997rodrigues@gmail.com, fernandesdacostadaniel@gmail.com, rosaclara@unifesspa.edu.br, nobregarh@ibb.unesp.br

* Corresponding author: diogenessilva@unifesspa.edu.br

that adapt according to the interactions among individuals and the changes in the surrounding habitats in which they live (Vazzoler 1996). Quirino et al. (2021), for example, described differences in the reproductive phases of two isolated populations of the same species (*Astyanax rivularis*) from the São Francisco River basin, which was linked with their habitats since one of the populations living in an area with low vegetation and a river with calm and crystalline waters, containing small well formations, while the second one was in current water and the well-established areas of arboreal vegetation. On the other hand, some characteristics, such as gamete morphology (Taborsky 1998) and gonadal structure (Siqueira-Silva et al. 2019), differ among species based on their evolutionary history and phylogenetic position (Parenti and Grier 2004).

The gonads (ovaries and testes) in fish are generally paired and elongated organs. Most of the species—such as the peacock bass *Cichla kelberi* (Siqueira-Silva et al. 2013) and the yellowtail tetra *Astyanax altiparanae* (Siqueira-Silva et al. 2017)—present tapered organs. However, in the South American catfishes *Pseudoplatystoma fasciatum* (Batlouni et al. 2006), *P. corruscans* (Brito and Bazzoli 2003), and *Steindachneridion parahybae* (Honji et al. 2013), the paired testes have surface fringes. Differing from the elongated pattern, some stingrays present almost spherical ovaries (Wourms 1977), and the Neotropical viviparous teleost fishes *Tomeurus gracilis* (Parenti et al. 2010) and *Poecilia mexicana* (Torres-Martínez et al. 2017) show spherical gonads. Shark species from the genus Scyliorhinus, Pristiophorus, Carcharhinus, Galeus, Mustelus, and Sphyrna may have asymmetrical ovaries with an atrophic and sterile left gonad (Wourms 1977).

Concomitantly with the macroscopic morphological diversity, internal and microscopic differences in fish gonadal structure are observed. In cartilaginous fish, such as the sharks from the Cetorhinus genus, the epithelium invaginate, forming tubules (Worms 1977), as well as the teleost ovaries that form ovarian lamellae (Parenti and Grier 2004). Testes in cartilaginous fish, such as the basking shark, are divided into several lobules, separated by connective tissue trabeculae (Matthews 1950), while in the teleost testes, the germinal epithelium can be distinguished in anastomosed tubules in the most primitive taxa and lobules in the most derived ones (Parenti and Grier 2004; Siqueira-Silva et al. 2019).

The germinal epithelium in both sexes comprises somatic and germ cells (Parenti and Grier 2004). In the testes, the somatic cells are named Sertoli cells (Grier 1993). In the ovaries, the somatic cells are called follicular cells (Grier 2000). The germ cells that constitute the testes are distinguished into spermatogonia, spermatocytes, and spermatid, while the oogonia and oocytes compose the ovaries. Those cells pass through mitotic and meiotic division and differentiation processes in order to form sperm and oocytes, respectively (Grier 2000).

The germ cells of fish may show morphological variations among the different species. Most of them, however, present ovoid oocytes showing a single small opening on their surface, called micropyle (Laale 1980; Ganeco and Nakaghi 2003). However, numerous micropyles were already described in a single oocyte in sturgeon species (Psenicka et al. 2011). In the Neotropical species, the yellowtail tetra *Astyanax altiparanae*, the oocytes have grooves, which hypothetically direct male gametes to the micropyle, facilitating fertilization (Santos et al. 2016). It is not different with male gametes; the spermatozoon is divided into a head, a midpiece, and usually a single flagellum (Nagahama 1983; Zhang et al. 2003). The head may be spherical, oval (Ginzburg and Billard 1972), or elongated (Dzyuba et al. 2019), while the midpiece has an ultrastructural pattern with a central flagellum and surrounding mitochondrial sheath. In some species, such as Neotropical Characidae *Oligosarcus paranensis* and *Roeboides descalvadensis,* the midpiece presents vesicles on its posterior region (Mirande et al. 2022). The flagellum tail exhibits a typically 9 + 2 axonemal pattern (Todd 1976). In spite of this common morphology for most species, some specificities can be observed in some species. For example, in sturgeons, the spermatozoon head presents an acrosome and sub-acrosome (Psenicka et al. 2007; Psenicka et al. 2008).

The spermatozoa morphology (Mattei 1991) directly influences the advances in the biotechnological approach applied to them (Asturiano et al. 2017). Cryopreservation, for example, has improved in the last decades mainly due to the spermatozoa's simple cellular structure, reduced length, and high cold resistance (Asturiano et al. 2017). However, cryopreservation protocols still face problems with the complexity of the freezing and thawing processes (osmotic changes, dehydration, and rehydration, cellular volume alterations, ice crystals formation, and cryoprotectant solution's toxicity, among others) (Cabrita et al. 2014). That complexity is still higher when oocytes are considered since their big cellular volume, chorion presence, low permeability to the cryoprotectant solutions, and high cold sensibility culminate in their cryopreservation failure (Asturiano et al. 2017).

This chapter will discuss morphological and structural particularities of the gonads and gametes of Neotropical fish species, pointing to the success of the protocol used in the cryopreservation. The special highlight of the Neotropical region is associated with the huge biodiversity of this region, which shelters the largest number of freshwater fish species in the world (Albert and Reis 2011). Reis et al. (2016) already reported the amount of 8,000 species for the region, and this number keeps increasing. Just in 2021, of the 212 newly described species, 94 were from South America (Edmondstone et al. 2022).

1. Testicular Morphology in Neotropical Fish

Testicular types of Neotropical fish generally follow the same pattern for the fish species. First described by Parenti and Grier (2004) and later confirmed by Grier et al. (2016), the testes in fish can be classified into two basic structures that are morphologically identified based on the germinal epithelium organization. In the lobular type, the germinal epithelium forms lobules that look like digitiform cords, ending in a bottom blind at the periphery of the testis (Figure 1); in the tubular anastomosed type, the germinal epithelium branched, forming loops with no defined beginning or end (Figure 1). These structures' arrangement show an evolutionarily well-conserved characteristic. Thus, since their description until the current knowledge, the type of testicle always depends on the phylogenetic position the target species occupy. Most primitive species belong to the Siluriformes (e.g., *Pseudoplatystoma fasciatum* (Batlouni et al. 2006), *Rhinelepis aspera,*

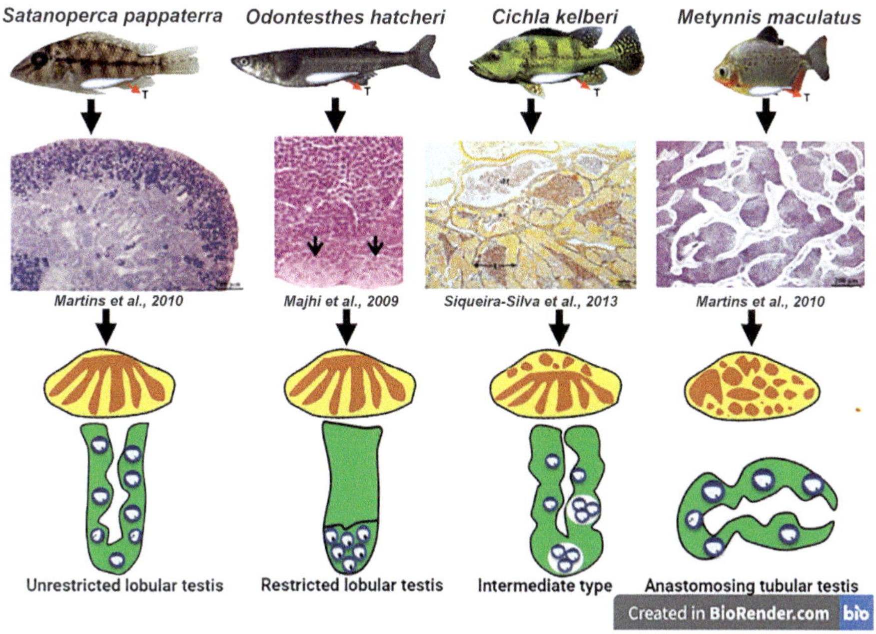

Figure 1. Illustrative diagram demonstrating the testicular types of Neotropical fish. In the unrestricted lobular type, spermatogonia can be distributed along the entire extension of the germinal epithelium, as found in *Satanoperca papaterra*. The second type is named restricted lobular. Note that *Odontesthes hatcheri* testis shows a lobule with a large cyst of spermatogonia (arrows) restricted to the periphery of the testes, in the blind bottom of the lobules. An intermediate type between the restricted and unrestricted can be found in *Cichla kelberi*. In this type, spermatogonia are mostly observed at the blind bottom of the lobules, forming a spermatogonial niche, but can also be distributed along the germinal epithelium. In the anastomosed tubular type, spermatogonia are distributed along the entire extension of the germinal epithelium, as observed for *Metynnis maculatus*. T: testis (Adapted from Siqueira-Silva et al. 2019).

Pimelodus maculatus and *Conorhynchos conirostris* (Melo et al. 2011a), and Characiformes orders (e.g., *Hoplias malabaricus*) (Bizott and Godinho 2007), *Leporinus macrocephalus* (Muñoz et al. 2011), *Astyanax altiparanae* (Siqueira-Silva et al. 2017), *A. bimaculatus* (Cordeiro et al. 2020), and *Copella arnoldi* (Cordeiro et al. 2022), for example, show the tubular anastomosed testicle type. Meanwhile, the most derived ones, the so-called *Neoteleost* orders, such as the Perciformes (e.g., The Cichla species *Cichla piquiti*) (Vieira et al. 2009), *C. kelberi* (Siqueira-Silva et al. 2013), Atheriniformes (e.g., *Odontesthes bonariensis*) (Majhi et al. 2009), Synbranchyformes and Cyprinodontiformes, such as in the oviparous fish *Crenichthys baileyi* (Uribe et al. 2018), Gymnotiformes, as in *Gymnotus carapo* (Vergílio et al. 2013), present the lobular type (Siqueira-Silva et al. 2019).

The testes of the lobular type can also be divided into two or three different types, depending on spermatogonial distribution along the germinal epithelium. The most traditional definition considers two types of lobular testicles. In the first type, spermatogonia can be found distributed along the entire extension of the germinal epithelium as in *Satanoperca pappaterra* (Martins et al. 2010); this arrangement is named spermatogonial unrestricted type (Figure 1). Thus, species containing this germinal epithelium structure are said to have testes of the unrestricted lobular type. The second type is named restricted lobular type, whose spermatogonia are restricted to the periphery of the testes, in the blind bottom of the lobules (Figure 1). This type was already described in the Patagonian pejerrey *Odontesthes hatcheri* (Majhi et al. 2009) and in the Rivulid *Hypsolebas sertanejo* (Evangelista et al. 2021). However, some authors consider an intermediate type between the restricted and unrestricted type, in which spermatogonia are mostly found at the blind bottom of the lobules, forming spermatogonial niches, but can also be observed distributed along the germinal epithelium (Siqueira-Silva et al. 2019), as observed for *Cichla kelberi* (Siqueira-Silva et al. 2013) (Figure 1). In the anastomosed tubular type, spermatogonia are always distributed along the entire extension of the germinal epithelium (Figure 1) (e.g., *Metynnis maculatus* and *Megalancistrus parananus*) (Martins et al. 2010; Parenti and Grier 2004; Siqueira-Silva et al. 2019).

Independently of the type of testes, germ cells develop during spermatogenesis supported by the somatic Sertoli cells. Those cells are in contact with the first spermatogonial type (A undifferentiated spermatogonia), and as soon they start mitotic division, the Sertoli cells enclose them inside cysts. Usually, all the spermatogenesis process occurs inside those cysts, which open to the lumen only when the spermatozoa are fully differentiated (Leal et al. 2009; Schulz et al. 2010), as in both the Trahira *Hoplias malabaricus* (Bizott and Godinho 2007), and the Jundiá *Rhandia quelen* (Lacerda et al. 2019). This way, the spermatogenesis in most of the species is named cystic. However, in some species, the spermatogenesis is not fully performed inside cysts; spermatocytes or spermatids, as in the Corydoradinae *Corydoras flaveolus, Corydoras aeneus, Sceleromystax lacerdai*, and *Aspidoras poecilus* (Spadella et al. 2006), are released inside the germinal epithelium lumen where they complete the process. This type of spermatogenesis is called semi-cystic since the process starts but does not finish inside the spermatogenetic cysts.

2. Ovary Morphology

Teleost fish display three types of ovarian organizations. For any kind, the ovaries are suspended in the coelomic cavity, attached to the dorsal body wall by connective tissue named mesovarium. The gymnovarian condition is found in basal animals, such as lungfish, sturgeon, and bowfin. In this type, mature oocytes are released into the coelomic cavity, enter the ostium, and then pass through the oviduct to the external environment. The secondary gymnovarian condition is similar to that; however, the eggs are shed into the peritoneal cavity and then spawned (Šmejkal et al. 2019). This type can be found in salmonids. Both ovarian types miss an oviduct (Figure 2). The third type, called cystovarian, is present in most teleost fishes, such as the Neotropical species. This ovarian type presents a cavity (ovarian lumen) where the mature oocytes are released. The ovarian lumen is directly connected to a common oviduct that opens externally via the urogenital papilla (McMillan 2007; Mokhtar 2017) (Figure 2).

The ovarian wall is composed of three layers formed by different tissues. The outermost is a thin connective tissue layer that continues from the mesothelium; the

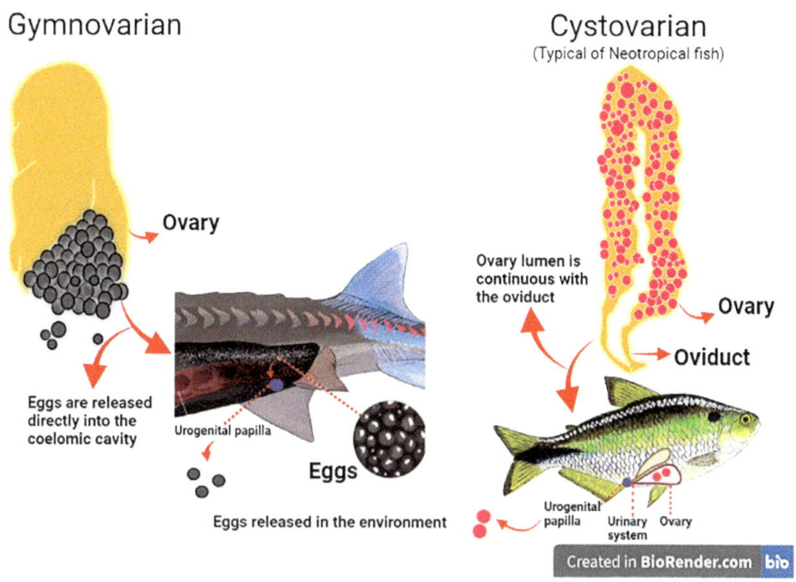

Figure 2. Diagram demonstrating the types of ovary in fish. Gymnovarian ovary type of sturgeons. Note that oocytes are released directly into the coelomic cavity. Fish image adapted from bluemarinefundation.com. The second type is named Cystovarian. In this type, the ovary lumen is continuous with the oviduct, as found in the Neotropical fish, such as *Astyanax altiparanae*. Fish image adapted from pescasgerais.com.br.

middle layer is named tunica albuginea, which is mainly constituted by collagenous and elastic fibers, and smooth muscle cells. During the breeding season, this layer becomes highly vascular in order to supply the vitellogenic oocytes, as can be seen in *Astyanax altiparanae* (Cassel et al. 2017) and the tripletail fish *Lobotes surinamensis* (Fagundes et al. 2021). The tunica albuginea is separated from the innermost layer, the germinal epithelium, by a basement membrane, which is formed as soon as the germinal epithelium develops, as shown by Mazzoni et al. (2015) in the Characidade *Gymnocorymbus ternetzi.* The germinal epithelium is composed of oogonia and oocytes in different stages of development scattered among somatic epithelial cells. This layer borders the ovarian lumen and is responsible for supporting oocyte development (Uribe et al. 2019).

Fish ovarian can be recognized as three different types, according to the pattern of oocyte development into the germinal epithelium (McMillan 2007) (Figure 3). In fish that spawn once in life, such as lampreys, some salmon species, and eels, the oocytes are all at the same development stage and thus are considered to own synchronous ovaries (Figure 3). This type of ovary was not identified in Neotropical fish until this moment. The other two types of ovaries are common for Neotropical fish. Species that spawn once per year during a breeding season are considered group-synchronous ovaries since, in the germinal epithelium, at least two populations of oocytes in different stages of development are found (e.g., oocytes in primary growth and vitellogenic oocytes) (Figure 3). This ovarian type is the most common among Neotropical fish and is usually found in species that perform the

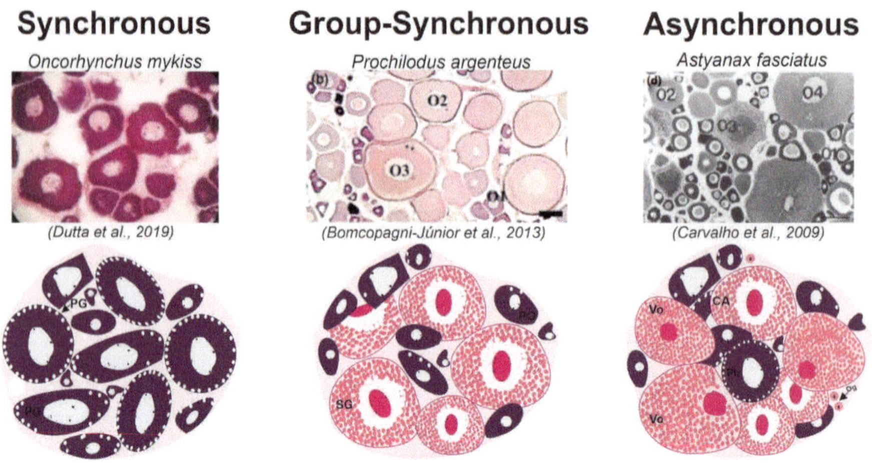

Figure 3. Diagram representing the types of oocyte development in fish. Synchronous development, in which the oocytes are all in the same developmental stage; group synchronous, in which the oocytes develop in successive groups, allowing repeated reproduction in the same reproduction season; and asynchronous, in which oocytes are in all developmental stages and then fish can be able to spawn during a long period. CA: Cortical alveolus oocytes; Og: Oogonia; PN: Perinucleolar oocytes; PG: Primary growth oocytes; SG: Secondary Growth oocytes: Vo: Vitellogenic oocytes (Created in CorelDRAW 2021).

reproductive migration for spawning in headwater rivers by releasing oocyte batches (Rizzo and Bazzoli 2020), as in the species *Prochilodus argenteus, Prochilodus costatus, Salminus franciscanus,* and *Pseudoplatystoma corruscans* (Sato 2003). In the third type, more common in species that live in lentic environments, female germ cells in all development stages are found at the same time in the ovarian germinal epithelium; because of this, the species presenting this type of ovarian can multiple spawn during a long period (Figure 3). This type of ovary has already been described in several Neotropical species, such as *Pygocentrus piraya, Serrasalmus brandtii, Hoplias malabaricus*, and *H. lacerdae* (Sato 2003).

3. Gametes Morphology

The detailed morphological characterization of males and females' gametes of Neotropical South American species have been well described. Those studies are represented by different species from several orders, including *Prochilodus scrofa* (Alexandrino et al. 1985; Borges Filho 1987; Kavamoto et al. 1999), *Prochilodus lineatus* (Egger et al. 2021), *Piaractus mesopotamicus* (Romagosa et al. 1993; Murgas et al. 2017), *Colossoma macropomum* (Maria et al. 2010; Pires et al. 2017), *Salminus maxillosus* = *S. braziliensis* (Veríssimo-Silveira et al. 2006; Murgas et al. 2017), *Astyanax scabripinnis* (Costa et al. 2019), *Astyanax altiparanae* (Santos et al. 2016), *Rhamdia quelen* (Sanches et al. 2011), species from *Brycon* genus (Romagosa et al. 1999; Cruz-Casallas et al. 2005; Ninhaus-Silveira et al. 2006; Veríssimo-Silveira et al. 2006; Galo et al. 2011; Viveiros et al. 2012), *Hemiodus* genus (Brandão et al. 2003), and others evaluated at the moment. Those morphological analyses are commonly used as an indicator of fish gamete quality.

It should be emphasized that the evaluation of fish sperm and eggs' characteristics, as well as their high quality during reproduction, is critical to the success of fertilization (Billard and Cosson 1992; Galo et al. 2011; Sanches et al. 2011; Figueroa et al. 2016; Lassen et al. 2021). However, fish egg and spermatozoa viability are very responsive to several factors, including nutrition, the physical and chemical parameters of water, temperature, age, cryopreservation methods, endocrine disrupting chemicals (EDCs), genetics, and stress situations that have an impact on the survivability of embryos/larvae in the short or long time (Mansour et al. 2006; Miliorini et al. 2011; Santos et al. 2016).

In this context, the morphological evaluation of gametes in Neotropical fish will open new opportunities to study academic science and aquaculture biotechnologies, including cell cryopreservation and germ cell transplantation to generate genetic resources of interest. Below, we review some of Neotropical South American species' main spermatozoa and egg morphology and the consequences of gametes morphological alterations on reproductive performance.

4. Types of Sperm

Since reproduction is one of the most important aspects of fish survival, studies have been describing spermatozoa morphology of Neotropical species. From the viewpoint of both academic research and applied practice, the aspects of spermatozoa structure can help researchers understand fish reproductive biology, being also of great importance for assessing the quality of the breeders (Carneiro-Leite et al. 2020), mainly from economically important species in South American countries, in order to both develop and improve artificial insemination or semen handling techniques and the development of conservation strategies (Caldas and Godoy 2019).

The characteristics of spermatozoa in teleost are defined in the process of spermatid differentiation, named spermiogenesis, and may be associated with fertilization strategies, reproductive habits, habitat, and oocyte morphology (Nagahama 1983; Burns et al. 2009; Murgas et al. 2017). In this sense, the shape of the head is related to the type of fertilization performed by the species (Billard 1970; Ginzburg 1972; Grier 1987; Jamieson 1991). The spherical head observed in *Prochilodus scrofa* spermatozoa is associated with external fertilization (also named aquasperm), as can be seen for several other Neotropical fish species, such as *P. lineatus, Brycon orbignyanus, S. brasiliensis* (Murgas et al. 2017), *Copella arnoldi* (Cordeiro et al. 2022), and most of the Characidae species (e.g., *Pristella maxillaris*, *Poptella compressa*, *Paracheirodon axelrodi*) (Mirande et al. 2022), while an elongated head may be associated with internal fertilization (introsperm), as observed in catfishes of the family Auchenipteridae (Parreira et al. 2009), which may be represented by *Trachelyopterus galeatus* (cumbá) (Melo et al. 2011a), and in the Siluriformes *Scoloplax distolothrix* (Spadella et al. 2006), other Neotropical fishes. The head morphology also determines spermatozoon behavior during motility.

Spermatozoa flagella can also present differences among species. The fire-eye-tetra *Moenkhausia oligolepis* has a single flagellum (Rodrigues et al. 2020), while both the Amazonian pirarucu *Arapaima gigas* (Kirschnik et al. 2022) and the protogynic species *Synbranchus marmoratus* (Lo Nostro et al. 2003) exhibit two flagella. The unique or paired flagella can exhibit on their extremity a structure similar to a keel (Alavi et al. 2009) or a pair of lateral fins along the entire flagellum extension, as in the peacock bass *Cichla kelberi* (Siqueira-Silva et al. 2012), in *A. gigas* (Kirschnik et al. 2022), and in *Moenkhausia pittieri* (Mirande et al. 2022). Those accessory structures for the flagellum are thought to be helpful during the fertilization process. The size of the spermatozoon flagellum can also vary according to the behavior and reproductive strategy of the species (Taborsky 1998). Non-territorial species that present external fertilization, such as the South American fishes *Moenkhausia oliglepis* (Rodrigues et al. 2020) and *Bryconops affinis* (Andrade et al. 2001) have a long flagellum, which can optimize their reproductive success, in terms of greater displacement toward the oocyte and, therefore, a greater possibility of fertilizing it (Taborsky 1998).

The first record of sperm morphology from a Neotropical teleost species is described by Alexandrino and collaborators (1985) for *Prochilodus scrofa* (Steindachner 1881). Popularly known as "Curimbatá," its sperm presents a spherical basophilic head without acrosome, a midpiece, and a single flagellum. According to Ginzburg (1972), the sperm of most teleosts do not present acrosome. This may be related to the presence of an egg micropyle in teleost eggs (Ginzburg and Billard 1972). The same statement was described in different species (Nagahama 1983), as well as in *P. mesopotamicus* (Pacu caranha, another Neotropical species) (Romagosa et al. 1993), and *Astyanax scabripinnis* (Costa et al. 2019) and to this date no acrosome was identified in Neotropical fish sperm.

Among other advantages, the sperm morphology description can help researchers understand problems in experimental studies, as in the case of Kavamoto and collaborators (1999), who demonstrated some abnormalities in *P. scrofa* sperm post-hormone induction. Overall, the results showed an elevated correlation between sperm head alterations and the time of hormonal treatment. The higher defects observed were bent tails and free sperm heads, while sperm motility resulted in a mean of 94.17%. In general, the morphological abnormalities of the midpiece and tail cause alterations in motility. Costa et al. (2022) found that head abnormalities in *Brycon hilarii* spermatozoa affected motility and movement kinetics and modified the pattern of sperm cell movement, sinuosity of movement, and vigor. Spermatozoa with circular or oscillatory movements decrease the fertilization rate (Kavamoto et al. 1999; Solis-Murgas et al. 2011).

Describing spermatozoa ultrastructure through spermiogenesis analysis is also essential for phylogeny studies (Quagio-Grassiotto et al. 2020; Mirande et al. 2022). According to the spermatid differentiation process, three types of spermatozoa are formed (Figure 4). Those formed through spermiogenesis Type I present the flagellar axis perpendicularly arranged in relation to the nucleus. This disposition resulted from a nuclear rotation of 90° upon the flagellum since the flagellum is synthesized parallel to the nucleus at the beginning of spermiogenesis (Mattei 1970) (Figure 4). From Basal to most derived Neotropical species, Type I spermiogenesis occurs, such as in the Silurifomes species *Diplomystes mesembrinus* (Quagio-Grassiotto et al. 2001a) and *Sorubim lima* (Quagio-Grassiotto and Carvalho 2000), the Perciformes peacock bass species *Cichla kelberi* (Siqueira-Silva et al. 2012) and *C. intermedia*

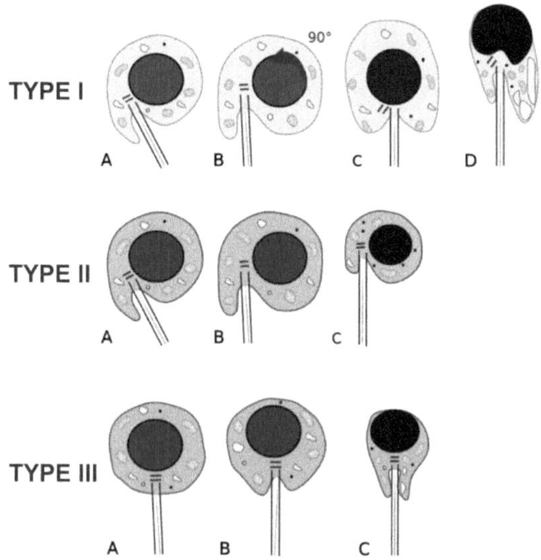

Figure 4. Diagram of longitudinal sections of the types of spermiogenesis. In spermiogenesis Type I, nuclear rotation is observed, and the spermatozoon nucleus is perpendicular to the flagellum; in spermiogenesis Type II, no nuclear rotation happens, and the nucleus remains parallel to the nucleus in the spermatozoon; in spermiogenesis Type III, the nucleus is perpendicular in relation to the flagellum axis even without nuclear rotation. (Adapted from Quagio-Grassiotto et al. 2020) (Created in Biorender.com and adjusted in CorelDRAW 2021.)

(Quagio-Grassiotto et al. 2003), and in the Characiformes *Hoplias malabaricus* (Quagio-Grassiotto et al. 2001b) and *Copella arnoldi* (Cordeiro et al. 2022). It is worth mentioning that of all Lebiasinidae species whose spermiogenesis was described (Santana et al. 2013), *Copella arnodi* is the only one in which spermiogenesis is from Type I; spermatozoa formed from spermiogenesis Type II present the flagellar axis parallel to the nucleus since in this type the nuclear rotation does not happen (Mattei 1970) (Figure 4). Its spermiogenesis type was described for *Gymnotus cf. anguillaris* (França et al. 2007) and species from the Lebiasinidae family (Santana et al. 2013). Finally, the spermatozoa formed from spermiogenesis Type III, as in the Siluriformes species *Pimelodus maculatus* and *Pseudoplatystoma fasciatum* (Quagio-Grassiotto and Oliveira 2008) possess the flagellum in medial position in relation to the nucleus, which does not rotate (Figure 4).

Spermatozoon morphology is also a powerful tool for application in biotechnologies involving artificial reproduction and spermatozoa preservation (Suquet et al. 1998). In this context, Murgas et al. (2017) evaluated *P. mesopotamicus* spermatozoa ultrastructure, which showed a spherical head without acrosome, as already observed before by Romagosa et al. (1993), a cylindrical midpiece and a single flagellum. Furthermore, morphometrical analysis of different portions of the spermatozoa showed the following measurements (μm): head length (2.19 ± 0.204), head width (1.33 ± 0.10), midpiece (1.61 ± 0.17), flagellum (26.46 ± 2.11), total length (29.86 ± 1.23), and proportion of flagellum (88.91 ± 6.86%). These spermatozoa biometry did not differ from those reported for other Neotropical species from the same order, the "piracanjuba" (*B. orbignyanus*), "dourado" (*S. brasiliensis*) (Veríssimo-Silveira et al. 2006) and the "tambaqui" (*C. macropomum*) (Maria et al. 2010; Pires et al. 2017). The midpiece is thin, long, and asymmetrical, but the average length of the spermatozoa midpiece for *S. braziliensis* is large compared to *P. mesopotamicus*. In addition, the mitochondria of the midpiece present proportional size, and differences in the quality of spermatozoa motility may appear (Veríssimo-Silveira et al. 2006; Murgas et al. 2017). Corroborating with Veríssimo-Silveira et al. (2006), the spermatozoa head of *P. mesopotamicus* presents a similar spherical core (Murgas et al. 2017). According to Viveiros et al. (2012), the "pirapitinga" (*Brycon nattereri*) spermatozoa show similar morphological and morphometric patterns to that reported for those Neotropical species (Veríssimo-Silveira et al. 2006; Murgas et al. 2017). However, three concentric membrane rings delimited by four membranes are observed near the end of their midpiece. In addition, the flagellum is perpendicular to the nucleus, showing a symmetrical shape of the spermatozoa (Viveiros et al. 2012).

Santos and collaborators (2016) described the spermatozoa morphology of another Neotropical Characiformes species, the yellowtail tetra *A. altiparanae*. The yellowtail tetra is a small characin that presents early sexual maturation, is simple to breed, and is useful for laboratory study and aquaculture (Rodrigues et al. 2015; Yasui et al. 2015; Yasui et al. 2022). Its sperm presents a spherical head, midpiece, and a single flagellum, as described before for other species (Fribourgh et al. 1970; Ohta et al. 1993; Maria et al. 2010). For this teleost, the total spermatozoa length is about 21.22 ± 4.29 µm. The spermatozoa head demonstrated a spherical conformation (diameter of 1.88 ± 0.08 µm), and the midpiece showed an average length of 0.75 ± 0.19 µm, while the flagellum presents 18.67 ± 4.32 µm of total length. *Astyanax altiparanae* demonstrated high fertilization rates (Yasui et al. 2015), which reports that the descriptive analysis of sperm characteristics contributes to the success of fertilization (Yasui et al. 2015; Santos et al. 2016).

Within this context, the morphological characterization of another Characiformes spermatozoon, the Yamú (*Brycon siebenthalae*), a fish with high-quality meat and great commercial importance for South American aquaculture was studied (Cruz-Casallas et al. 2005). It had no acrosome, and the head region was approximately spherical, while the midpiece was almost cylindrical and presented three ovoid mitochondria and a very long flagellum. Its morphology was very similar to other *Brycon* species spermatozoa (Romagosa et al. 1999; Cruz-Casallas et al. 2005; Ninhaus-Silveira et al. 2006; Veríssimo-Silveira et al. 2006; Viveiros et al. 2012; Murgas et al. 2017), except by the chromatin density pattern. The spermatozoa of *Brycon* species are classified as aquasperm type (Jamieson 1991), in which spermatozoa usually presents a small head with a round-shaped nucleus containing homogeneous and highly dense chromatin, midpiece and cytoplasmic channel of different sizes (Ninhaus-Silveira et al. 2006).

The effects of external factors, such as seasonality, on *P. lineatus* post-thaw sperm morphology, were evaluated (Egger et al. 2021). Following the methodology described by Miliorini et al. (2011) to classify sperm morphological damages on this species, two categories were considered. The primary damages considered were head and midpiece degeneration, tail stump, fractured tail, strongly coiled tail, macrocephaly, and microcephaly, and the secondary alterations evaluated were free normal head, simple net tail, and proximal and distal droplet. The researchers reported that midpiece degeneration was the most significant damage in post-thaw spermatozoa. Figueroa et al. (2019) consider that alterations in the midpiece negatively affect the success of sperm fertilization due to its association with mitochondrial dysfunction. Structural deformation of spermatozoa during freezing and thawing may affect the number of mitochondria in the midpiece. Loss of mitochondria may happen due to midpiece swells, pushing away the mitochondria or inducing structural damage inside mitochondria (Zhang et al. 2003), affecting the cellular energy supply and decreasing flagellar movement. Similar alteration was also observed in the semen of *Macrozarces americanus* (Yao et al. 2000). In "Jundiá" (*R. quelen*), another Neotropical fish, the sperm morphology was evaluated before and after cryopreservation. According to Sanches et al. (2011), impairment of sperm motility and morphological damages were observed post-cryopreservation. The most observed damages were tail disorders (distally curled tail, strongly curled tail, broken tail, and folded tail) and macrocephaly, following the classification described before by Miliorini et al. (2011). After freezing and thawing spermatozoa of *B. orbignyanus*, Galo et al. (2011) observed degenerated tails and free heads. Those described morphological alterations are sufficient to cause circular and oscillatory movements, making fertilization ineffective and thus indicating a substantial fragility of the spermatozoa toward the cryopreservation process.

Considering that sperm morphology is a criterion for species identification, conservation, and breeding success, more efforts must be employed in the structure description of Neotropical fish species spermatozoa.

5. Female Gametes

Considering that fecundity depends largely on fish size, oocyte diameter, and reproductive strategy, the morphological analysis of fish eggs (number, shape, total size, micropyle shape, and zona radiata thickness) is necessary to elucidate the reproductive dynamics in teleost fish.

In general, teleost fish eggs are surrounded by the zona radiata, an acellular layer that shows morphological changes according to species (Rizzo et al. 2002). In the Neotropical Curimatidae species, *Curimatella lepidura* and *Steindachnerina elegans,* the zona radiata presents two layers (Perini et al. 2013), while three layers are observed in the Cichlid *Laetacara araguaiae* (Santos-Silva et al. 2016). During vitellogenesis, the zona radiata participate in the transference of vitellogenin to oocytes (Shabanipour and Heidari 2017), as observed in *L. araguaiae* (Santos-Silva et al. 2016). In the Neotropical species, *Orthospinus franciscensis*, *Tetragonopterus chaleuss*, and *Hemigramus marginatus,* along with other Characidae species (De Carvalho et al. 2009) the zona radiata, are thin and undulated. The *Hemiodus* species, *H. unimaculatus* and *H. ternetzi,* showed a thinner zona radiata compared with *H. ternetzi.* However, the zona radiata of the last species presented grooves with pores or channels filled with filaments that could be associated with adhesiveness (Brandão et al. 2003).

Moreover, the zona radiata plays an important role in controlling interactions between the egg and aquatic environments (Kaviani et al. 2013). According to Yamagami et al. (1992), it undergoes extensive molecular modifications after spawning to form the chorionic structure that reflects adaptations to environmental conditions. For example, pelagic eggs usually present a thin chorion, while demersal eggs have a thicker and more complex chorion structure (Stehr and Hawkes 1979). *Rhinelepis aspera* has demersal eggs, which are spawned on rock substrates. Its vitellogenic oocytes have a thick zona radiata as described in some catfishes (Suzuki et al. 2000). On the other hand, species such as *Conorhynchos conirostris*, *Pimelodus maculatus*, and *Pseudoplatystoma corruscans* have pelagic eggs that disperse freely and then present thin zona radiata (Sato and Godinho 2003; Godinho et al. 2007). The zona radiata from species living in seasonal environments, with constant environmental changes, or in fast waters, are usually thicker to increase their strength against the mechanical shock during spawn (Melo et al. 2011b). The Amazonian species *Serrapinus kriegi*, for example, lives in regions of lotic streams and presents a thick zona radiata (8.4 millimeter; Felicio et al. 2021), ensuring its survival in a stressful environment.

Comparing eggs of 23 Characiformes and nine Siluriformes migratory and sedentary Neotropical species, Rizzo et al. (2002) observed the eggs from sedentary species present morphologically more elaborated zona radiata, owning apparatus to assure egg attachment to the substrate. In this sense, they described the degree of egg adhesiveness, classifying different eggs into adhesive, semi-adhesive, and non-adhesive. *Brycon affinis* have adhesive eggs, which may result from morphological alterations in the zona radiata layer. It presents globular projections formed by the synthesis of substances produced by follicular cells (Andrade et al. 2001; Rizzo et al. 2002; Sato 2003; Weber et al. 2013). Species that present adhesive eggs show a prismatic follicular cell layer and a large number of organelles for mucous substance synthesis; however, in *A. bimaculatus*, the eggs have squamous follicular cells and a smooth zona radiata with pore-channels (Rizzo et al. 2002; Weber et al. 2013). Egg adhesiveness may correspond to the reproductive strategies of the females (Melo et al. 2011b). Usually, species producing small vitellogenic oocytes in diameter have no adhesive eggs and lack parental care; however, species with large vitellogenic oocytes show an adhesive disk around the micropyle and exert parental care (Sato 2003; Perini et al. 2010; Melo et al. 2011b).

Some other morphological specializations on the zona radiata were also attributed to the place of spawning (Rizzo et al. 2002). The jelly coat, for example, is a mucous layer that covers the zona radiata of sedentary species, mainly from Siluriformes, Perciformes, Cypriniformes, Cyprinodontiformes, Acipenseriformes, and Dipnoi eggs (Kemp 1994; Riehl and Patzner 1998; Rizzo et al. 2002; Perini et al. 2010; Zelazowska 2010). Weber et al. (2013) described a jelly coat in Characiformes egg. It was in *Astyanax bimaculatus*, which shows a narrow jelly coat (0.04 ± 0.01 mm), compared with the jelly coat described in *R. quelen* (0.51 mm) (Sato 2003). The jelly coat is usually thick in species that leave their eggs in rocky areas, working as a protection against mechanical shocks (Groot and Alderdice

1985). However, species that lay eggs near the surface or in marginal vegetation present eggs with a thin jelly coat to facilitate gas exchange (Riehl 1996).

Another specialized structure present in the zona radiata of teleost fish eggs is the micropyle. It is a small opening, composed of the vestibule, and the micropillar channel, which consists of a connection between the inner and outer external environment of the oocyte, participating in the fertilization process (Ricardo et al. 1996; Rizzo and Godinho 2003; Ganeco and Nakaghi 2003; Zelazowska 2010; Yanagimachi et al. 2013). It is the only entry site for spermatozoa (Yanagimachi et al. 2013). The disposition of the micropyle morphology on the oocyte surface has potential phylogenetic value and shows morphological differences among fish species, and micropyle-spermatozoa attraction would be specie-specific (Li et al. 2000; Yanagimachi et al. 2013). The micropyle in Curimatidae species *Curimatella lepidura* and *Steindachnerina elegans* showed a funnel-shaped, similar to other Characiformes species (Alexandre et al. 2010; Isaú et al. 2011). In contrast, Melo et al. (2011b) described different micropyle morphologies among Siluriform species. In *O. franciscensis*, *T. chalceus*, and *H. marginatus,* the micropyle is ovoid and has folds, a deep vestibule, and short micropylar channels. However, the micropyle of *B. affinis* and *T. guentheri* is funnel-shaped, forming no folds; the vestibule is shallow, with an elongated micropylar channel. Santos et al. (2016) showed two morphological moments of *A. altiparanae* micropyle. The non-hydrated oocytes showed a spherical conformation with a diameter of 695.119 ± 3.20 μm, while hydrated oocytes showed a micropyle surrounded by an extensive region with grooves, which hypothetically guide the spermatozoa for the single micropyle channel. In addition, the authors observed the formation of the fertilization cone on the micropyle, which works as a physical barrier to the other spermatozoa, preventing polyspermy. Other teleosts with external fertilization also demonstrated this mechanism (Kudo et al. 1980). The micropyles in the Characiformes *P. lineatus, S. brasiliensis, P. mesopotamicus*, and *B. orbignyanus* were similarly small and composed of the vestibule and a mycropilar channel (Murgas et al. 2017). However, in *B. orbignyanus*, ridges, and folds were observed around the micropylar vestibule. In general, the oocyte morphology description is essential to the biological understanding of the Neotropical species or the phylogeny of *incertae sedis* groups.

6. Perspective on the Neotropical Fish Gametogenesis Studies

As already stated, the Neotropical region is a constant source for discovering new fish species. However, human actions have negatively impacted their habitat, and several species were already extinct from nature or are threatened. Thus, understanding the reproductive biology of fish species is the first tool in order to think of conservation strategies. In this sense, morphological characterization of fish gametes could be an important ally in developing genetic fish cryobanks through cryopreservation techniques. Moreover, it can help to comprehend the phylogenetic relationships among them.

On the other hand, morphological gonadal characterization can elucidate the reproductive peak of a target species. This knowledge could then be used in reproductive biotechnologies, such as the hypofization technique, aiming their reproduction *ex-situ*. Thus, studying Neotropical fish gametogenesis is an opportunity to increase and provide new advances in fish biology.

References

Alavi, S.H., M. Rodina, A.T. Viveiros, J. Cosson, D. Gela, S. Boryshpolets et al. 2009. Effects of osmolality on sperm morphology, motility and flagellar wave parameters in Northern pike (*Esox lucius* L.). Theriogenology 72: 32–43.

Albert, J.S. and R.E. Reis. 2011. Introduction to neotropical freshwaters. pp. 3–20. *In*: Albert, J.S. and R.E. Reis [eds.]. Historical Biogeography of Neotropical Freshwater Fishes. University of California Press Berkeley, Berkeley, USA.

Alexandre, J.S., A. Ninhaus-Silveira, R. Veríssimo-Silveira, H. Buzollo, J.A. Senhorini and M.P. Chaguri. 2010. Structural analysis of the embryonic development in *Brycon cephalus* (Günther, 1869). Zygote 18: 173–183.

Alexandrino, A.C., M.T. Phan and E.F.G. Pinheiro. 1985. Caracterização macroscópica das gônadas do Curimbatá, *Prochilodus scrofa* (Steindachner, 1881), durante o ciclo reprodutivo. Bol. Zool. 9: 159–175.

Andrade, R.F., N. Bazzoli, E. Rizzo and Y. Sato. 2001. Continuous gametogenesis in the neotropical freshwater teleost, *Bryconops affinis* (Pisces: Characidae). Tissue Cell 33: 524–532.

Arcifa, M.S., T.G. Northcote and O. Froehlich. 1991. Interactive ecology of two cohabiting characin fishes (*Astyanax fasciatus* and *Astyanax bimaculatus*) in an eutrophic Brazilian reservoir. J. Trop. Ecol. 7: 257–268.

Asturiano, J.F., E. Cabrita and Á. Horváth. 2017. Progress, challenges and perspectives on fish gamete cryopreservation: a mini-review. Gen. Comp. Endocrinol. 245: 69–76.

Balon, E.K. 1977. Early ontogeny of *Labeotropbeus* Ahl, 1927 (Mbuna, Cichlidae, Lake Malawi), with a discussion on advanced protective styles in fish reproduction and development. Environ. Biol. Fishes 2: 147–176.

Batlouni, S.R., E. Romagosa and M.I. Borella. 2006. The reproductive cycle of male catfish *Pseudoplatystoma fasciatum* (Teleostei, Pimelodidae) revealed by changes of the germinal epithelium: An approach addressed to aquaculture. Anim. Reprod. Sci. 96: 116–132.

Bazzoli, N. and H.P. Godinho. 1994. Cortical alveoli in oocytes of freshwater neotropical teleost fish. Ital. J. Zool. 61: 301–308.

Billard, R. 1970. La spermatogènese de *Poecilia reticulate*. IV. La spermiogenese. Etude ultrastructurale. – Annals of Biological Animal Biochemica Biophysics 10: 493–510.

Billard, R. and M.P. Cosson. 1992. Some problems related to the assessment of sperm motility in freshwater fish. J. Exp. Zool. 261: 122–131.

Bizzott, P.M. and H.P. Godinho. 2007. Morphometric evaluation of the spermatogenesis in trahira *Hoplias malabaricus* (Bloch) (Characiformes, Erythrinidae). Rev. Bras. Zool. 24: 541–544.

Borges Filho, O.F. 1987 Caracterização dos estádios de maturação e correlação com avaliações histoquímico-enzimáticas e ultraestruturais das células endócrinas testiculares, durante o ciclo reprodutivo de *Prochilodus scrofa* Steindachner, 1881. Ph.D. Thesis, University of São Paulo, São Paulo.

Brandão, C.A.S., M.F.M. Valentim and E. Pellegrini-Caramaschi. 2003. Ovary maturation stages and oocyte features in three species of the neotropical fish *Hemiodus* (Muller, 1842). Braz. Arch. Biol. Technol. 46: 433–441.

Brito, M.F.G. and N. Bazzoli. 2003. Reproduction of the surubim catfish (Pisces, Pimelodidae) in the São Francisco River, Pirapora Region, Minas Gerais, Brazil. Arq. Bras. Med. Vet. Zootec. 55: 624–633.

Burns, J.R., I. Quagio-Grassiotto and B.G.M. Jamieson. 2009. Ultrastructure of spermatozoa: Ostariophysi. pp. 287–388. *In*: Jamieson, B.G.M. [ed.]. Reproductive Biology and Phylogeny of Fishes (Agnathans and Bony Fishes). CRC Press, Florida, USA.

Cabrita, E., S. Martínez-Páramo, P.J. Gavaia, M.F. Riesco, D.G. Valcarce, C. Sarasquete et al. 2014. Factors enhancing fish sperm quality and emerging tools for sperm analysis. Aquaculture 432: 389–401.

Caldas, J.S. and L. Godoy. 2019. Sperm characterization of the endangered Amazonian fish *Hypancistrus zebra*: Basic knowledge for reproduction and conservation strategies. Anim. Reprod. Sci. 204: 117–124.

Carneiro-Leite, L., C. Bashiyo-Silva, Y.A.A. Oliveira, L.P. Borges, M.P. Sanchez, L.G.D. Silva et al. 2020. Seminal characteristics and sensitivity of *Astyanax lacustris* (Characiformes: Characidae) sperm to cryoprotective solutions based on dimethylsufoxide and methylglicol. Neotrop. Ichthyol. 18: e200039.

Cassel, M., C. Chehade, G.S. Branco, D. Caneppele, E. Romagosa and M.I. Borella. 2017. Ovarian development and the reproductive profile of *Astyanax altiparanae* (Teleostei, Characidae) over one year: Applications in fish farming. Theriogenology 98: 1–15.

Cordeiro, J.G., J. Rodrigues, M.S. Rodrigues, R.M. Bertolini, R.H. Nóbrega, G.S. Yasui et al. 2020. Reproductive cycle of the tetra *Astyanax bimaculatus* (Characiformes: Characidae) collected in Amazonian streams. Zygote 28: 37–44.

Cordeiro, J.G., J. Rodrigues, R. Santos, M.S. Rodrigues, R.H. Nóbrega, F.F.S. Siqueira et al. 2022. Reproductive biology of the Amazonian amphibian fish the splash tetra *Copella arnoldi* with emphasis to histological characterization. Acta Zool. 104: 444–457.

Costa, B.B., L.S. Marques, P.G. Lassen, R.B. Rodrigues and D.P. Streit Jr. 2019. Cryopreservation-induced morphological changes in the sperm of South American silver catfish (*Rhamdia quelen*). J. Appl. Ichthyol. 35: 987–993.

Costa, B.B., J.A. Povh, E.A. Sanches, L.N. Spica, R.B. Rodrigues, N.S. Teixeira et al. 2022. Characterization of sperm quality in *Brycon hilarii*: How does morphology affect sperm movement? Theriogenology Wild 1: 100007.

Cruz-Casallas, P.E., D.A. Lombo-Rodríguez and Y.M. Velasco-Santamaría. 2005. Milt quality and spermatozoa morphology of captive *Brycon siebenthalae* (Eigenmann) broodstock. Aquac. Res. 36: 682–686.

De Carvalho, P.A., A.L. Paschoalini, G.B. Santos, E. Rizzo and N. Bazzoli. 2009. Reproductive biology of *Astyanax fasciatus* (Pisces: Characiformes) in a reservoir in southeastern Brazil. J. Appl. Ichthyol. 25: 306–313.

Dzyuba, V., S. Sampels, A. Ninhaus-Silveira, M. Kahanec, R. Veríssimo-Silveira, M. Rodina et al. 2019. Sperm motility and lipid composition in internally fertilizing ocellate river stingray *Potamotrygon motoro*. Theriogenology 130: 26–35.

Edmondstone, M.R.J., H.C. Patricio and M. Baltzer. 2022. New Species 2021: The Freshwater Fish Species Described in 2021 (Report 1). Shoal. http://shoalconservation.org/wp-content/uploads/2022/03/NEW-SPECIES-2021-v.12-1.pdf.

Egger, R.C., N.C. Motta, T.S. França, A.V. Oliveira, W.F. Fernandes-Braga, J. Alvarez-Leite et al. 2021. Sperm cryopreservation of *Prochilodus lineatus* throughout the same reproductive season. Aquac. Res. 52: 6453–6463.

Evangelista, M.M., E. Romagosa, D.H. Siqueira-Silva, G.S. Yasui, T. Fujimoto and J.A. Senhorini. 2021. Reproductive strategies and chromosomal aberrations affect survival in the Rivuliid fish *Hypsolebias sertanejo*. Zygote 29: 20–26.

Fagundes, K.R.C., R.M. Honji, M.C. Escalante-Rojas, C.E.D.O. Garcia and R.G. Moreira. 2021. Seasonal morphophysiological changes during the reproductive cycle of the tripletail fish *Lobotes surinamensis* (Perciformes: Lobotidae) in the western South Atlantic. Environ. Biol. Fishes 104: 779–796.

Felicio, G.R., J.G. Cordeiro, B.P. Dutra-Costa, C. Maximino, G.S. Branco, P.P. Quirino et al. 2021. Gonadal characterization of the Amazonian fish *Serrapinnus kriegi* (Characidae: Cheirodontinae). Braz. J. Biol. 83: e249158.

Figueroa, E., I. Valdebenito and J.G. Farias. 2016. Technologies used in the study of sperm function in cryopreserved fish spermatozoa. Aquac. Res. 47: 1691–1705.

Figueroa, E., M. Lee-Estevez, I. Valdebenito, I. Watanabe, R. Oliveira, J. Romero et al. 2019. Effects of cryopreservation on mitochondrial function and sperm quality in fish. Aquaculture 511: 1–13.

França, G.F., C. Oliveira and I. Quagio-Grassiotto. 2007. Ultrastructure of spermiogenesis and spermatozoa of *Gymnotus cf. anguillaris* and *Brachyhypopomus cf. pinnicaudatus* (Teleostei: Gymnotiformes). Tissue cell 39: 131–139.

Fribourgh, J.H., D.E. McClendon and B.L. Soloff. 1970. Ultrastructure of the goldfish, *Carassius auratus* spermatozoon. Copeia 2: 274–9.

Fricke, R., W.N. Eschmeyer and R. Van der Laan. 2022. Eschmeyer's Catalog of Fishes: Genera, Species, References. (http://researcharchive.calacademy.org/research/ichthyology/catalog/fishcatmain.asp). Electronic version accessed 30 dec 2022.

Galo, J.M., D.P. Streit-Junior, R.N. Sirol, R.P. Ribeiro, M. Digmayer, V.X.L. Andrade et al. 2011. Spermatic abnormalities of piracanjuba *Brycon orbignyanus* (Valenciennes, 1849) after cryopreservation. Braz. J. Biol. 71: 693–699.

Ganeco, L.N. and L.S.O. Nakaghi. 2003. Morfologia da micrópila e da superfície dos ovócitos de piracanjuba, *Brycon orbignyanus* (Osteichthyes, Characidae), sob microscopia eletrônica de varredura. Acta Sci.: Biol. Sci. 25: 227–231.

Ginzburg, A.S. 1972. Fertilization in fishes and the problem of polyspermy. Springfield, Department of Commerce, National AM Technical Information Service, Israel.

Ginsburg, A.S. and R. Billard. 1972. Ultrastructure du spermatozoïde d'Anguille. J. Microsc. 14: 50–51a.

Godinho, A.L., B. Kynard and H.P. Godinho. 2007. Migration and spawning of female surubim (*Pseudoplatystoma corruscans*, Pimelodidae) in the São Francisco River, Brazil. Environ. Biol. Fishes 80: 421–433.

Grier, H.J. 1981. Cellular organization of the testis and spermatogenesis in fishes. Am. Zool. 21: 345–357.

Grier, H.J. 1987. The mechanism of tubule elongation during testicular recrudescence in the redfish, *Sciaenops ocellatus* (Perciformes). pp. 285–291. In Proceedings of the Vth congress of European ichthyologists, Stockholm 1985. Swedish Museum of Natural History.

Grier, H.J. 1993. Comparative organization of Sertoli cells including the Sertoli cell barrier. pp. 703–739. *In*: Russel, L.D. and M.D. Griswold [eds.]. The Sertoli cell. Cache River Press, Clearwater, USA.

Grier, H.J. 2000. Ovarian germinal epithelium and folliculogenesis in the common snook, *Centropomus undecimalis* (Teleostei: Centropomidae). J. Morphol. 243: 265–281.

Grier, H.J., M.C. Uribe, F.L. Lo Nostro, S.D. Mims and L.R. Parenti. 2016. Conserved form and function of the germinal epithelium through 500 million years of vertebrate evolution. J. Morphol. 277: 1014–1044.

Groot, E.P. and D.F. Alderdice. 1985. Fine structure of the external egg membrane of five species of Pacific salmon and steelhead trout. Can. J. Zool. 63: 552–566.

Honji, R.M., D. Caneppele and R.G. Moreira. 2013. Caracterização macroscópica das gônadas durante a reprodução induzida em cativeiro do surubim-do-paraíba. Pesqui. Agropecu. Bras. 48: 1110–1114.

Isaú, Z.A., E. Rizzo, T.B. Amaral, N.M.N. Mourad and A.T.M. Viveiros. 2011. Structural analysis of oocytes, post-fertilization events and embryonic development of the Brazilian endangered teleost *Brycon insignis* (Characiformes). Zygote 21: 85–94.

Jamieson, B.G.M. 1991. Fish Evolution and Systematics: Evidence from Spermatozoa. Cambridge: Cambridge University Press.

Kavamoto, E.T., V.H. Barnabe, B.E.S. Campos and E.F.A. Talmelli. 1999. Anormalidades morfológicas nos espermatozoides do curimbatá, *Prochilodus scrofa* (Steindachner, 1881) (Osteichthyes, Characiformes, Prochilodontidae). Bol. Inst. Pesca 25: 61–66.

Kaviani, E.F., N. Shabanipour and S.B. Mirnategh. 2013. Light and electron microscope structural study of the zona radiata in the oocyte of zebrafish (*Danio rerio*). Microscopy 62: 377–382.

Kemp, A. 1994. Pathology in eggs, embryos, and hatchlings of the Australian lungfish, *Neoceratodus forsteri* (Osteichthyes: Dipnoi). Copeia 1994: 935–43.

Kirschnik, L.N.G., J.T. Lopes, A.F. Lima and L.S. Torati. 2022. The spermatozoa ultrastructure of the giant Amazon pirarucu *Arapaima gigas* (Schinz, 1822). *In*: Conferencia Latinoamericana sobre Cultivo de Peces Nativos, 7, 2022, Belo Horizonte. Anais. Belo Horizonte: UFMG.

Kudo, S. 1980. Sperm penetration and the formation of a fertilization cone in the common carp egg. Dev. Growth Differ. 22: 403–414.

Laale, H.W. 1980. The perivitelline space and egg envelopes of bony fishes: a review. Copeia 1980: 210–226.

Lacerda, S.M.S.N., E.R.M. Martinez, I.L.D.D. Mura, L.B. Doretto, G.M.J. Costa, M.A. Silva et al. 2019. Duration of spermatogenesis and identification of spermatogonial stem cell markers in a Neotropical catfish, Jundiá (*Rhamdia quelen*). Gen. Comp. Endocrinol. 273: 249–259.

Lassen, P.G., M.E.A. Canozzi, B.B. da Costa, J.O.J. Barcellos and D.P. Streit. 2021. Morphological abnormalities in cryopreserved semen from South American neotropical fish: A meta-analysis. Aquac. Res. 52: 3077–3086.

Leal, M.C., E.R. Cardoso, R.H. Nóbrega, S.R. Batlouni, J. Bogerd, L.R. França et al. 2009. Histological and stereological evaluation of zebrafish (*Danio rerio*) spermatogenesis with an emphasis on spermatogonial generations. Biol. Reprod. 81: 177–187.

Li, Y.H., C.C. Wu and J.S. Yang. 2000. Comparative ultrastructural studies of the zona radiata of marine fish eggs in three genera in Perciformes. J. Fish Biol. 56: 615–621.

Lo Nostro, F.L., H. Grier, F.J. Meijide and G.A. Guerrero. 2003. Ultrastructure of the testis in *Synbranchus marmoratus* (Teleostei, Synbranchidae): the germinal compartment. Tissue Cell 35: 121–132.

Lubzens, E., G. Young, J. Bobe and J. Cerdà. 2010. Oogenesis in teleosts: how fish eggs are formed. Gen. Comp. Endocrinol. 165: 367–389.

Majhi, S.K., R.S. Hattori, S.M. Rahman, T. Suzuki and C.A. Strüssmann. 2009. Experimentally induced depletion of germ cells in sub-adult Patagonian pejerrey (*Odontesthes hatcheri*). Theriogenology 71: 1162–1172.

Mansour, N., M. McNiven and F. Richardson. 2006. The effect of dietary supplementation with blueberry, a-tocopherol or astaxanthin on oxidative stability of Arctic char (*Salvelinus alpinus*) semen. Theriogenology 66: 373–82.

Maria, A.N., H.C. Azevedo, J.P. Santos, C.A. Silva and P.C.F. Carneiro. 2010. Semen characterization and sperm structure of the Amazon tambaqui *Colossoma macropomum*. J. Appl. Ichth. 26: 779–783.

Martins, Y.S., D.F. De Moura, G.B. Santos, E. Rizzo and N. Bazzoli. 2010. Comparative folliculogenesis and spermatogenesis of four teleost fish from a Reservoir in south-eastern Brazil. Acta Zool. 91: 466–473.

Martins, Y.S., F.P. Arantes, Y. Sato, J.E. dos Santos, E. Rizzo and N. Bazzoli. 2012. Comparative analysis of gonadal morphology in six fish species of the *Incertae Sedis* genera in Characidae of occurrence in the São Francisco River Basin, Brazil. Acta Zool. 93: 48–56.

Mattei, X. 1970. Spermiogenèse comparée des poisons. pp. 57–69. *In*: Baccetti, B. (ed.). Comparative Spermatology. Academic Press, New York, USA.

Mattei, X. 1991. Spermatozoon ultrastructure and its systematic implications in fishes. Can. J. Zool. 69: 3038–3055.

Mattei, X., B. Marchand and Y. Quilichini. 2019. A biflagellate spermatozoon in the African bonytongue *Heterotis niloticus* (Teleostei, Osteoglossidae). J. Fish Biol. 94: 335–338.

Matthews, L.H. 1950. Reproduction in the basking shark, *Cetorhinus maximus* (Gunner). Philos. Trans. R. Soc. Lond., B, Biol. Sci. 234: 247–316.

Mazzoni, T.S., H.J. Grier and I. Quagio-Grassiotto. 2015. The basement membrane and the sex establishment in the juvenile hermaphroditism during gonadal differentiation of the *Gymnocorymbus ternetzi* (Teleostei: Characiformes: Characidae). Anat. Rec. 298: 1984–2010.

McMillan, D.B. 2007. Fish Histology: Female Reproductive Systems. Springer Science and Business Media, Dordrecht, Netherlands.

Miliorini, A.B., L.D.S. Murgas, P.V. Rosa, G. Oberlender, G.J.M. Pereira and D.V. Costa. 2011. A morphological classification proposal for curimba (*Prochilodus lineatus*) sperm damages after cryopreservation. Aquac. Res. 42: 177–187.

Mirande, J.M., C.M. Baicere-Silva, J.C. Santana and I. Quagio-Grassiotto. 2022. Sperm phylogeny of Characidae (Teleostei, Characiformes). Zool. Scri. 52: 117–135.

Melo, R.M.C., F.P. Arantes, Y. Sato, J.E. Santos, E. Rizzo and N. Bazzoli. 2011a. Comparative morphology of the gonadal structure related to reproductive strategies in six species of neotropical catfishes (Teleostei: Siluriformes). J. Morphol. 272: 525–535.

Melo, R.M.C., C.M. Ferreira, R.K. Luz, Y. Sato, E. Rizzo and Bazzoli, N. 2011b. Comparative oocyte morphology and fecundity of five characid species from São Francisco River basin, Brazil. J. Appl. Ichth. 27: 1332–1336.

Menezes, N.A., P.A. Buckup, J.L. Figueiredo and R.L. Moura. 2003. Catálogo das espécies de peixes marinhos do Brasil (Vol. 1). Museu of Zoology University of São Paulo, São Paulo.

Mokhtar, D.M. 2017. Fish Histology: From Cells to Organs. Apple Academic Press, New York.

Muñoz, M.E., S.R. Batlouni, I.B.F. Vicentini and C.A. Vicentini. 2011. Testicular structure and description of the seminal pathway in *Leporinus macrocephalus* (Anostomidae, Teleostei). Micron. 42: 892–897.

Murgas, L.D.S., M.S. Paulino, P.C. Palhares, A.B. Miliorini, E. Alves and V. Oliveira Felizardo. 2017. Ultrastructural and morphometric analysis of gametes in neotropical teleost fishes. J. Fish. Sci. 11: 56–61.

Nagahama, Y. 1983. The functional morphology of teleost gonads. Fish Physiol. 9: 223–275.

Narahara, M.Y., E.F.A. Talmelli and F.M.S. Braga. 1993. Mudanças morfológicas dos testículos de pacu, *Piaractus mesopotamicus* (Holmberg, 1887), em condições de confinamento. Rev. UNIMAR. 15: 1–17.

Ninhaus-Silveira, A., F. Foresti, R. Veríssimo-Silveira and J.A. Senhorini. 2006. Seminal analysis, cryogenic preservation, and fertility in matrinxã fish, *Brycon cephalus* (Günther, 1869). Braz. Arch. Biol. Techn. 49: 651–659.

Ohta, T., K.H. Kato, T. Abe and T. Takeuchi. 1993. Sperm morphology and distribution of intramembranous particles in the sperm heads of selected freshwater teleosts. Tissue Cell 25: 725–735.

Parenti, L.R. and H.J. Grier. 2004. Evolution and phylogeny of gonad morphology in bony fishes. Integr. Comp. Biol. 44: 333–348.

Parenti, L.R., F.L. LoNostro and H.J. Grier. 2010. Reproductive histology of *Tomeurus gracilis* Eigenmann, 1909 (Teleostei: Atherinomorpha: Poeciliidae) with comments on evolution of viviparity in atherinomorph fishes. J. Morphol. 271: 1399–1406.

Parreira, G.G., H. Chiarini-Garcia, R.C. Melo, F.O. Vieira and H.P. Godinho. 2009. Spermatozoon and its relationship with the ovarian lamellae in the internally inseminating catfish *Trachelyopterus galeatus*. Microsc. Res. Tech. 72: 889–897.

Perini, V.R., Y. Sato, E. Rizzo and N. Bazzoli. 2010. Biology of eggs, embryos and larvae of *Rhinelepis aspera* (Spix and Agassiz, 1829) (Pisces: Siluriformes). Zygote 18: 159–71.

Perini, V.R., Y. Sato, E. Rizzo and N. Bazzoli. 2013. Comparative analysis of the oocytes and early development of two species of Curimatidae teleost fish. Anat. Histol. Embryol. 42: 40–47.

Pires, L.B., E.A. Sanches, E. Romagosa, R.A.C. Corrêa Filho, D.P. Streit Junior, R.A.R. Nass et al. 2017. Semen characteristics of *Colossoma macropomum* from three successive sample collections in the same reproductive cycle. Aquac. Res. 48: 5104–5110.

Psenicka, M., S.H. Alavi, M. Rodina, D. Gela, J. Nebesarova and O. Linhart. 2007. Morphology and ultrastructure of Siberian sturgeon (*Acipenser baerii*) spermatozoa using scanning and transmission electron microscopy. Biol. Cell. 99: 103–115.

Psenicka, M., S.M.H. Alavi, M. Rodina, Z. Cicova, D. Gela, J. Cosson et al. 2008. Morphology, chemical contents and physiology of chondrostean fish sperm: a comparative study between Siberian sturgeon (*Acipenser baerii*) and sterlet (*Acipenser ruthenus*). J. Appl. Ichth. 24: 371–377.

Psenicka, M., V. Kaspar, S.M.H. Alavi, M. Rodina, D. Gela, P. Li et al. 2011. Potential role of the acrosome of sturgeon spermatozoa in the fertilization process. J. Appl. Ichth. 27: 678–682.

Quagio-Grassiotto, I. and E.D. Carvalho. 2000. Ultrastructure of *Sorubim lima* (Teleostei, Siluriformes, Pimelodidae) spermiogenesis. J. Submicrosc. Cytol. Pathol. 32: 629–633.

Quagio-Grassiotto, I., C. Oliveira and A.E. Gosztonyi. 2001a. The ultrastructure of spermigenesis and spermatozoa in *Diplomystes mesembrinus* (Teleostei, Siluriformes, Diplomystidae). J. Fish Biol. 58: 1623–1632,

Quagio-Grassiotto, I., J.N.C. Negrão, E.D. Carvalho and F. Foresti. 2001b. Ultrastructure of spermatogenic cells and spermatozoa in *Hoplias malabaricus* (Teleostei, Characiformes, Erythrinidae). J. Fish Biol. 59: 1494–1502.

Quagio-Grassiotto, I., F.N. Antoneli and C. Oliveira. 2003. Spermiogenesis and sperm ultrastructure in *Cichla intermedia* with some considerations about Labroidei spermatozoa (Teleostei, Perciformes, Cichlidae). Tissue and Cell 35: 441–446.

Quagio-Grassiotto, I. and C. Oliveira. 2008. Sperm ultrastructure and a new type of spermiogenesis in two species of Pimelodidae, with a comparative review of sperm ultrastructure in Siluriformes (Teleostei: Ostariophysi). Zool. Anz. 247: 55–66.

Quagio-Grassiotto, I., C.M. Baicere-Silva, J.C.O. Santana and J.M. Mirande. 2020. Spermiogenesis and sperm ultrastructure as sources of phylogenetic characters. The example of characid fishes (Teleostei: Characiformes). Zool. Anz. 289: 77–86.

Quirino, P.P., D.H. Siqueira-Silva, M.S. Rodrigues, A.P. Santos-Silva, M.L.R. Delgado, J.A. Senhorini et al. 2021. Gonadal morphology and difference in reproductive development of two isolated populations of *Astyanax rivularis* (Teleostei, Characidae). J. Fish Biol. 99: 1719–1728.

Reis, R.E., J.S. Albert, F. Di Dario, M.M. Mincarone, P. Petry and L.A. Rocha. 2016. Fish biodiversity and conservation in South America. J. Fish Biol. 89: 12–47.

Ricardo, M.C.P., C.A. Aguiar, E. Rizzo and N. Bazzoli. 1996. Morfologia da micrópila e da célula micropilar em teleósteos neotropicais de água doce (Morphology of the micropyle and micropylar cell in neotropical freshwater teleosts). Arq. Bras. Med. Vet. Zoot. 48: 17–24.

Riehl, R. 1996. The ecological significance of the egg envelope in teleosts with special reference to limnic species. Limnogica 26: 183–9.

Riehl, R. and R.A. Patzner. 1998. Mini review: the modes of egg attachment in teleost fishes. Ital. J. Zool. 65: 415–420.

Rizzo, E., Y. Sato, B.P. Barreto and H.P. Godinho. 2002. Adhesiveness and surface patterns of eggs in neotropical freshwater teleosts. J. Fish Biol. 61: 615–632.

Rizzo, E.Y. and H.P. Godinho. 2003. Superfície de ovos de peixes Characiformes e Siluriformes (Surface of fish eggs in Characiformes and Siluriformes). pp. 115–132. *In*: Godinho, H.P. and A.L. Godinho [eds.]. Águas, peixes e pescadores do São Francisco das Minas Gerais. PUC Minas, Belo Horizonte, Brazil.

Rizzo, E., H.P. Godinho and Y. Sato. 2003. Short-term storage of oocytes from the neotropical teleost fish *Prochilodus marggravii*. Theriogenology 60: 1059–1070.

Rizzo, E. and N. Bazzoli. 2020. Reproduction and embryogenesis. pp. 287–313. *In*: Baldisserotto, B., E.C. Urbinati and J.E.P. Cyrino [eds.]. Biology and Physiology of Freshwater Neotropical Fish. Academic Press, Cambridge, UK.

Rodrigues, M.D.S., D.H. Siqueira-Silva, P.P. Quirino, A. Ninhaus-Silveira and R. Verissimo-Silveira. 2015. Spermatogenesis in the yellowtail tetra *Astyanax altiparanae*: a histological analyses with emphasis to spermatogonial and spermatid types. Bol. Inst. Pesca 41: 697–705.

Rodrigues, J., R.S. Santos, J.G. Cordeiro, M. Leite, H.S.T.O. Oliveira, H. Tercya et al. 2020. Seminal characterization of the Amazonian fire-eye tetra *Moenkhausia oligolepis* (Günther, 1864). Zygote 28: 453–458.

Romagosa, E., H.M. Godinho and F.M.S. Braga. 1993. Mudanças morfológicas dos ovários de pacu, *Piaractus mesopotamicus* em condições de confinamento. Bol. Inst. Pesca. 20: 67–80.

Romagosa, E., M.Y. Narahara, M.I. Borella, S.F. Parreira and N. Fenerich-Verani. 1999. Ultrastructure of the germ cells in the testis of matrinxã, *Brycon cephalus* (Teleostei, Characidae). Tissue and Cell 31: 540–544.

Sampaio, K.H. 2006. Superfície ovocitária e desenvolvimento inicial de quatro espécies de peixes de interesse comercial da bacia do Rio São Francisco. Dissertation, University of Minas Gerais, Belo Horizonte.

Sanches, E.A., G. Neumann, C.P.R. Toledo, R.A. Bombardelli, P.A. Piana and E. Romagosa. 2011. Temperature and storage period over spermatic parameters of jundiá, *Rhamdia quelen* (Quoy and Gaimard, 1824). Aquac. Res. 44: 534–541.

Santana, J.C.D.O., A.L. Netto-Ferreira, D. Calcagnotto and I. Quagio-Grassiotto, 2013. Sperm characteristics as additional evidence of close relationship between Lebiasina and Piabucina (Characiformes: Lebiasinidae: Lebiasininae). Neotrop. Ichthyol. 11: 573–579.

Santos, M.P., G.S. Yasui, P.L.P. Xavier, N.S.M. Adamov, N.F. Nascimento, T. Fujimoto et al. 2016. Morphology of gametes, post-fertilization events and the effect of temperature on the embryonic development of *Astyanax altiparanae* (Teleostei, Characidae). Zygote 24: 795–807.

Santos-Silva, A.P., D.H. Siqueira-Silva, A. Ninhaus-Silveira and R. Veríssimo-Silveira. 2016. Oogenesis in *Laetacara araguaiae* (Ottoni and Costa, 2009) (Labriformes: Cichlidae). Zygote 24: 502–10.

Sato, Y., N. Fenerich-Verani, A.P.O. Nuñer, H.P. Godinho and J. Verani. 2003. Padrões reprodutivos de peixes da bacia do São Francisco. Águas, peixes e pescadores do São Francisco das Minas Gerais. pp. 229–274. *In*: Godinho, H.P. and A.L. Godinho [eds.]. Water, Fish and Fishermen of Sáo Francisco, Minas Gerais, Belo Horizonte: Editora PUC, Belo Horizonte, Brazil.

Sato, Y. and H.P. Godinho. 2003. Migratory fishes of the São Francisco river. pp. 195–232. *In*: Carolsfeld, J. and B. Harvey [eds.]. Migratory Fishes of South America: Biology, Fisheries and Conservation Status. IDRC, Victoria, Canada.

Schulz, R.W., L.R. França, J.J. Lareyre, F. LeGac, H. Chiarini-Garcia, R. Nobrega et al. 2010. Spermatogenesis in fish. Gen. Comp. Endocrinol. 165: 390–411.

Shabanipour, N. and B. Heidari. 2017. A histological study of the zona radiata during late oocyte developmental stages in the Caspian Sea mugilid, *Liza aurata* (Risso 1810). Braz. J. Morphol. Sci. 21: 191–195.

Siqueira-Silva, D.H., R.S. Costa, A. Ninhaus-Silveira, C.A. Vicentini and R. Veríssimo-Silveira. 2012. Ultrastructural analysis of spermiogenesis in the neotropical cichlid *Cichla kelberi* Kullander and Ferreira, 2006 (Perciformes: Cichlidae). J. Appl. Ichth. 28: 878–882.

Siqueira-Silva, D.H., C.A. Vicentini, A. Ninhaus-Silveira and R. Veríssimo-Silveira. 2013. Reproductive cycle of the Neotropical cichlid yellow peacock bass *Cichla kelberi*: A novel pattern of testicular development. Neotrop. Ichthyol. 11: 587–596.

Siqueira-Silva, D.H., A.P. Santos-Silva, C. Bashiyo-Silva, M.S. Rodrigues, C.A. Vicentini, A. Ninhaus-Silveira et al. 2017. Testicular structure and seminal pathway in the yellowtail tetra *Astyanax altiparanae* (Characiforms: Characidae). Anat. Histol. Embryol. 46: 342–346.

Siqueira-Silva, D.H., M.S. Rodrigues and R.H. Nóbrega. 2019. Testis structure, spermatogonial niche and Sertoli cell efficiency in Neotropical fish. Gen. Comp. Endocrinol. 273: 218–226.

Solis-Murgas, L.D., V.O. Felizardo, M.R. Ferreira, E.S. Andrade and G.C. Veras. 2011. Importância da avaliação dos parâmetros reprodutivos em peixes nativos. Rev. Bras. Reprod. Animal. 35: 186–191.

Šmejkal, M., P. Blabolil, D. Bartoň, J. Duras, L. Vejřík, Z. Sajdlova et al. 2019. Sex-specific probability of PIT tag retention in a cyprinid fish. Fish. Res. 219: 105325.

Spadella, M.A., C. Oliveira and I. Quagio-Grassiotto. 2006. Spermiogenesis and introsperm ultrastructure of *Scoloplax distolothrix* (Ostariophysi: Siluriformes: Scoloplacidae). Acta Zool. 87: 341–348.

Stehr, C.M. and J.W. Hawkes. 1979. The comparative ultrastructure of the egg membrane and associated pore structures in the starry flounder, *Platichthys stellatus* (Pallas), and pink salmon, *Oncorhynchus gorbuscha* (Walbaum). Cell Tissue Res. 202: 347–356.

Suquet, M., C. Dreanno, G. Dorange, Y. Normant, L. Quemener, J.L. Gaignon et al. 1998. The ageing phenomenon of turbot spermatozoa: effects on morphology, motility and concentration, intracellular ATP content, fertilization, and storage capacities. J. Fish Biol. 52: 31–41.

Suzuki, H.I., A.A. Agostinho and K.O. Winemiller. 2000. Relationship between oocyte morphology and reproductive strategy in loricariid catfishes of the Parana′ river, Brazil. J. Fish Biol. 57: 791–807.

Svensson, P.A., J.D. Blound, E. Forsgren and T. Tamudsen. 2010. Female ornamentation and egg carotenoids of six sympatrics gobies. J. Fish Biol. 75: 2777–2787.

Taborsky, M. 1998. Sperm competition in fish 'bourgeois' males and parasitic spawning. Ecol. Evol. 13: 222–227.

Todd, P.R. 1976. Ultrastructure of the spermatozoa and spermiogenesis in New Zealand freshwater eels (Anguillidae). Cell Tissue Res. 171: 221–232.

Torres-Martínez, A., A. Hernández-Franyutti, M.C. Uribe and W.M. Contreras-Sánchez. 2017. Ovarian structure and oogenesis of the extremophile viviparous teleost *Poecilia mexicana* (Poeciliidae) from an active sulfur spring cave in Southern Mexico. J. Morphol. 278: 1667–1681.

Uribe, M.C., H.J. Grier and L.R. Parenti. 2018. Testicular structure and spermatogenesis of the oviparous goodeids *Crenichthys baileyi* (Gilbert, 1893) and *Empetrichthys latos* Miller, 1948 (Teleostei, Cyprinodontiformes). J. Morphol. 279: 1787–1797.

Uribe, M.C., G. De la Rosa Cruz, A. García-Alarcón, J.C. Campuzano-Caballero and M.G. Guzmán-Bárcenas. 2019. Structures associated with oogenesis and embryonic development during intraovarian gestation in viviparous teleosts (Poeciliidae). Fishes 4: 35.

Valdebenito, I., P. Gallegos and B. Effer. 2015. Gamete quality in fish: Evaluation parameters and determining factors. Zygote 23: 177–197.

Vazzoler, A.E.A.M. 1996. Biologia da reprodução de peixes teleósteos: Teoria e Prática. EDUEM - Editora da Universidade Estadual de Maringá, São Paulo.

Vergílio, C.D.S., R.V. Moreira, C.E. Carvalho and E.J. Melo. 2013. Characterization of mature testis and sperm morphology of *Gymnotus carapo* (Gymnotidae, Teleostei) from the southeast of Brazil. Acta Zool. 94: 364–370.

Veríssimo-Silveira, R., P. Gusmão-Pompiani, C.A. Vicentini and I. Quagio-Grassiotto. 2006. Spermiogenesis and spermatozoa ultrastructure in *Salminus* and *Brycon*, two primitive genera in Characidae (Teleostei: Ostariophysi: Characiformes). Acta Zool. 87: 305–313.

Vieira, A.B., R. Melo, G.B. Santos and N. Bazzoli. 2009. Reproductive biology of the peacock bass *Cichla piquiti* (Perciformes: Cichlidae), an exotic species in a Neotropical reservoir. Neotrop. Ichthyol. 7: 745–750.

Viveiros, A.T., A.N. Maria, T.B. Amaral, L.H. Orfão, Z.A. Isau and R. Veríssimo-Silveira. 2012. Spermatozoon ultrastructure and sperm cryopreservation of the Brazilian dry season spawner fish pirapitinga, *Brycon nattereri*. Aquac. Res. 43: 546–555.

Weber, A., F. Arantes, Y. Sato, E. Rizzo and N. Bazzoli. 2013. Oocyte adhesiveness and embryonic development of *Astyanax bimaculatus* (Linnaeus, 1758) (Pisces: Characidae). Zygote 21: 198–202.

Wourms, J.P. 1977. Reproduction and development in chondrichthyan fishes. Am. Zool. 17: 379–410.

Yamagami, K., T.S. Hamazaki, S. Yasumasu, K. Masuda and I. Iuchi. 1992. Molecular and cellular basis of formation, hardening and breakdown of the egg envelope in fish. Int. Rev. Cytol. 136: 51–92.

Yanagimachi, R., G. Cherr, T. Matsubara, T. Andoh, T. Harumi, C. Vines et al. 2013. Sperm attractant in the micropyle region of fish and insect eggs. Biol. Reprod. 88: 1–11.

Yao, Z., L.W. Crim, G.F. Richardson and C.J. Emerson. 2000. Motility, fertility and ultrastructural changes of ocean pout (*Macrozoarces americanus* L.) sperm after cryopreservation. Aquaculture 181: 361–375.

Yasui, G.S., J.A. Senhorini, E. Shimoda, M. Pereira-Santos, L.S.O. Nakaghi, T. Fujimoto et al. 2015. Improvement of gamete quality and its short-term storage: an approach for biotechnology in laboratory fish. Animal 9: 464–70.

Yasui, G.S., N.F. Nascimento, M. Pereira-Santos, A.P.S. Silva, G.C.Z. Coelho, J.A. Visintin et al. 2022. Establishing a model fish for the Neotropical region: The case of the yellowtail tetra *Astyanax altiparanae* in advanced biotechnology. Front. Genet. 13: 903990.

Zelazowska, M. 2010. Formation and structure of egg envelopes in Russian sturgeon *Acipenser gueldenstaedtii* (Acipenseriformes: Acipenseridae). J. Fish Biol. 76: 694–706.

Zhang, Y.Z., S.C. Zhang, X.Z. Liu, Y.Y. Xu, C.L. Wang, M.S. Sawant et al. 2003. Cryopreservation of flounder (*Paralichthys olivaceus*) sperm with a practical methodology. Theriogenology 60: 989–996.

Chapter 6

Quali-Quantitative Analysis of Semen of Neotropical South American Species

Eduardo A. Sanches[1,*] and *Victor Gallego*[2]

INTRODUCTION

1. Why Spermatozoa Are Important?

Fish reproduction success occurs after the fusion of haploid gametes (egg and sperm) for the formation of a zygote (2n), and the presence of quality gametes is necessary to promote fecundation and subsequent normal embryonic development (Bobe and Labbé 2010). In that sense, it seems essential to define what gamete quality means, and, regarding male gametes, sperm quality could be defined as the ability of the spermatozoa to fertilize the oocyte (Valdebenito et al. 2015). Spermatozoa results from a maturation process undergoing several haploid cells created to transmit paternal genetic information to the next generation. It is important to remark that, throughout this maturation process, several factors will affect the final cell (spermatozoa) and, thus, the fertilization success. The most important factors are described in depth in Chapter 7 and can range from the reproductive age of the breeders to the hormonal treatment applied to achieve sexual maturation. Therefore, to achieve good quality spermatozoa, it is essential to consider all these factors because slight variations in any of them could negatively affect the sperm quality.

[1] Universidade Estadual Paulista "Julio de Mesquita Filho" - UNESP, FCAVR and CAUNESP. Av. Nelson Brihi Badur 430, Vila Tupy, 11900-00, Registro, São Paulo, Brazil.

[2] Universitat Politècnica de València Instituto de Ciencia y Tecnología Animal Camino de Vera s/n. València, Spain.

Email: vicgalal@upvnet.upv.es

* Corresponding author: eduardo.sanches@unesp.br

However, spermatozoa must reach their goal (the oocyte) and be successful in many different aquatic environments, and a series of fish reproductive strategies and tactics are employed and verified in nature in different periods (Vazzoler 1996), mainly in the environments found in South America that present an ichthyological diversity with more than 5,570 species (Fricke et al. 2023) being 4,967 freshwater distributed in different habitats with specific characteristics (Cassemiro et al. 2023). Due to their different reproductive strategies, species have evolved and adapted their reproductive structures, such as the sperm cells to find greater reproduction efficiency. Thus, different spermatozoa shapes are verified, and this results in different swimming/movement behaviors in the aquatic habitat (Viveiros et al. 2014), varying according to the environment in which spawning occurs, something already observed since the 1930s for South American fish (Ihering and Azevedo 1935).

The sperm movement behavior in relation to the environment is related to sperm biology; thus, the way of swimming and displacement in water depends on the structure and components of the cell (Viveiros et al. 2014). However, numerous other factors directly influence sperm behavior (Kowalski and Cejko 2019), such as the composition of the aquatic environment because the fish spermatozoa only become activated after contact with an activating solution that triggers the beginning of movement. This occurs due to the changes to which the cells are exposed after collection because, until then, they are immobile in the testes and diluted in seminal plasma, which has a composition that maintains inactivation (Alavi and Cosson 2005).

The sperm activation process occurs when the semen is diluted in water or the activating solution used; under these circumstances, ATP activation occurs, and the flagellar beat starts from a cascade of events that provide movement (Alavi and Cosson 2006). From this, the spermatozoa will swim in an aquatic environment with the objective of finding the micropyle (channel used for fertilization) during the short time of movement to effect fertilization. In this sense, many activating solutions with the aim of providing greater sperm vigor (velocity) and movement time were successfully employed for South American fish species (Viveiros and Godinho 2009), but still without a defined standard so that it can be efficiently applied by producers. Allied with this, many times, the lack of adequate equipment and technical conditions in the production companies of young forms limit the evaluations in the field because, as there is still a favorable availability of broodstock, few are concerned with evaluating the sperm quality, even knowing it is crucial for reproduction. In this sense, more refined analyses are restricted to the laboratories of universities and research institutions; however, this scenario is gradually changing with the possibility of sperm cryopreservation of some species used in genetic improvement, hybrid production, and asynchrony of males and females. Currently, some farms in South America already have basic equipment for assessing sperm quantity and quality, contributing to a more rational use of gametes.

In addition, it should be noted that knowledge of sperm characteristics is of fundamental importance for the reproductive success of the species and different evaluation techniques can be applied for its measurement (Cosson et al. 2008; Cabrita et al. 2014). Thus, this chapter approaches the main techniques used to measure the sperm quantity and quality of South American fish, not restricting it only to the most refined techniques that can be used in research laboratories but also bringing a practical approach, which can be applied by producers who need to improve techniques on their properties.

2. Sperm Quantity Analysis

Quantitative analyses are often ignored in fish breeding laboratories, as many producers have not yet visualized the applicability of this type of analysis, and some questions such as "Why quantify the sperm produced by males?" and "What is the importance of this theme in fish reproduction?" are still performed. These questions often arise from different actors in the production chain, including producers, students, technicians, and researchers, and the answer may be through other questions, such as: "Have you ever thought about using semen in a rational way, without loss?" and "Did you know that the genetic material at hand can be optimized, used in other reproductions, and even commercialized?" are just two that open up a range of possibilities based on quantification. These analyses can be summarized only in two, measuring the semen volume collected and sperm concentration. Putting these two measures together, we have the total number of sperm cells collected from the male, and this makes it possible to estimate the number of oocytes that can be fertilized by this fish.

2.1 Sperm Volume

The sperm or semen volume is a parameter that can be measured using graduated syringes, graduated test tubes, such as "falcon" tubes, and micropipettes of variable volume immediately after collection. However, the volume of semen collected does not guarantee that all the semen produced by the fish was collected since the usual methods of collection through abdominal massage in the cranium-tail direction do not remove all semen present in the testicular lumen (Ferreira et al. 2001). This occurs because the fish have different testicle shapes and often have fringes, which limit the complete collection, such as *Rhamdia quelen* testicles (Reidel et al. 2010), either prevent the release or provide little semen in the collection, as is the case of *Pimelodus britski* (Damasceno et al. 2017), *Zungaro jahu* (Nogueira et al. 2012), *Pimelodus maculatus* (Sato et al. 1999) and *Sorubim lima* (Shibatta et al. 2011). In addition, the testicles are found in the lower part of the belly of the fish, below the ribs, as occurs in *Leporinus macrocephalus* (Streit Jr et al. 2003), *Leporinus elongatus* (Streit Jr et al. 2008), *Colossoma macropomum* (Farias et al. 1999), *Prochilodus lineatus* (Streit Jr et al. 2004), *Brycon amazonicus* (Ninhaus-Silveira et al. 2006), *Brycon orbignyanus* (Murgas et al. 2004), among many other species of scales, something that does not prevent collection but makes it difficult to remove the total amount of semen present in the lumen.

Generally, measurements of semen volume are performed in milliliters (mL) or microliters (μL) and can be related to the individual (mL/fish) or by weight (mL/g of fish or mL/kg of fish), the latter also known as relative seminal volume (Bombardelli et al. 2006). Despite the quantification being something simple to carry out, many factors directly interfere with the volume and the sperm concentration of South American species, such as the use of different hormone inducers (Bernardes Jr et al. 2017; Martins et al. 2017), time of collection during the reproductive period (Pires et al. 2017), performance or not of successive seminal collections (Spica et al. 2021), nutritional status of breeders (Tessaro et al. 2012), fish age (Chalde et al. 2016), and presence of environmental contaminants in water (Assis et al. 2021), among others.

The amount of semen released by South American fish is very variable (Table 1) and occurs between individuals of the same species reared under the same conditions (Paulino et al. 2011), such as the large difference in production verified for *Rhinelepis aspera*, an armored catfish native to the Prata basin, where fish weighing 0.8 at 1.6 kg produced 0.6 at 44.2 mL of semen after manual massage (Sanches et al. 2011a). In addition, using inducers allows the increase of seminal fluid and consequent greater volume of semen when compared to non-induced fish (Maria et al. 2012; Araújo et al. 2014). However, this behavior is not always evident, such as the volume of semen collected from *Steidachneridion parahybae*, an endemic catfish to the Paraíba do Sul River, where hormone induction by carp pituitary extract (CPE) did not provide a significant increase in semen volume (Caneppele et al. 2015).

Despite this, the measurement of seminal volume is still not usual in field situations, even with the reproduction routine of most South American fish species being by manual extrusion and "dry" fertilization (Reynalte-Tataje et al. 2013). In this procedure, the female gametes (eggs) are first collected in a clean and dry basin, and then the male is extruded above the egg mass, promoting direct contact of the gametes, which are then homogenized before the addition of water, which promotes fertilization (Zaniboni Filho and Nuñer 2004). This method is usually done in a hurry to avoid loss of post-collection egg quality (Sanches et al. 2011b), and the producers do not "lose one's time" in collecting and measuring semen volume before, although it is a practical procedure. Although efficient, some flaws can be minimized during "dry" fertilization, such as adding the appropriate amount of semen, improving fertilization rates from the best sperm:egg:water ratio (Sanches et al. 2016; Souza et al. 2022). In addition, another possibility that few producers have used is to collect the semen before the eggs (Sanches et al. 2011b) and storage in Styrofoam boxes containing ice (Sanches et al. 2013a), methodology that can be easily used by many laboratories and allows a more accurate estimate of semen to be used depending on the number of eggs released by the females (Sanches et al. 2009, 2016), therefore, there is a need to estimate sperm concentration.

2.2 Sperm Concentration

The sperm concentration is a parameter that can be used to estimate the number of sperm cells per milliliter (sperm/mL). This estimate is also called sperm density and is usually measured using cell counts in counting chambers, such as Makler, Fertility, Fuchs-Rosenthal, and Neubauer hematimetric (hemacytometer), the most usual (CBRA 1998); however, other methodologies can also be used, such as spermatocrit (Sanches et al. 2011c), spectrophotometer, computerized, and flow cytometry (Cuevas-Uribe and Tiersch 2011).

Table 1. Semen quantity parameters (volume and sperm concentration) reported in several Neotropical fish species over the last years. For each species, the article from which the quantity parameters were obtained, the body measure (if available) and the hormonal therapy (if applied), are indicated.

Species/Common Name	Body Measure	Hormonal Therapy	Volume (mL)	Density (x10^9spz/mL)	References
Astyanax altiparanae = A. lacustris	5.52 ± 0.50 g, 6.24 ± 0.18 cm	5.0 mg/kg CPE		0.13 ± 0.07	(Rocha et al. 2020)
		3.0 mg/kg CPE		0.62 ± 0.39	(Yasui et al. 2015)
	7.45 ± 0.16 g, 8.40 ± 0.05 cm	5.0 mg/kg CPE		4.17 ± 0.16	(Pinheiro et al. 2020)
Brycon amazonicus = B. cephalus	1.31 ± 0.69 kg	2.0 mg/kg CPE	16.5 ± 7.0	16.3	(Oliveira and Villacorta 2021)
	0.37 ± 0.06 kg	1+1/3 pellet/kg mGnRH + metoclopramide	4.7 ± 1.5	8.1 ± 3.4	(Bashiyo-Silva et al. 2016)
	0.98 ± 0.08 kg	2.5 mg/kg CPE	8.75 ± 2.88	10.9	(Souza et al. 2022)
Brycon orbignyanus	0.9 ± 0.1 kg	0.25 ml/kg sGnRHa + domperidone	9.1 ± 5.3	13.4 ± 2.1	(Viveiros et al. 2015b)
	1.16 ± 0.12 kg	5.0 mg/kg CPE	7.6 ± 1.1	5.88 ± 1.2	(Palhares et al. 2020)
	1.3 ± 0.5 kg	1.0 mg/kg CPE		5.7 ± 3.4	(Chiacchio et al. 2017)
	1.1 ± 0.8 kg	3.0 mg/kgCPE	5.0	7.1 ± 5.6	(Nascimento et al. 2018)
	1.0 ± 0.1kg	1.0 mg/kg CPE	7.9 ± 3.2	5.68 ± 0.93	**(França et al. 2020)**
		Untreated	5,43 ± 0,73	18.31 ± 3.34	(Oliveira et al. 2015)
Colossoma macropomum	7.0 ± 1.0 kg	0.25 + 2.5 mg/kg CPE	12.6 ± 4.7	7.9 ± 2.4	(Maria et al. 2012) (Vieira et al. 2011) (Martins et al. 2017) (Pires et al. 2017)
		Untreated	0.5 ± 0.6	15.5 ± 4.6	
	5.41 ± 0.29 kg	2.0 mg/kg CPE	5.05 ± 2.00	22.93 ± 9.79	
	5.45 ± 0.34 kg	Untreated	0.55 ± 0.50	40.46 ± 18.31	
	6.2 ± 1.5 kg	0.3 pellet mGnRH + metoclopramide	3.9 ± 0.5	10.7 ± 2.7	
		Untreated	0.9 ± 0.4	7.94 ± 2.0	
	6.4 ± 1.5 g	2.5 mg/kg CPE	6.7	2.7	

Leiarius marmoratus	1.56 ± 0.16 kg	3.0 mg/kg CPE	0.66 ± 0.28	2.51 ± 2.16	(Borges et al. 2020)
	2.54 kg	2.5 mg/kg CPE	1.53	3.68	(Araújo et al. 2014) (Mira et al. 2010) (Spica et al. 2021)
		0.6 pellet mGnRH + metoclopramide	1.70	3.72	
		Untreated	0.0		
	1.41 ± 0.07 kg	0.2 + 0.5 + 2.0 mg/kg CPE	2.84 ± 0.60	1.87 ± 0.27	
		0.25 mL sGnRHa + domperidona	1.59 ± 0.40	6.30 ± 1.03	
		0.9% NaCl	0.15 ± 0.10	9.46	
	3.24 ± 1.24 kg	0.9 + 2.7 mg/kg CPE	0,59 at 1.38	0.35 at 1.98	
Megaleporinus obtusidens	0.62 ± 0.16 kg	60 µg/kg gonadorelin	≈ 4.8 ml/kg	≈ 25	(Bernardes Jr et al. 2017) (Bernardes Jr et al. 2018)
		4.0 mg/kg CPE	≈ 3.5 ml/kg	≈ 25	
		Untreated	≈ 0.5 mL/kg	≈ 35	
	0.79 ± 0.18 kg	60 µg/kg gonadorelin		28 ± 3	
Piaractus mesopotamicus	3.00 ± 0.99 kg	2.5 mg/kg CPE	13.64 ± 6.82	35	(Sanches et al. 2011d) (Teodozia et al. 2020) (Galo et al. 2019) (Kuradomi et al. 2016)
	2.80 ± 0.40 kg		12.03 ± 1.62	53	
		2.0 mg/kg CPE		10.48 ± 5.38	
		2.5 mg/kg CPE		21.09	
	2.38 ± 0.22 kg	10 µg/kg buserelin acetate	6.33 ± 2.49	69.4 ± 3.4	
Piaractus brachypomus		4.0 mg/kg CPE	7.44 ± 2.18	18.58 ± 4.90	(Martínez et al. 2019) (Nascimento et al. 2010) (Ramirez-Merlano et al. 2011)
			5.94 ± 1.86	15.81 ± 5.26	
	3.5 ± 0.5 kg	5.0 mg/kg CPE	1.8 ± 0.6	55.5 ± 2.3	
	5.0 ± 0.6 kg	4.0 mg/kg CPE	13 ± 1	11 ± 1	
Prochilodus magdalenae	0.21 ± 0.02 kg	4.8 mg/kg CPE	2.3 ± 0.4	18.8 ± 7.3	(Atencio et al. 2021)
	0.17 ± 0.02 kg	4.0 mg/kg CPE	1.3 ± 0.4	18.5 ± 3.4	(Atencio et al. 2013)
	0.25 ± 0,01 kg	4.5 mg/kg CPE	1.05 ± 0.1	17.37 ± 0.43	(Martínez et al. 2012)

Table 1 contd. ...

...Table 1 contd.

Species/Common Name	Body Measure	Hormonal Therapy	Volume (mL)	Density (x10⁹spz/mL)	References
Prochilodus lineatus	0.55 ± 0.25 kg	0.5 + 5.0 mg/kg CPE	1.4 ± 0.7	79.5 ± 2.4	(Andrade et al. 2014)
		0.25 + 0.5 mL/kg buserelin acetate	1.2 ± 0.4	77.1 ± 25.6	(Sanches et al. 2015a)
		60 + 60 µg/kg gonadorelin	1.4 ± 1.0	86.1 ± 16.5	(Paula et al. 2019)
	0.51 ± 0.10 kg	3.0 mg/kg CPE	1.6 ± 1.1	28.3 ± 2.1	(Romagosa et al. 2010)
	2.00 ± 0.80 kg	0.5 + 5.0 mg/kg CPE	6.41 ± 3.17	31.31 ± 4.86	
	0.41 ± 0.13 kg	0.5 + 5.0 mg/kg CPE	0.51 ± 0.35	29.5	
Pseudoplatystoma reticulatum		Untreated		2.01	(Streit Jr et al. 2012)
		2.5 mg/kg CPE		1.00	
Pseudoplatystoma magdaleniatum		0.4 mL sGnRHa + domperidona	6.1 ± 4.3	55.9 ± 22.9	(Herrera-Cruz et al. 2019)
Pseudoplatystoma mataense	1.50 ± 0.20 kg	4.0 mg/kg CPE	10.14 ± 1.83	35.18 ± 3.8	(Ramirez-Merlano et al. 2011)
Rhamdia quelen	0.32 ± 0.07 kg	2.5 mg/kg CPE	5.0	60.8 ± 13.4	(Sanches et al. 2013a)
	0.76 ± 0.17 kg	3.0 mg/kg CPE	13 ± 4	59 ± 21	**(França et al. 2022)**
	0.25 ± 0.06 kg	3.0 mg/kg CPE	6.23 ± 1.30	34 ± 3	(Neumann et al. 2019)
	0.14 ± 0.02 kg	2.5 mg/kg CPE	3.47 ± 0.45	1.34 ± 0.13	(Rodrigues et al. 2022)
	0.12 ± 0.02 kg		2.28 ± 0.66	1.88 ± 0.02	
	0.13 ± 0.03 kg		6.00 ± 2.29	1.19 ± 0.04	
Steindachneridion parahybae	0.54 ± 0.04 kg	Untreated	17.32 ± 2.20	9.24 ± 0.88	(Caneppele et al. 2015)
	0.52 ± 0.03 kg	3.0 mg/kg CPE	18.91 ± 1.97	8.75 ± 0.58	
	0.64 ± 0.10 kg	Untreated		35.40 ± 0,25	(Sanches et al. 2013b)
	0.72 ± 0.1 kg	Untreated	16.8 ± 8.6	9.3 ± 5.9	(Sanches et al. 2015b)
	1.66 ± 0.55 kg	Untreated	26.80 ± 7.07	17.0	(Sanches et al. 2016)

The estimate using the cell count in a Neubauer hematimetric chamber is more widespread among laboratories that work with seminal analyses of South American fish (Viveiros et al. 2014a). For this purpose, the semen must be previously diluted (1:500; 1:1000, or 1:2000) in a 4.63% buffered saline formalin solution (20% of buffer solution [$Na_2HPO_42H_2O$ – 3.1% + KH_2PO_4 – 1.27%] + 30% of saline solution [NaCl – 18.02%] + 12.5% of formalin [P.A. – 37%] + 37.5% of distilled water), or formalin citrate 2.94% modified (96% of sodium citrate [$Na_3C_6H_5O_7$ – 3%] + 4% of formol [P.A. – 37%]), according Mies-Filho (1987). From that point, the fixed material is added to the chamber for performed count. It should be noted that the 4.63% saline buffered formalin solution can also be used as a fixative for cells that will be submitted to evaluate morphological changes (Streit Jr et al. 2004).

The Neubauer chamber is composed of two fields, and each field contains 25 squares equivalent to 1 mm² with a depth of 0.1 mm; thus, the volume of the chamber corresponds to 0.1 mm³. It is recommended to count five random squares (diagonal or on the extremities) of each counting field. The summatory of cells counted in these 10 squares is used to calculate the concentration in cells/mL, according to the formula, below adapted from Herman et al. 1994):

$$\text{SC(SPZ/mL)} = \left(\frac{\sum \text{SPZ}}{\text{c.s.}}\right) \times \frac{\text{t.s.} \times \text{dilution} \times 1{,}000}{\text{c.v.}}$$

where: SC = sperm concentration; SPZ = sperm; ∑SPZ = total number of sperm COUNT; c.s. = COUNT squares (=10 for two field); t.s. = total squares in one field of 1 mm² (= 25); c.v. = chamber volume (= 0.10 mm³); diluition = semen dilution factor; 1,000 = constant for conversion to mL.

The spectrophotometer method uses a photocolorimeter, a device that allows measuring the amount of a light beam that passes through a standardized volume of semen diluted in an established proportion (Cuevas-Uribe and Tiersch 2011). The part of the light beam that reaches the photoelectric cell has an inverse relationship to the concentration of the sample under examination. The device is calibrated against a sample of diluted semen of known concentration through previous determination by direct count.

The evaluation by the spermatocrit method is based on measuring the percentage of sperm cells present in the total volume of semen (Shimoda et al. 2007). The measurement is performed using micro-hematocrit capillary tubes filled with approximately 70% of semen, sealed at one end, and subjected to centrifugation in a hematocrit tubes centrifuge, and then the cell mass percentage is measured in the semen through a hematimetric ruler. This measure presents proportional linear correlations with concentration determined by the Neubauer chamber counting method in some species of fish (Viveiros et al. 2014a). For being a method that is easy to apply in commercial systems and field research, it is welcome in sperm analysis routines; however, the application is recommended based on prior knowledge of the literature on the species in question (Sanches et al. 2011c).

Another procedure is the computerized analysis using CASA, where the sperm concentration can be determined by means of specific software through previous dilution of the semen in specific chambers. In this system, the software processes the images captured during the analysis and determines the amount of sperm. The use of CASA systems presents several advantages, such as the speed and versatility of the analyses in an objective way; however, they need validation of the methods to be efficient according to the developers' recommendations (Gallego and Asturiano 2018).

The choice of method for evaluating sperm concentration is conducted according to the laboratory routine and the availability of equipment available in each laboratory so the efficiency and reliability of the results are crucial for these analyses. Wide variations in the sperm concentration of fish are reported in the literature (see Table 1), and their knowledge is important in reproduction laboratories because from these data, gametes, and reproducers can be optimized, in addition to increasing the fertilization percentage through the application of inseminating doses of the South American species (Bombardelli et al. 2006; Sanches et al. 2009). In addition, they are fundamental in sperm cryopreservation programs, as in these programs, one must have knowledge of the quantity and sperm that are being frozen to be used in fertilization (Adames et al. 2015; Neumann et al. 2019; França et al. 2022).

Despite this, due to the diversity of South American species studied so far, we have compiled information (Table 1) on some species that have commercial and/or conservation importance that have been studied; however, we decided to use more up-to-date references from the compilation of data already carried out by Viveiros and Godinho (2009). Evidently, other works not mentioned in the table are of great importance for understanding the patterns presented by the native species of South America.

3. Sperm Quality Analysis

To begin with, we can assert that any quantifiable parameter directly correlated with the fertilization capacity could potentially be used as a sperm quality marker, and over the years, a relatively high number of sperm parameters have been used to assess sperm quality. These sperm biomarkers have so far been documented in scientific papers, and several traits of the fish/sperm itself, such as osmolality, pH and chemical composition of the seminal plasma, enzymatic activity, ATP concentration, spermatocrit and sperm density, sperm motility, or/and sperm morphology and ultrastructure have been linked to the ability of sperm to fertilize the ova [reviewed by Kowalski and Cejko (2019) and Cabrita et al. (2014)]. However, despite it having been reported that it is the set of sperm biomarkers (as a whole) that contribute to determining sperm quality, sperm motility is currently considered the most useful parameter (by itself) for assessing sperm quality in fish. Nowadays, sperm motion values from more than 300 fish species have been reported in more than 1,500 scientific articles covering a wide range of topics, from molecular biology to ecology issues (Gallego and Asturiano 2019).

3.1 Sperm Motility in Fish: How to Measure

Sperm motility assessment can be performed in two different ways in the lab: (i) the subjective way, in which a technician evaluates sperm motility through a simple observation under the microscope, and (ii) the objective way, in which a sophisticated software records a video-track of sperm cells for calculate the trajectories and their kinetic characteristics. Although the subjective evaluation method has been the most used technique in history, some problems have emerged from this technique (Cosson 2021). The first drawback is focused on the limitation of the human eye, through which we can only provide a simple evaluation of spermatozoa, giving (i) the percentage of motile cells and (ii) the motility duration (how long the sperm move). In addition, this type of evaluation will depend on the observer's experience and several aspects (e.g., sperm density, the velocity of cells, drift, etc.), which can induce overestimated or underestimated assessments (Gallego et al. 2018). Therefore, the low reproducibility of this subjective motility evaluation, which can result in variations of 30–60% from the same sample, often makes it difficult to interpret and compare the results intra- and interlaboratory (Rosenthal et al. 2010).

On the other hand, the gradual appearance of computer-assisted sperm analysis (CASA) systems has made it possible to estimate a higher number of sperm motion parameters not given by subjective evaluation, providing the scientific community with new useful tools to be applied in multidisciplinary studies (Gallego and Asturiano 2018). These systems, which are the evolution of multiple photomicrography exposure and video-micrography techniques for spermatozoa track, represent an objective, sensitive, and accurate technique for obtaining sperm kinetic features. CASA systems were first used in the 1970s in mammalian sperm, and only in the 1990s modern CASA systems have been adapted for fish spermatozoa studies. Differences in the size and the biology of fish and mammalian spermatozoa may explain the delay in the release of adequate tools for measuring sperm motility in fish. To date, these systems have been used and validated in a wide range of animal groups, such as marine invertebrates, birds, marine mammals, reptiles, and even insects (van der Horst 2020).

The most used parameters for fish sperm analysis were revised by Kowalski and Ciejko (2019). In general, the percentages of motile (TM) and progressive motile spermatozoa (PM) can provide a general overview of the quality of the sperm sample. TM means any spermatozoa showing any movement, while PM is determined by the spermatozoa, which swim in an essentially straight line. However, the most useful parameters for some authors are the spermatozoa velocities (Figure 1). In this respect, VCL (curvilinear velocity) is defined as the actual velocity along the real spermatozoon trajectory, and VSL (straight-line velocity), means the straight-line distance between the start and end points of the track. Finally, VAP (average path velocity) is the velocity along a derived smooth path, and using VAP, it is possible to generate other related parameters such as the percentage of fast (FA), medium (ME), and slow (SL) spermatozoa, which can provide useful information for assessing the kinetic evolution of spermatozoa over the whole swimming period (Blanes-García et al. 2021).

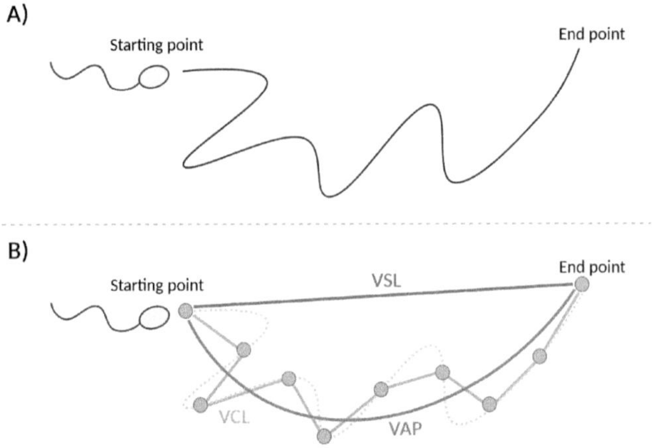

Figure 1. Diagram of some sperm kinetic parameters of fish sperm: (A) real trajectory of the spermatozoa; (B) spermatozoa velocities assessed by a CASA system. (VCL: Curvilinear velocity; VSL: Straight-line velocity; VAP: Averaged path velocity.)

3.2 Sperm Motility in South American Tropical Fish

Since this date, a high number of sperm kinetic parameters from more than 40 Neotropical species have already been reported in more than 200 scientific articles, covering a wide range of topics such as sperm physiology, sperm storage, broodstock management, and ecotoxicology studies. However, five of these species represent more than 50% of published papers, where the "tambaqui" (*C. macropomum*), the "curimba" (*P. lineatus*), and the "jundiá" (*R. quelen*) have been the most studied Neotropical species. Table 2 summarizes the most important sperm motion parameters (motility, curvilinear velocity, straight-line velocity, average path velocity, and sperm motility duration) reported in several Neotropical fish species over the last few years. Regarding motility values, fresh sperm from almost all species listed in Table 2 showed high motility initial values over 80% of motile cells, and even many exceeded 90%. Only a few species (*S. brasiliensis, R. voulezi, B. hilarii*, and *J. multidentata*) showed motility values around 70% after the activation process. Regarding sperm velocities, only data from fish sperm assessed by a CASA system are available. In that sense, values range widely from species showing from 60–65 µm/s in *P. mesopotamicus* (Nascimento et al. 2010) or *P. promelas* (Frankel et al. 2018) to 240–280 µm/s in *P. lineatus* (Viveiros et al. 2019), or *B. orbignyanus* (Chiacchio et al. 2017). However, despite these huge differences (4-fold higher), that contrast could be due to the technical factors of the CASA system with which the samples were analyzed (Gallego et al. 2013). In that sense, it is important to highlight that motility parameters were highly affected by the species and the CASA used for analyses, so special care should be taken regarding recording conditions and the quality of video recordings (Boryshpolets et al. 2013). Regarding the duration of the sperm movement, data from Neotropical species also present a wide dispersion, and we can find values ranging from 25–30 s in *R. voulezi* or *B. cephalus* (Diemer et al. 2014; Bashiyo-Silva et al. 2016) to even 612 s in *P. scrofa* (Coser et al. 1984). The

Table 2. Sperm motion parameters (Motility), Curvilinear Velocity (VCL), Rectilinear Velocity (VSL), Average Path Velocity (VAP), Sperm Motility Duration (SMD) reported in several Neotropical fish species over the last years. For each species, the article from which the motion parameters were obtained and the topic on which the article had focused are indicated (ECO: Ecotoxicology; CRYO: Cryobiology; BROOD: Broodstock management; STO: Short-Term Storage; DES: Descriptive Studies.)

Neotropical Fish Species	Motility (%)	VCL (µm/s)	VSL (µm/s)	VAP (µm/s)	SMD (s)	Topic	Reference
Astyanax altiparanae	90 ± 2	65 ± 1	42 ± 2	44 ± 2		ECO	(Assis et al. 2021)
Brycon amazonicus (*siebenthalae*)	88 ± 9				41 ± 7	CRYO	(Cruz-Casallas et al. 2005; Oliveira et al. 2018)
Brycon henni	88 ± 21	108 ± 37	59 ± 23	77 ± 29		CRYO	(Restrepo-Betancur et al. 2017)
Brycon insignis	98 ± 4	122 ± 26	87 ± 22	103 ± 26		CRYO	(Viveiros et al. 2012a)
Brycon nattereri	99 ± 4					CRYO	(Viveiros et al. 2012b)
Brycon hilarii	68 ± 5	125 ± 5	71 ± 3	105 ± 6		BROOD	(Costa et al. 2022)
Brycon opalinus	94 ± 1				99 ± 2	CRYO	(Orfão et al. 2011)
Brycon orbignyanus	92–94	239–243	154–156	207–213	48 ± 1	BROOD	(Chiacchio et al. 2017; Palhares et al. 2021)
Brycon cephalus	81–100				32 ± 8	BROOD	(Bashiyo-Silva et al. 2016)
Brycon vermelha	90–95				46 ± 1	DES	(Faustino et al. 2015)
Brycon lundii	80–85					STO	(Marques and Godinho 2004)
Brycon orthothaenia	90 ± 10				39 ± 6	CRYO	(Melo et al. 2018)
Colossoma macropomum	98 ± 4	165 ± 4	115 ± 4	128 ± 5		BROOD	(Pires et al. 2019)
Jenynsia multidentata	68 ± 6	85–88	81–83			DES	(Roggio et al. 2014)
Leporinus elongatus	85–95				56 ± 18	STO	(Streit Jr et al. 2008)
Leporinus macrocephalus	79 ± 9					CRYO	(Ribeiro and Godinho 2003)
Leporinus obtusidens	93 ± 2	160–170			216 ± 103	CRYO	(Júnior et al. 2017; Taitson et al. 2008)

Table 2 contd. ...

...Table 2 contd.

Neotropical Fish Species	Motility (%)	VCL (µm/s)	VSL (µm/s)	VAP (µm/s)	SMD (s)	Topic	Reference
Leporinus friderici	85–90					STO	(Marques and Godinho 2004)
Piaractus brachypomus	86 ± 1				45 ± 1	CRYO	(Ramirez-Merlano et al. 2011)
Piaractus mesopotamicus	95 ± 3	60	33	46	50–60	CRYO	(Nascimento et al. 2010; Kuradomi et al. 2016)
Pimephales promelas	80–83	65–70	85–90	50–55		ECO	(Frankel et al. 2018)
Prochilodus marggravii	90–95					STO	(Marques and Godinho 2004)
Prochilodus brevis	99 ± 1	98 ± 8	44 ± 9	69 ± 9		CRYO	(Almeida-Monteiro et al. 2020)
Prochilodus lineatus	94±7	283±50	124±26		96 ± 2	BRO	(Viveiros et al. 2019)
Prochilodus magdalenae	95–100	90–110	45–55	55–60		ECO	(Sierra-Marquez et al. 2019)
Prochilodus mariae	95 ± 3				52 ± 5	CRYO	(Medina-Robles et al. 2021)
Prochilodus scrofa	82–87				612 ± 78	BRO	(Coser et al. 1984; Kavamoto et al. 1997)
Prochilodus vimboides	93–5				98–134	STO	(França et al. 2020)
Rhamdia quelen	71 ± 12	89 ± 22	42 ± 6	56 ± 8	48 ± 2	STO	(Sanches et al. 2013b)
Rhamdia voulezi	67 ± 6				25 ± 2	BRO	(Diemer et al. 2014)
Salminus brasiliensis	> 70				300 ± 84	BRO	(Coser et al. 1984)
Steindachneridion melanodermatum	99 ± 1	185 ± 14	49 ± 5	87 ± 4	154 ± 72	STO	(Marcos et al. 2015; Santana et al. 2019)
Steindachneridion parahybae	89 ± 7	107 ± 15	77 ± 22	83 ± 22		STO	(Sanches et al. 2015b)
Steindachneridion scriptum	97±2				78±4	CRYO	(Pereira et al. 2019)

differences found between these species could be due to the reproductive biology of each species, as sperm kinetic parameters (motility, speed, and longevity) are usually closely linked to the reproductive behaviour of the species (Gallego et al. 2014). In addition, these differences could also be due to how the sperm is activated. In that sense, Viveiros and Godinho (2009) published a nice revision about some Brazilian fish species, reporting that various activation media have been used to induce sperm motility and evaluate fresh sperm quality, and they can generate disparate results when activating the same sample, as occur in species such as *P. lineatus*, *L. elongatus*, or *B. amazonicus*.

On the other hand, sperm motility has been widely used for assessing sperm storage techniques and methods. In that sense, this topic has been the most investigated field using sperm motility as a research tool, and more than 100 scientific publications reporting kinetic spermatozoa parameters have contributed to the discovery and improvement of sperm storage protocols in many Neotropical fish species. First, and for keeping the sperm by either short-, medium, or long-term period, it is mandatory to design an "extender" in which we can dilute the original sperm sample to keep the sperm quality over time. As in the rest of the fish, the addition of extenders stabilises sperm conditions and suppresses the initiation of motility during storage, overcomes the problem of desiccation, and reduces bacterial growth (Viveiros et al. 2014b). In that sense, the original sperm sample collected is usually too concentrated (in the order of billions of spermatozoa), so the task of lowering that initial concentration by using an extender is essential both to keep the sperm immotile and for keeping the sperm motion parameters as high as possible. As a rule, a proper extender must be iso-osmotic and have a chemical composition like the seminal plasma. Regarding Neotropical fish, using the extender Beltsville thawing solution (BTS), which was developed for pigs, has shown satisfactory results for fish sperm in several Neotropical species. For example, in *P. lineatus*, sperm cooled in BTS yielded a motility rate above 60% after four days, while no motility was observed for undiluted control samples (Orfão et al. 2010). This same extender (BTS) has also been successfully applied in other species such as *C. macroporum* (Garcia et al. 2016), *A. lacustris* (Machado et al. 2022), *L. marmoratus* (Gheller et al. 2019), and *B. orbignyanus* (Viveiros et al. 2015a). However, the variety of extenders is great, and in the literature, we can find different diluents, such as the Hank's balanced salt solution (HBSS) successfully applied in *C. macroporum* (Garcia et al. 2016) or the "powdered coconut water" extender solution (ACP), which was developed and successfully adapted for Neotropical species (Zanandrea et al. 2014; Linhares et al. 2015). However, it is important to remark that instead of diluting the original sperm sample, it is also possible to apply a chilled storage method to fresh (undiluted) sperm for a "short" time without compromising its quality (motility). For example, the study carried out by Marques and Godinho (2004), in which they analyzed six Neotropical characiformes fishes, reported that the shortest sperm storage duration (7 h) was recorded for *L. friderici* (reaching motilities about 30%), whereas the longest duration (20 h) was obtained with the sperm of *P. lineatus* (reaching motilities about 60%).

On the other hand, and as mentioned at the beginning of the chapter, it is noteworthy that most of the parameters evaluated using CASA systems have been correlated positively with spermatozoa fertilization potential; thus, CASA is a very useful tool for assessing sperm quality in fish reproduction research. Regarding Neotropical fish, sperm motion parameters have been correlated with fertilization and hatching rates. For example, in *P. lineatus*, there were positive correlations between fertilization rates and all sperm velocities, with the highest correlation ($r = 0.8$) for VCL (Viveiros et al. 2010). In the tambaqui, *C. macroporum*, a study carried out by Maria et al. (2015) reported that all sperm motility parameters presented a strong positive correlation with fertilization rates ($r = 0.89$, 0.86, 0.79, 0.78, 0.69 for TM, PM, VCL, VAP, VSL, respectively). Finally, Araújo (2011) reported in *S. parahybae* a direct proportional linear correlation between VCL and VAP and the fertilization rates. In that study, Pearson's linear correlation indicated that MOT, VCL, and STR were good indicators of sperm quality and for predicting fertilization rates.

Finally, and in addition to individually assessing the results of the kinetic parameters (e.g., motility, velocity, etc.), another approach currently used is the study of fish sperm subpopulations. The appearance of sperm subpopulations means that sperm is not homogenous throughout the whole population of an ejaculate and that spermatozoa with various kinematic properties exist within one sample (Martínez-Pastor 2022). The number and type of subpopulations could vary inter and intra-species, but the scarce reports published on fish show that sperm samples usually display three or four sperm subpopulations (Marinović et al. 2021). In addition to the number of subpopulations, their biological meaning is even more important and could be used for assessing different techniques and processes, such as cryopreservation protocols, toxicity tests, etc. Some authors have correlated the different subpopulations with crucial parameters such as the fertilization rates (Beirão et al. 2011), and it can be clearly seen as samples containing higher percentages of subpopulations with fast and liner spermatozoa show higher fertilization rates than other ones. Regarding Neotropical species, there is only one study focused on spermatozoa subpopulations (Gallego et al. 2017). In that sense, the assessment of the kinetic data obtained with a CASA system (applying a two-step cluster analysis) yielded in tambaqui three different subpopulations in fresh sperm: SP1, considered as a slow nonlinear subpopulation; SP2, considered as a fast nonlinear subpopulation; finally, SP3, considered as a fast linear subpopulation. In addition, this study also yielded that SP2 and SP3 (the fast-spermatozoa subpopulations) showed a high-positive correlation with the fertilization rates ($r = 0.93$). In addition, the positive significant correlations found in curvilinear velocity ($r = 0.78$), straight-line velocity ($r = 0.57$), and average velocity ($r = 0.75$) indicate that sperm kinetic features seem to be a key factor in the fertilization process in tambaqui, as occur in other fish species. In this sense, it would be important to highlight that further research in Neotropical fish needs a deeper data analysis, where a broader use of CASA systems and in-depth statistical analysis (such as subpopulations studies) would allow progress and achieve more relevant results both in aquaculture applied industries such research centers.

3.3 Emerging Tools for Fish Sperm Quality Analysis

In addition to motility as an easy and useful tool for assessing sperm quality in fish, a relatively high number of sperm quality biomarkers have been successfully applied in fish (Cabrita et al. 2014). In that sense, it is important to highlight that these biomarkers add more relevant information about the "general quality" of the gamete (spermatozoa), giving, for example, if the cell shows DNA damage even though it can move for fertilizing the ova.

One of the most studied effects generated by the cryopreservation process is the cryodamage on nuclear DNA, which has been analyzed using various methods (e.g., Comet assay). For example, (Paula et al. 2019) carried out a trial testing several cooling rates in *P. lineatus* sperm, and the authors reported that DNA damage was significantly higher in all treatments when compared to fresh sperm. In addition, data also revealed significant differences in DNA integrity between the two cooling rates (15°C/min and 30°C/min) applied during the freezing process. In another study carried out by Atencio-Garcia et al. (2023) in *Sorubim cuspicaudus*, different cryo-extenders were tested, assessing the damage suffered by cells on mitochondria, plasmatic membranes, and DNA fragmentation. Data showed that in post-thawed sperm, cell damage increased to rise 13-to-18.5-fold times (regarding DNA integrity) than in fresh sperm. In addition, the study's results show that DNA fragmentation suffered by the spermatozoa during the cryopreservation process (pre-freezing, freezing, and thawing) also affected its fertilizing capacity. Finally, in another study carried out in *P. magdalenae* (Martínez et al. 2012), authors reported that the DNA integrity of frozen-thawed sperm was significantly higher than that found in fresh sperm. In that sense, DNA damage varied according to DMSO and glucose concentration, and two cryoprotectants had an interactive effect.

Regarding the topic of this book, fish gamete cryopreservation, it is important to note that "supplementary" quality biomarkers acquire a special relevance since, after the freezing and thawing process, the spermatozoa may have suffered genetic damage that could influence the correct fish embryo development (Figueroa et al. 2016). Those kinds of DNA tools, which range from simple techniques to very sophisticated approaches involving molecular methods, have been scarcely applied over the last years on Neotropical fish species but must be gradually and regularly applied in the very near future with the aim of improving the cryopreservation techniques in South American fish species. This "new" generation of tools will improve the knowledge on sperm quality assessment, complementing the information provided by the traditional and mostly used sperm motion assessment, and the identification of predictive estimators or markers of sperm quality would have major applications in research, fish farms, and biotechnological industries.

Acknowledgements

During the edition of this chapter, V. Gallego was granted a "Ramón y Cajal" contract (RYC2021-031558-I) funded by the Ministerio de Ciencia e Innovación (Spain) and the NextGenerationEU (European Union); and a "Beatriz Galindo" aid

(BG22/00024; for the attraction of research talent – 2022) funded by the Ministerio de Universidades (Spain).

References

Adames, M.S., C.P.R. Toledo, G. Neumann, A.H. Buzzi, C.N. Buratto, P.A. Piana et al. 2015. Optimization of the sperm:oocyte ratio and sperm economy in the artificial reproduction of *Rhamdia quelen* using fructose as a sperm motility modulator. Anim. Reprod. Sci. 161: 119–128.

Alavi, S.M.H. and J. Cosson. 2005. Sperm motility in fishes (I). Effects of temperature and pH: A review. Cell Biol. Int. 29: 101–110.

Alavi, S.M.H. and J. Cosson. 2006. Sperm motility in fishes (II). Effects of ions and osmolality: A review. Cell. Biol. Int. 30: 1–14.

Almeida-Monteiro, P.S., R. Pinheiro, M.S. Oliveira-Araújo, Y.S. Sales, R.V. Nascimento, L.T. Nunes et al. 2020. Sperm vitrification of *Prochilodus brevis* using Powder Coconut Water (ACP-104) in association with different cryoprotectant concentrations. Aquac. Res. 51: 4565–4574.

Andrade, E.S., A.F.S. Carvalho, M.R. Ferreira, F.G. Paula, F.S. Rodrigues and V.O. Felizardo. 2014. Indutores hormonais na reprodução artificial de curimba (*Prochilodus lineatus*) R. Bras. Rep. Anim. 38: 230–236.

Araújo, R.V. 2011. Motilidade, velocidade e fertilidade do sêmen de surubim-do-paraíba *Steindachneridion parahybae* (Siluriformes) criopreservado em diferentes diluidores. Lavras, 93f. Tese (Doutorado) no Programa de Pós-Graduação em Zootecnia, área de concentração em Produção Animal, Universidade Federal de Lavras. Lavras.

Araújo, J., D.P. Streit Jr, J.S. Ribeiro, E. Martins, F.N. Souza, C.A.L. Oliveira et al. 2014. Ovopel and carp pituitary extract as spawning inducers in males of the Amazon catfish *Leiarius marmoratus* (Gill, 1970). Braz. Arch. Bio. Technol. 57: 882–886.

Assis, C.B., G.S. Branco, R.G. Moreira and J.P. Pinheiro. 2021. Aluminum at environmental concentrations affects the sperm quality of the freshwater teleost *Astyanax altiparanae*: An *in vitro* study. Comp. Bio. Physiol. C: Toxicol. and Pharmacol. 243: 108983.

Atencio, V.J., E.J. Perez, J.A. Espinosa and S.C. Pardo. 2013. Evaluation of dimethylacetamide as cryoprotectant for cryopreservation of sperm of bocachico *Prochilodus magdalenae*. Arch. Med. Vet. 45: 151–158.

Atencio-García, V., S.S. Cabrales-Hessen and J.A. Espinosa-Araujo. 2021. Efecto de etilenglicol y leche en polvo en la criopreservación de semen de bocachico *Prochilodus magdalenae*. R. Col. Biotec. 23: 25–35.

Atencio-García, V., D. Padilla-Izquierdo, J. Robles-González, M. Pietro-Guevara, S. Pardo-Cardoso and J. Espinosa-Araujo. 2023. Damage to *Sorubim cuspicaudus* sperm cryopreserved with ethylene glycol. Animals 13: 235.

Bashiyo-Silva, C., R.D.S. Costa, D.D.C. Ribeiro, J.A. Senhorini, R. Veríssimo-Silveira and A. Ninhaus-Silveira. 2016. Hormonal induction of *Brycon cephalus* (Characiformes, Characidae) to spermiation using D-Ala6, pro9net-mGnRH + metoclopramide. Zygote 24: 319–325.

Beirão, J., E. Cabrita, S. Pérez-Cerezales, S. Martínez-Páramo and M.P. Herráez. 2011. Effect of cryopreservation on fish sperm subpopulations. Cryobiology 62: 22–31.

Bernardes Júnior, J.J., R.A. Bombardelli and A.P. Nuñer. 2017. Gonadorelin increases semen production and does not affect its quality in *Leporinus obtusidens*. Anim. Reprod. Sci. 185: 154–160.

Bernardes Júnior, J.J., J.E. Jimenez, R.A. Bombardelli and A.P. Nuñer. 2018. Changes in external osmolality and ionic composition affect *Megaleporinus obtusidens* sperm motility. Anim. Reprod. Sci. 190: 63–74.

Blanes-García, M., P. Risueño, L. Pérez, J.F. Asturiano and V. Gallego. 2021. Sperm quality parameters of Iberian toothcarp (*Aphanius iberus*) and Valencia toothcarp (*Valencia hispanica*): New conservation tools from a gamete perspective. Aquaculture 530: 735819.

Bobe, J. and C. Labbé. 2010. Egg and sperm quality in fish. Gen. Comp. Endocrinol. 165: 535–548.

Bombardelli, R.A., E.F. Mörschbächer, R. Campagnolo, E.A. Sanches and M.A. Syperreck. 2006. Insemination dose for artificial fertilization of grey jundia oocytes, *Rhamdia quelen*. R. Bras. Zootec. 35: 1251–1257.

Borges, A.M., K.O. Araujo, I. Pivato and R.D. Navarro. 2020. Ultrastructure and sperm cryopreservation of the amazon catfish (*Leiarius marmoratus*) in captivity. Arq. Bras. Med. Vet. Zootec. 72: 253–262.

Boryshpolets, S., R.K. Kowalski, G.J. Dietrich, B. Dzyuba and A. Ciereszko. 2013. Different computer-assisted sperm analysis (CASA) systems highly influence sperm motility parameters. Theriogenology 80: 758–765.

Brasil-Sato, M.C. and G.C. Pavanelli. 1999. Ecological and reproductive aspects of *Neoechinorhynchus pimelodi* and *Pimelodus maculatus* of the São Francisco River, Brazil. R. Bras. Zootec. 16: 73–82.

Cabrita, E., S. Martínez-Páramo, P.J. Gavaia, M.F. Riesco, D.G. Valcarce, C. Sarasquete et al. 2014. Factors enhancing fish sperm quality and emerging tools for sperm analysis. Aquaculture 432: 389–401.

Caneppele, D., E.A. Sanches and E. Romagosa. 2015. Sperm production of *Steidachneridion parahybae* (Steindachner 1877) and the effect of hormonal induction throughout one reproductive cycle. J. Appl. Ichthyol. 31: 54–61.

Cassemiro, F.A., J.S. Albert, A. Antonelli, A. Menegotto, R.O. Wüest, F. Cerezer et al. 2023. Landscape dynamics and diversification of the megadiverse South American freshwater fish fauna. Proc. Natl. Acad. Sci. USA 120: e2211974120.

Chalde, T., A. Gárriz, E.A. Sanchez and L.A. Miranda. 2016. Influence of pejerrey *Odontesthes bonariensis* (Valenciennes, 1835) broodstock age on gamete quality, reproductive performance and plasma sex steroid levels during the spawning season. Aquac. Res. 47: 969–982.

Chiacchio, I.M., I.L. Almeida, M.C. Leal and A.T.M. Viveiros. 2017. Sperm quality and its freezing ability throughout the spawning season in *Prochilodus lineatus* and *Brycon orbignyanus*. Theriogenology 90: 284–288.

Coser, A.M., H. Godinho and D. Ribeiro. 1984. Cryogenic preservation of spermatozoa from *Prochilodus scrofa* and *Salminus maxillosus*. Aquaculture 37: 387–390.

Cosson, J., A.L. Groison, M. Suquet, C. Fauvel, C. Dreanno and R. Billard. 2008. Marine fish spermatozoa: Racing ephemeral swimmers. Reproduction 136: 277–294.

Cosson, J. 2021. A 40-years journey with fish spermatozoa as companions as I personally experienced it. Fish Physiol. Biochem. 47: 757–765.

Costa, B.B., J.A. Povh, E.A. Sanches, L.N. Spica, R.B. Rodrigues, N. Teixeira et al. 2022. Characterization of sperm quality in *Brycon hilarii*: How does morphology affect sperm movement? Theriogenology Wild 1: 100007.

Cruz-Casallas, P.E., D.A. Lombo-Rodríguez and Y.M. Velasco-Santamaría. 2005. Milt quality and spermatozoa morphology of captive *Brycon siebenthalae* (Eigenmann) broodstock. Aquac. Res. 36: 682–686.

Cuevas-Uribe, R. and T.R. Tiersch. 2011. Estimation of fish sperm concentration by use of spectrophotometry. pp. 162–200. *In*: Tiersch, T.R. and C.C. Green [eds.]. Cryopreservation in Aquatic Species. Word Aquaculture Society, Baton Rouge, USA.

Damasceno, D.Z., R.A. Krause and M.S. Adames. 2017. Induced spermiation of *Pimelodus britskii* (Teleostei: Pimelodidae) during the reproductive period. Aquac. Res. 48: 862–874.

Diemer, O., F. Bittencourt, L.G. Barcellos, W.R. Boscolo, A. Feiden and E. Romagosa. 2014. Lysine in the diet of *Rhamdia voulezi* male broodstocks confined in net cages. Aquaculture 434: 93–99.

Farias, J.O., S.F. Nunes, M.A.M. Carvalho and C.C.M. Salgueiro. 1999. Avaliação "*in vitro*" e "*in vivo*" do sêmen de Tambaqui (*Colossoma macropomum*) conservado a temperatura ambiente e criopreservado em água de coco. R. Cient. Prod. Anim. 1: 44–58.

Faustino, F., R.C. Silva, C.C. Hilbig, L.C. Makino, J.A. Senhorini, A. Ninhaus-Silveira et al. 2015. Spermatozoon ultrastructure and semen parameters of *Brycon vermelha* (Characiformes, Characidae). Anim. Reprod. Sci. 157: 17–23.

Figueroa, E., I. Valdebenito and J.G. Farias. 2016. Technologies used in the study of sperm function in cryopreserved fish spermatozoa. Aquac. Res. 47: 1691–1705.

França, T.S., N.C. Motta, R.C. Egger, A.V. Oliveira and L.D.S. Murgas. 2020. Impact of activation solutions on fresh and frozen-thawed sperm motility and fertilization success for two species of migratory freshwater fishes. Theriogenology 149: 6–15.

França, T.S., I.C. Gomes, E.A. Sanches, M.P. Atehortúa, N.S. Teixeira, R.B. Rodrigues et al. 2022. Post-thaw dilution of *Rhamdia quelen* sperm improves the reproductive success. Anim. Reprod. Sci. 243: 107018.

Frankel, T., L. Yonkos, F. Ampy and J. Frankel. 2018. Exposure to levonorgestrel increases nest acquisition success and decreases sperm motility in the male fathead minnow (*Pimephales promelas*). Environ. Toxicol. Chem. 37: 1131–1137.

Fricke, R., W.N. Eschmeyer and R. van der Laan. 2023. Eschmeyer's Catalog of Fishes: Genera, Spices, References. http://researcharchive.calacademy.org/research/ichthyology/catalog/fishcatmai Electronic Version accessed January 31 2023.

Gallego, V., P.C.F. Carneiro, I. Mazzeo, M.C. Vílchez, D.S. Peñaranda, C. Soler et al. 2013. Standardization of European eel (*Anguilla anguilla*) sperm motility evaluation by CASA software. Theriogenology 79: 1034–1040.

Gallego, V., L. Pérez, J.F. Asturiano and M. Yoshida. 2014. Sperm motility parameters and spermatozoa morphometric characterization in marine species: A study of swimmer and sessile species. Theriogenology 82: 668–676.

Gallego, V., S.S. Cavalcante, R.Y. Fujimoto, P.C.F. Carneiro, H.C. Azevedo and A.N. Maria. 2017. Fish sperm subpopulations: Changes after cryopreservation process and relationship with fertilization success in tambaqui (*Colossoma macropomum*). Theriogenology 87: 16–24.

Gallego, V., J.G. Herranz-Jusdado, C. Rozenfeld, L. Pérez and J.F. Asturiano. 2018. Subjective and objective assessment of fish sperm motility: when the technique and technicians matter. Fish. Physiol. Biochem. 44: 1457–1467.

Gallego, V. and J.F. Asturiano. 2018. Sperm motility in fish: Technical applications and perspectives through CASA-Mot systems. Reprod. Fertil. Dev. 30: 820–832.

Gallego, V. and J.F. Asturiano. 2019. Fish sperm motility assessment as a tool for aquaculture research: a historical approach. Rev. Aquac. 11: 697–724.

Galo, J.M., D.P. Streit Jr, C.A. Oliveira, J.P. Povh, D.C. Fornari, M. Digmayer et al. 2019. Quality of fresh and cryopreserved semen and their influence on the rates of fertilization, hatching and quality of the larvae of *Piaractus mesopotamicus*. Braz. J. Biol. 79: 438–445.

Garcia, R.F., A.C.N. Vasconcelos, J.A. Povh, E.R. Oberst, L.R. Eloy and D.P. Streit Jr. 2016. Different extenders solutions for tambaqui semen cooling. Pesqui. Agropecu. Bras. 51: 780–784.

Gheller, S.M.M., C.D. Corcini, C.R.C. Brito, I.B. Acosta, G.C. Tavares, S.L. Soares et al. 2019. Use of trehalose in the semen cryopreservation of Amazonian catfish *Leiarius marmoratus*. Cryobiology 87: 74–77.

Herrera-Cruz, E., J. Aristizabal-Regino, A.E. Yepes-Blandón, J. Espinosa-Araujo and V. Atencio-Garcia. 2019. Evaluation of three cryoprotectants to preserve striped catfish (*Pseudoplatystoma magdaleniatum*) semen. R. Col. Biotec. 21: 55–62.

Herman, H.A., J.R. Mitchell and G.A. Doak. 1994. The artificial insemination and embryo transfer of dairy and beef cattle (including techniques for goats, sheep, horses and swine). A Handbook and Laboratory Manual for Students, Herd Operators and Workers in the AI Field. 8 ed. Intersate Publishers Inc, Danville, USA.

Horst, G. van der. 2020. Computer Aided Sperm Analysis (CASA) in domestic animals: Current status, three D tracking and flagellar analysis. Anim. Reprod. Sci. 220: 106350.

Ihering, R. and P. Azevedo. 1935. Experiências com o esperma de curimatã (*Prochilodus*) dos açudes salgados da Parahyba. A. Acad. Bras. Scienc. 7: 19–27.

Kavamoto, E.T., C.S.R. Mainardes-Pinto, T. Andrade-Talmelli, F. Laine and B.E. Campos. 1997. Sperm production of "curimbatá" *Prochilodus scrofa* Steindachner, 1881. B. Inst. Pesca. 24: 73–78.

Kowalski, R. and B.I. Cejko. 2019. Sperm quality in fish: Determinants and affecting factors. Theriogenology 135: 94–108.

Kuradomi, R.Y., T.G. Souza, F. Foresti, R.W. Schulz, J. Bogerd, R.G. Moreira et al. 2016. Effects of re-stripping on the seminal characteristics of pacu (*Piaractus mesopotamicus*) during the breeding season. Gen. Comp. Endocrinol. 225: 162–173.

Linhares, F.R.A., C.S.B. Salmito-Vanderley, M.A. Carvalho, R.R. Pinheiro, F.C.E. Oliveira and J.F. Nunes. 2015. Cinética e morfologia de espermatozoides de carpa comum criopreservados em água de coco em pó ACP-104. Arq. Bras. Med. Vet. Zootec. 67: 1313–1320.

Machado, G.J., N.C. Motta, J.A. Tello-Ballinas, C.C. Martínez-Chávez, C.A. Martínez-Palácios and L.D.S. Murgas. 2022. Short-term sperm storage protocol for an endangered Mexican fish acúmara (*Algansea lacustris*). Theriogenology Wild 1: 100008.

Marcos, R.M., G. Neumann, C.P.R. de Toledo, J.M. Sena, G. Baumgartner and R.A. Bombardelli. 2015. Seminal and spermatic characteristics of fresh semen and the effects of sperm cooling in *Steindachneridion melanodermatum* (Garavello, 2005). Semin. Cienc. Agrar. 36: 4493–4505.

Maria, A.N., H.C. Azevedo, J.P. Santos and P.C.F. Carneiro. 2012. Hormonal induction and semen characteristics of tambaqui *Colossoma macropomum*. Zygote 20: 39–43.

Maria, A.N., A.C.M. Carvalho, R.V. Araújo, J.P. Santos, P.C.F. Carneiro and H.C. Azevedo. 2015. Use of cryotubes for the cryopreservation of tambaqui fish semen (*Colossoma macropomum*). Cryobiology 70: 109–114.

Marinović, Z., I. Šćekić, J. Lujić, B. Urbányi and Á. Horváth. 2021. The effects of cryopreservation and cold storage on sperm subpopulation structure of common carp (*Cyprinus carpio* L.). Cryobiology 99: 88–94.

Marques, S. and H.P. Godinho. 2004. Short-term cold storage of sperm from six neotropical characiformes fishes. Bra. Arch. Bio Tech. 47: 799–804.

Martínez, J.G., V.A. García and S.P. Carrasco. 2012. DNA fragmentation and membrane damage of bocachico *Prochilodus magdalenae* sperm following cryopreservation with dimethylsulfoxide and glucose. Neotrop. Ichth. 10: 577–586.

Martínez, R.O.S., V.M.M. Robles and P.E.C. Casallas. 2019. Effect of two semen collection in a breeding season on the seminal quality of red-bellied pacu (*Piaractus brachypomus*). Rev. Inv. Vet. Peru. 30: 1184–1195.

Martínez-Pastor, F. 2022. What is the importance of sperm subpopulations? Anim. Reprod. Sci. 246: 106844.

Martins, E.D., D.P. Streit Jr, J.S. Abreu, R.A. Corrêa-Filho, C.A. Oliveira, N.M. Lopera-Barrero et al. 2017. Ovopel and carp pituitary extract for the reproductive induction of *Colossoma macropomum* males. Theriogenology 98: 57–61.

Medina-Robles, V.M., L.Y. Sandoval-Vargas, D. Guaje-Ramírez, L.C. Marín-Cossio, I. Valdebenito and P.E. Cruz-Casallas. 2021. Cryopreservation of coporo (*Prochilodus mariae*) milt using three permeating cryoprotectant agents and two freezing systems. Aquac. Res. 52: 6760–6769.

Melo, F.C., F.C. Melo and H.P. Godinho. 2018. A protocol for cryopreservation of spermatozoa of the fish *Brycon orthotaenia*. Anim. Reprod. 3: 380–385.

Mies-Filho, A. 1987. Inseminação artificial. 6 ed. Sulina, Porto Alegre, Brazil.

Mira, L.T., R.M. Medina and C.P. Cruz. 2010. Evaluation of three protocols for hormonal induction of spermiation in *Leiarius marmoratus*. Rev. Univ. Cordoba 15: 2070–2077.

Murgas, L.D., A.B. Miliorini, R.T. Franciscatto and A.N. Maria. 2004. Spermatic viability of piracanjuba (*Brycon orbignyanus*) semen cooled at 4°C. Rev. Bra. Zoot. 33: 1361–1365.

Nascimento, A.F., A.N. Maria, N.O. Pessoa, M.A. Carvalho and A.T.M. Viveiros. 2010. Out-of-season sperm cryopreserved in different media of the Amazonian freshwater fish pirapitinga (*Piaractus brachypomus*). Anim. Reprod. Sci. 118: 324–329.

Nascimento, A.F., A.C.S. Gonçalves, R.V., Neto, M.C. Leal, A.T.M. Viveiros, A.F. Nascimento et al. 2018. Extender composition, osmolality, cryoprotectant and equilibration time effects on fresh sperm motility of two Characiformes fish: piracanjuba (*Brycon orbignyanus*) and streaked prochilod (*Prochilodus lineatus*). Anim. Reprod. 9: 103–110.

Neumann, G., P.V. Sanches and R.A. Bombardelli. 2019. Effects on fertility of motile sperm to egg ratio with use of cryopreserved *Rhamdia quelen* semen at different post-activation times. Anim. Reprod. Sci. 201: 84–92.

Ninhaus-Silveira, A., F. Foresti, R. Veríssimo-Silveira and J.A. Senhorini. 2006. Seminal analysis, cryogenic preservation, and fertility in matrinxã fish, *Brycon cephalus* (Günther, 1869). Braz Arch. Biol. Technol. 49: 651–659.

Nogueira, L.B., P.G. Azevedo, M.R. Canelhas, A.G. Bedore, J.M. Lopes and H.P. Godinho. 2012. Induced spawning and early ontogeny in hatchery-reared catfish *Zungaro jahu* (Siluriformes: Pimelodidae. Neotrop. Ichthyol. 10: 89–98.

Oliveira, D.J., F.Y. Ashikaga, F. Foresti and J.A. Senhorini. 2015. Indução a reprodução artificial e caracterização espermática da piracanjuba *Brycon orbignyanus* (Bryconidae, Characiformes), espécie em perigo de extinção. Evol. Conserv. Biodivers. 5: 11–19.

Oliveira, R.G., A.F. Oliveira and Villacorta M.A. Correa. 2018. Sperm abnormalities in matrinxã *Brycon amazonicus* after hormonal treatment with carp pituitary extract and Ovaprim™. Zygote 26: 279–285.

Oliveira, R.G. and M.A. Villacorta-Correa. 2021. Inseminating dose for the artificial fertilization of *Brycon amazonicus* (Teleostei: Characidae). Zygote 29: 252–255.

Orfão, L.H., A.N. Maria, A.F. Nascimento, Z.A. Isaú and A.T.M. Viveiros. 2010. Sperm fertility of the subtropical freshwater streaked prochilod *Prochilodus lineatus* (Characiformes) improved after dilution and cold storage. Aquac. Res. 41.

Orfão, L.H., A.F. Nascimento, F.M. Corrêa, J. Cosson and A.T.M. Viveiros. 2011. Extender composition, osmolality and cryoprotectant effects on the motility of sperm in the Brazilian endangered species *Brycon opalinus* (Characiformes). Aquaculture 311: 241–247.

Palhares, P.C., I.D.L. Assis, J.G. Souza, T.D. França, R.C. Egger, D.A. Paula et al. 2020. Effect of melatonin supplementation to a cytoprotective medium on post-thawed *Brycon orbignyanus* sperm quality preserved during different freezing times. Cryobiology 96: 159–165.

Palhares, P.C., I.D.L. Assis, G.J. Machado, R.M. Freitas, M.B. Freitas, D.A. Paula et al. 2021. Sperm characteristics, peroxidation lipid and antioxidant enzyme activity changes in milt of *Brycon orbignyanus* cryopreserved with melatonin in different freezing curves. Theriogenology 176: 18–25.

Paulino, M.S., A.B. Miliorini, L.D.S. Murgas, F.S. Lima and V.O. Felizardo. 2011. Reproductive performance of pacu, piracanjuba and curimba induced by busereline extract. Bolet. Instit. Pes. 37: 39–45.

Paula, D.A., L.D.S. Murgas, T.F.D. Castro, I. de Lima Assis, R.V. Neto and S. Marcussi. 2019. Effects of cooling rates on the quality of *Prochilodus lineatus* (Valenciennes, 1836) sperm. Reprod. Domest. Anim. 54: 1034–1043.

Pereira, J.R., F.A. Pereira, C.T. Perry, D.M. Pires, J.R. Muelbert, J.R. Garcia et al. 2019. Dimethylsulfoxide, methanol and methylglycol in the seminal cryopreservation of Suruvi, *Steindachneridion scriptum*. Anim. Reprod. Sci. 200: 7–13.

Pinheiro, J.P., C.B. Assis, E.A. Sanches and R.G. Moreira. 2020. Aluminum, at an environmental concentration, associated with acidic pH and high water temperature, causes impairment of sperm quality in the freshwater teleost *Astyanax altiparanae* (Teleostei: Characidae. Environ. Pollut. 262: 114252.

Pires, L.B., E.A. Sanches, E. Romagosa, R.A. Corrêa Filho, D.P. Streit Jr, R.A. Nass et al. 2017. Semen characteristics of *Colossoma macropomum* from three successive sample collections in the same reproductive cycle. Aquac. Res. 48: 5104–5110.

Pires, L.B., E.A. Sanches, E. Romagosa, R.A. Corrêa Filho, N.M. Lopera-Barrero and J.A. Povh. 2019. Sperm quality of *Colossoma macropomum* after room-temperature and cold storage. J. App. Ichthyol. 35: 747–753.

Ramirez-Merlano, J.A., V. Medina-Robles and P. Cruz-Casallas. 2011a. Stational variation on seminal characteristics in bagre rayado *Pseudoplatystoma metaense* (Telostei, pimelodidae). Rev. Univ. Cordoba. 16: 2336–2348.

Ramirez-Merlano, J.A., Y.M. Velasco-Santamaría, V.M. Medina-Robles and P. Cruz-Casallas. 2011b. Cryopreservation effects on the sperm quality of cachama blanca *Piaractus brachypomus* (Cuvier 1818). Aquac. Res. 42: 738–745.

Reidel, A., W.R. Boscolo, A. Feiden and E. Romagosa. 2010. The effect of diets with different levels of protein and energy on the process of final maturation of the gametes of *Rhamdia quelen* stocked in cages. Aquaculture 298: 354–359.

Reynalte-Tataje, D.A., C.A. Lopes, S. Ávila-Simas, J.R. Garcia and E. Zaniboni-Filho. 2013. Artificial reproduction of neotropical fish: extrusion or natural spawning? Nat. Sci. 5: 1–6.

Restrepo-Betancur, G., J.D. Páez and L.A. Chacón. 2017. Evaluation of two cryprotectans and three curves of programable freezing in semen cryopreservation of *Brycon henni* (Pisces: Characidae). Rev. Inv. Vet. Peru. 28: 597–605.

Ribeiro, R.I. and H.P. Godinho. 2003. Testicular sperm cryopreservation of the teleost "piau-açu" *Leporinus macrocephalus*. Arq. Bras. Med. Vet. Zootec. 55: 75–79.

Rocha, N.A., G.M. Schade, A.M. Alves, C.S. Silva, J.M. Nakirimoto and L.S. Lauri. 2020. Acute exposure to hyperosmotic conditions reduces sperm activation by urine in the yellowtail tetra *Astyanax altiparanae*, a freshwater teleost fish. Braz. J. Vet. Res. Ani. Sci. 57.

Rodrigues, M.L., L.G. Barcellos, E.B. Moro, B.S. Sosa, R.L.M. Gomes and F. Bittencourt. 2022. Gonad development and sperm characteristics of male silver catfish (*Rhamdia quelen*) fed diets with different oil sources. Braz. J. Develop. 8: 24032–24051.

Roggio, M.A., M.E. Teves, A.C. Hued, L.C. Giojalas and M.A. Bistoni. 2014. Spermatozoa characterization in the one-sided livebearing *Jenynsia multidentata* (Cyprinodontiformes: Anablepidae). Rev. Biol. Trop. 62: 997–1006.

Romagosa, E., B.E. Souza, E.A. Sanches, D.M. Baggio and R.A. Bombardelli. 2010. Sperm motility of *Prochilodus lineatus* in relation to dilution rate and temperature of the activating medium. J. Appl. Ichthyol. 26: 678–681.

Rosenthal, H., J.F. Asturiano, O. Linhart and A. Horvath. 2010. Foreword on the biology of fish gametes: summary and recommendations of the Second International Workshop, Valencia, Spain, 2009. J. Appl. Ichthyol. 26: 621–622.

Sanches, E.A., R.A. Bombardelli, D.M. Baggio and B.E. Souza. 2009. Insemination dose for artificial fertilization of dourado oocytes. R. Bras. Zootec. 38: 2091–2098.

Sanches, E.A., R.A. Bombardelli, D.M. Baggio, R.M. Sykora and A.M.M. Xavier. 2011a. Características seminais do cascudo-preto (*Rhinelepis aspera*). R. Bras. Rep. Anim. 35: 357–362.

Sanches, E.A., D.M. Neumann, R.A. Baggio, R.A. Bombardelli, P. Piana and E. Romagosa. 2011b. Time and temperature on the storage of oocytes from jundia catfish, *Rhamdia quelen*. Aquaculture 319: 453–458.

Sanches, E.A., R.M. Marcos, D.M. Baggio, L. Tessaro, R.E. Balen and R.A. Bombardelli. 2011c. Sperm concentration estimate of fish semen using spermatocrit method. R. Bras. Zootec. 40: 1163–1167.

Sanches, E.A., D.M. Baggio, P.A. Piana, B.E. de Souza and R.A. Bombardelli. 2011d. Artificial fertilization of oocytes and sperm activation in pacu: Effects of the spermatozoa:oocyte ratio, water volume, and in natura semen preservation. Rev. Bra. Zoot. 40: 1–6.

Sanches, E.A., G. Neumann, C.P.R. de Toledo, R.A. Bombardelli, P.A. Piana and E. Romagosa. 2013a. Temperature and storage period over spermatic parameters of jundiá, *Rhamdia quelen* (Quoy & Gaimard, 1824). Aquac. Res. 44: 534–541.

Sanches, E.A., R.M. Marcos, R.Y. Okawara, D. Caneppele, R.A. Bombardelli and E. Romagosa. 2013b. Sperm motility parameters for *Steindachneridion parahybae* based on open-source software J. Appl. Ichthyol. 29: 1114–1122.

Sanches, E.A., G. Neumann, C.P.R. de Toledo and R.A. Bombardelli. 2015a. Effects of water pH on gamete activation, embryonic development, and larval normality in *Prochilodus lineatus*. Semin. Ciênc. Agrár. 36: 2871–2880.

Sanches, E.A., R.Y. Okawara, D. Caneppele, C.P.R. de Toledo, R.A. Bombardelli and E. Romagosa. 2015b. Sperm characteristics of *Steindachneridion parahybae* (Steindachner, 1877) throughout 112 h of storage at four temperatures. J. Appl. Ichthyol. 31: 79–88.

Sanches, E.A., D. Caneppele, R.Y. Okawara, D.Z. Damasceno, R.A. Bombardelli and E. Romagosa. 2016. Inseminating dose and water volume applied to the artificial fertilization of *Steindachneridion parahybae* (Steindachner, 1877) (Siluriformes: Pimelodidae): Brazilian endangered fish. Neot. Ichthyol. 14: 140158.

Sant'Anna, J.F.M., A. Moraes-Neto, J.P. Castro, T.S. Sczepanski, J.A. Pohv, N. Lopera-Barrero et al. 2019. Quantitative and qualitative variables of semen from surubim do Iguaçu, *Steindachneridion melanodermatum* Garavello, 2005 (Siluriformes: Pimelodidae). Braz. J. Bio. 79: 1–5.

Shibatta, O.A., J.L. Novelli, J.H.P. Dias, S.G.C. Britto and M.C. Filho. 2011. Reprodução em cativeiro do jurupecê *Sorubim lima* (Siluriformes, Pimelodidae) por meio de indução hormonal. Semin. Ciênc. Agrár. 32: 363–372.

Shimoda, E., D.R. Andrade, J. Vidal, H.P. Yasui and J.F.S. Silva. 2007. Use of spermatocrit to estimate sperm concentration of semen from piabanha (*Brycon insignis*). Braz. J. Vet. Res. Ani. Sci. 44: 19–24.

Sierra-Marquez, L., J. Espinosa-Araujo, V. Atencio-Garcia and J. Olivero-Verbel. 2019. Effects of cadmium exposure on sperm and larvae of the neotropical fish *Prochilodus magdalenae*. Comp. Biochem. Physiol. C: Toxic. Pharmacol. 225: 108577.

Souza, A.R., Y.S. Mori, S.A.P. Andrade, J.S.I. Valandro, D. Caneppele and E.A. Sanches. 2022. Ratio of sperm, oocytes and water for artificial fertilization of *Brycon insignis* (Characiformes: Characidae). Aquac. Research. 53: 1276–1283.

Spica, L.N., E.A. Sanches, D.P. Streit Jr, R.A. Correa, L.D. Brasileiro and R.B. Rodrigues. 2021. Successive seminal collections from *Leiarius marmoratus* during the reproductive period. Ani. Reprod. Sci. 234: 106852.

Streit Jr, D.P., G.V. Moraes, R.P. Ribeiro, W.D.C. Caçador, E.S. Sakaguti, J.A. Povh et al. 2003. Estudo comparativo da indução hormonal da espermiação em piavuçu (*Leporinus macrocephalus*) com extrato de hipófise de frango, coelho e carpa. Acta Sci. 25: 261–2666.

Streit Jr, D.P., G. Moraes, R.P. Ribeiro, J.A. Povh, E.D. Souza and C.A.L. Oliveira. 2004. Avaliação de diferentes técnicas para coloração de sêmen de peixes. Arq. Ciênc. Vet. Zool. UNIPAR. 7: 157–162.

Streit Jr, D.P., R.N. Sirol, R.P. Ribeiro, G.V. Moraes, L.D.M. Vargas and A.L. Watanabe. 2008. Qualitative parameters of the piapara semen (*Leporinus elongatus* Valenciennes, 1850). Braz. J. Biol. 68: 373–377.

Streit Jr, D.P., R.N. Sirol, R.P. Ribeiro, M. Digmayer, J.M. Galo and J.A. Povh, 2012. Seminal parameters of *Pseudoplatystoma reticulatum* farmed broodstock pre and post-hormonal induction. Rev. Bras. Reprod. Anim. 36: 188–193.

Taitson, P.F., E. Chami and H.P. Godinho. 2008. Gene banking of the neotropical fish *Leporinus obtusidens* (Valenciennes, 1836): A protocol to freeze its sperm in the field. Anim. Reprod. Sci. 105: 283–291.

Teodozia, E.R., M.S.O. Araujo, J.T. Lopes, A.R. Montenegro and C.S.B. Salmito-Vanderley. 2020. Semen cryopreservation in *Piaractus mesopotamicus*: the effect of different diluents and freezing equipment. R. Cienc. Agron. 51: 20196754.

Tessaro, L., C.P.R. Toledo, G. Neumann, R.A. Krause, F. Meurer and R.A. Bombardelli. 2012. Growth and reproductive characteristics of *Rhamdia quelen* males fed on different digestible energy levels in the reproductive phase. Aquaculture 326–329: 74–80.

Vazzoler, A.E. 1996. Biologia de reprodução de peixes teleósteos: teoria e prática. EDUEM, Maringá, Brazil.

Valdebenito, I., P.C. Gallegos and B.R. Effer. 2015. Gamete quality in fish: evaluation parameters and determining factors. Zygote 23: 177–197.

Vieira, M.J., M.A. Carvalho, C.S.B. Salmito-Vanderley, C.C. Salgueiro, A.T.M. Viveiros, A.N. Moura et al. 2011. Características Do Sêmen De Tambaqui (*Colossoma Macropomum*) Em Latitude Equatorial. Arch. Zoot. 60: 1263–1270.

Viveiros, A.T.M. and H.P. Godinho. 2009. Sperm quality and cryopreservation of Brazilian freshwater fish species: A review. Fish Physiol Biochem. 35: 137–50.

Viveiros, A.T.M., A.F. Nascimento, L.H. Orfão and Z.A. Isaú. 2010. Motility and fertility of the subtropical freshwater fish streaked prochilod (*Prochilodus lineatus*) sperm cryopreserved in powdered coconut water. Theriogenology 74: 551–556.

Viveiros, A.T.M., Z.A. Isaú, D. Caneppele and M.C. Leal. 2012a. Sperm cryopreservation affects postthaw motility, but not embryogenesis or larval growth in the Brazilian fish *Brycon insignis* (Characiformes). Theriogenology 78: 803–810.

Viveiros, A.T.M., A.N. Maria, T.B. Amaral, L.H. Orfão, Z.A. Isau and R. Veríssimo-Silveira. 2012b. Spermatozoon ultrastructure and sperm cryopreservation of the Brazilian dry season spawner fish pirapitinga, *Brycon nattereri*. Aquac Res. 43: 546–555.

Viveiros, A.T.M., L.H. Orfão and M.C. Leal. 2014a. Biologia e Conservação de Espermatozoide. pp. 307–327. *In*: Baldisserotto, B., J.E.P. Cyrino and E.C. Urbinati [eds.]. Biologia e Fisiologia de Peixes Neotropicais de Água Doce. FUNEP, Jaboticabal, Brazil.

Viveiros, A.T.M., T.R. Taffarel and M.C. Leal. 2014b. Osmolality and composition of the extender during the cold storage of *Prochilodus lineatus* (Characiformes: Prochilodontidae) sperm. Neotrop. Ichthyol. 12: 643–648.

Viveiros, A.T.M., A.F. Nascimento, M.C. Leal, A.C.S. Gonçalves, L.H. Orfão and J. Cosson. 2015a. Methylglycol, methanol and DMSO effects on post-thaw motility, velocities, membrane integrity and mitochondrial function of *Brycon orbignyanus* and *Prochilodus lineatus* (Characiformes) sperm. Fish. Physiol. Biochem. 41: 193–201.

Viveiros, A.T.M., A.C. Gonçalves, A.F. Nascimento and M.C. Leal. 2015b. Fresh, equilibrated and post-thaw sperm quality of *Brycon orbignyanus* (Valenciennes, 1850) and *Prochilodus lineatus* (Valenciennes, 1837) treated with either salmon GnRHa and domperidone or pituitary extract. Neotrop. Ichthyol. 13: 157–164.

Viveiros, A.T.M., I.M. Chiacchio, I.L. Almeida and M.C. Leal. 2019. Seminal plasma features of *Prochilodus lineatus* and *Brycon orbignyanus* throughout two consecutives spawning seasons. Mol. Reprod. Dev. 86: 776–785.

Yasui, G.S., J.A. Senhorini, E. Shimoda, M. Pereira-Santos, L.S.O. Nakaghi, T. Fujimoto et al. 2015. Improvement of gamete quality and its short-term storage: an approach for biotechnology in laboratory fish. Animal 9: 464–470.

Zanandrea, A.C., M. Weingartner and E. Zaniboni-Filho. 2014. Use of ACP-104 at different dilutions for sperm cryopreservation of dourado, *Salminus brasiliensis*. J. World. Aquac. Soc. 45: 82–87.

CHAPTER 7

Maintenance and Management of Breeders of Neotropical South American Species in Laboratories

Sergio Ricardo Batlouni,[1,*] *Ruy Alberto Caetano Corrêa Filho,*[2] *Rafael Yutaka Kuradomi,*[3] *Louise Nex Spica,*[2] *Mariana Roza De Abreu,*[1] *Lucas de Oliveira Brasileiro,*[2] *Naiara Cristina Motta,*[3] *Laíza Maria De Jesus-Silva,*[1] *Susana Amaral Teixeira*[2] and *Jayme Aparecido Povh*[2,*]

INTRODUCTION

Global fish production has experienced significant growth in recent years, primarily driven by the expansion of aquaculture. Among the most cultivated fish species worldwide, two carp species, grass carp (*Ctenopharyngodon idellus*) and silver carp (*Hypophthalmichthys molitrix*), along with Nile tilapia (*Oreochromis niloticus*), hold prominent positions. These species, along with four other carp species, collectively contribute to 58.6% of global fish production (FAO 2022). Notably, some of these species have also gained prominence in fish farming practices in South American

[1] Universidade Estadual Paulista – Centro de Aquicultura da UNESP, Jaboticabal, 14884-900, Brazil,
[2] Universidade Federal do Mato Grosso do Sul – Faculdade de Medicina Veterinária e Zootecnia da UFMS, Campo Grande, 79070-900, Brazil,
[3] Instituto de Ciências Exatas e Tecnologia – Universidade Federal do Amazonas/ICET-UFAM, Itacoatiara, 69103-128 – Brazil,
Emails: marirozaabreu@gmail.com, laizamjs@gmail.com, louise.spica@ufms.br, lucaslob_@hotmail.com, ruy.filho@ufms.br, susana.teixeira@ufms.br, rkuradomi@ufam.edu.br, naiara.motta.ufv@gmail.com
* Corresponding authors: sergio.ricardo@unesp.br; jayme.povh@ufms.br

countries, such as Brazil, where Nile tilapia is the predominant species produced despite its exotic origin (PEIXE BR 2023).

Given that the production of exotic fish is not permitted in certain regions of South American countries, the lack of technological packages for the development of native aquaculture is ultimately a major obstacle to increasing the production of these fish and, consequently, impacting aquaculture in these areas. Insufficient production of juveniles in terms of quantity and quality is identified as one of the main bottlenecks in native fish production. Therefore, enhancing existing protocols related to breeder maintenance and management is essential to foster increased production of these fish species.

While it may appear straightforward, maintaining breeders of Neotropical South American species encompasses crucial considerations to effectively achieve desired objectives, such as genetic conservation, production, or the establishment of new broodstocks. Understanding and monitoring gonadal differentiation and initial gonadal maturation are essential for guiding the formation of a suitable broodstock. Furthermore, the production system, management practices, nutrition, and water quality significantly influence breeder management. Thus, knowledge about these factors is of utmost importance.

Establishing a well-formed broodstock becomes futile if breeders are not managed appropriately. Moreover, selecting the most suitable breeding system is crucial, as certain species may not adapt well to particular breeding systems. Recent research has also explored alternative reproductive protocols to the traditional approach using carp pituitary extract, as well as the potential for reusing breeders within the same reproductive period. These studies provide alternative options that, depending on the species, can contribute to more effective breeder management. Consequently, this chapter presents important aspects regarding the maintenance and management of breeders of South American Neotropical fish species.

1. Maintenance of Breeders

1.1 Broodstock Formation

The formation of a broodstock necessitates careful attention to genetic aspects, primarily aimed at avoiding (i) increased inbreeding, (ii) loss of genetic variability, and (iii) genetic contamination of the stock through the introduction of hybrids. Achieving such control in producing South American Neotropical fish is challenging, particularly due to the absence of well-established conservation and selective breeding programs. Although some institutions, such as Embrapa Fisheries and Aquaculture, have undertaken programs for native fish species like tambaqui (*Colossoma macropomum*) (Shiotsuki et al. 2019; Povh et al. 2022), the reality of native fish breeding lags far behind desirable standards, especially when compared to well-maintained breeding programs for Nile tilapia in Brazil, where broodstocks are kept under great care. These programs have yielded significant genetic advancements in the growth rate of Nile tilapia, as reported by Yoshida et al. (2021).

To compensate for the lack of conservation and selective breeding programs, many producers of migratory fish (rheophilic species) resort to empirical fish reproduction techniques they deem suitable. These techniques include choosing fish based solely on certain morphological traits, without considering genetic or productive performance criteria, as well as employing crosses of fish from different locations and species, aiming to exploit potential improvements resulting from heterosis in productive traits. In some cases, fish that are readily available and prepared for breeding are used. However, adopting these practices can lead to serious issues, such as reduced productivity due to increased inbreeding and genetic contamination of productive or natural stocks by hybrid fish.

Mating closely related breeders can result in high levels of inbreeding in the offspring, leading to deformities in larvae/juveniles (Alexandru et al. 2014), compromised productive performance (Nunes 2019), and poor reproductive performance (Silva et al. 2020). The impact of inbreeding on reproduction is often overlooked or difficult to identify in productive environments, given the high prolificacy of fish species. Failures in reproduction are often attributed to other factors that can also influence reproductive traits, such as nutrition (Bombardelli et al. 2017), population density, and rearing conditions (Yoshida et al. 2015). While studies on the effects of inbreeding on production and reproduction are more prevalent for Nile tilapia and rainbow trout (*Oncorhynchus mykiss*) (Sarder et al. 1999; Tave 1999), limited research exists on Neotropical South American fish. Therefore, conducting such studies is crucial to emphasize the significance of conservation and selective breeding programs and prevent losses in commercial production.

1.1.1 Formation Focused on Genetic Conservation Programs

Genetic conservation programs play a crucial role in ensuring a secure and diverse source of genetic resources for various purposes, such as (i) restocking programs, (ii) selective breeding programs, (iii) and replenishing broodstocks for commercial juvenile producers. Maintaining genetic variability is a key concern in genetic conservation programs as it enables fish stocks to adapt to environmental changes, including shifts in water quality and the presence of pathogens. This concern becomes even more significant when focusing on restocking programs and replacing producers' stocks (Povh et al. 2008), as the juveniles are destined for different production environments. Furthermore, genetic variability is essential for achieving genetic progress in breeding programs that involve selection or crosses since it allows the exploitation of genetic diversity within groups. Therefore, the formation of broodstocks for genetic conservation programs should prioritize adequate genetic variability, which can be achieved by maintaining multiple stocks of breeders, different families within each stock, and an appropriate number of breeders within each family.

In the context of genetic conservation programs aiming to support restocking programs, which involve introducing fish into areas where natural stocks have become extinct or diminished, it is necessary to form different stocks using fish captured from different locations, such as rivers. Each stock should have a sufficient number of breeders to represent the genes present in natural stocks, ensuring the preservation of genetic variability. However, the lack of knowledge about the genetic variability of natural stocks and the cost of genotyping a large number of captured fish pose initial challenges in forming broodstocks for restocking programs.

The formation of broodstocks for restocking programs involves additional challenges beyond obtaining the necessary authorizations from environmental agencies (Vidal and Batlouni 2023). Various zootechnical and biological issues must be addressed (Iwersen et al. 2019). Wild-collected breeders face acclimatization and adaptation difficulties, particularly related to feeding. These fish may not readily adapt to commercial feeds, and many species struggle to adjust to the lentic and high-density environments commonly found in fish farms. It is recommended that these fish undergo a quarantine period and remain in fish farming for at least one year without undergoing hormonal induction to allow for adaptation (Vianna et al. 2013). Failure to consume the provided feed can negatively affect gonadal development and gamete quality in native fish. Some fish farmers have reported an alternative approach of inducing reproduction immediately after capture (e.g., first hormonal application shortly after capture and again at the destination fish farm) when breeders are at the appropriate stage for induced reproduction, as an adaptation period would not be necessary in this case. However, this hypothesis requires further scientific research support.

In order to prevent inbreeding issues, careful mate selection should be practiced to avoid mating closely related animals. One of the most delicate steps in this process is to obtain wild broodstock from the river that is to be restocked (Vianna et al. 2013). Following this stage, the fish should be carefully marked under anesthesia, avoiding low-temperature months to minimize stress and the risk of diseases. Many native species with valuable zootechnical traits, such as mapará (*Hypophtalmus marginatus*) (Hainfellner et al. 2019; Pozzer et al. 2023), and species of the genus *Brycon*, such as piracanjuba (*Brycon orbignyanus*) (Lopera-Barrero et al. 2014), are extremely sensitive to handling, requiring special attention during rearing and breeding.

1.1.2 Formation Focused on Production

In forming broodstocks for production purposes, it is ideal to obtain breeders from a selective breeding program, which allows for genetic improvements in productive and reproductive traits, ultimately leading to increased productivity. In the absence of a selective breeding program, efforts should be made to mate unrelated or closely related breeders.

When focusing on fish farming, precautions must be taken to avoid mating between related individuals and the subsequent production of inbred progenies. Inbreeding can lead to the expression of diseases associated with deleterious recessive genes and inbreeding depression, which negatively impacts productive traits, especially reproductive traits. Inbred progenies generally have decreased adaptive value, i.e., they are less vigorous and display lower resistance to stress, environmental challenges, and pathogens, in addition to poorer reproductive performance (Kincaid 1983; Gjedrem 2005). In a selective breeding program, inbreeding control (often neglected in fish) is carefully managed throughout the generations of selection, resulting in families that are better adapted to specific production conditions (Araújo et al. 2022) or possess desirable traits (Garcia et al. 2017), thereby avoiding issues in productive or reproductive performance.

As previously mentioned, inbreeding control is essential but complex to implement in practice. It involves (i) maintaining an adequate effective number of broodstock, (ii) a sufficient number of families, (iii) efficient genealogy control, (iv) a breeding system that enables directed mating, (v) avoidance of mating related breeders, and (vi) the availability of separate tanks for rearing different progenies until the fish can be identified with a microchip, after which they can be stored together until reaching reproductive age. Marking the fish enables genealogy monitoring and inbreeding control.

An interesting alternative for fish tagging involves using microchips or transponders, which are small identification devices implanted under the skin of the fish. This method allows for long-term identification throughout the productive life of the fish (Figure 1), as described by Streit Jr et al. (2012). By identifying the fish and controlling their genealogy, it becomes possible to plan matings between unrelated individuals. Furthermore, tagged fish can be genotyped, providing information for more effective control of genetic kinship and inbreeding. Accurately estimating the stock size for each variety is crucial and requires careful calculation. For example, if it is desired to maintain the inbreeding coefficient of the stock to change by less than 0.50% per generation (ΔF), and each family is to consist of one male and one female in a 1:1 ratio, a minimum of 50 families with one male and one female each are needed ($\Delta F = 1/N_m + 1/N_f$). From each family, one male and one female are selected to replace the broodstock and form the families of the next generation. The selected males rotate among families in a systematic sequence that spans 50 generations until a descendant returns to the initial family (Gjedrem 2005).

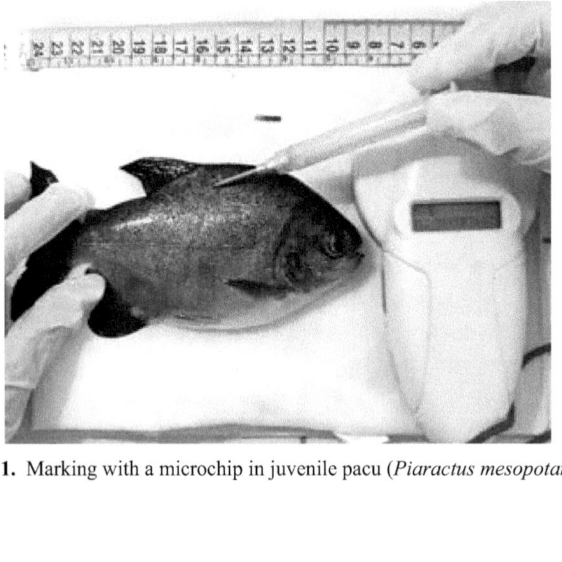

Figure 1. Marking with a microchip in juvenile pacu (*Piaractus mesopotamicus*).

1.1.3 Renewal of Stocks With Juveniles

When forming broodstocks from locally produced or obtained young individuals, selecting fish from different progenies resulting from the mating of various males and females is crucial. Each breeder should participate in a single mating group, such as a pair or trio, to generate different progenies. Once the origin or genealogy of the breeders is known, mating between individuals should be carefully planned to ensure the pairing of genetically diverse males and females. Alternative approaches in forming distinct progenies include exchanging fish between fish farms or using fish captured from rivers. These strategies allow the production of unrelated breeding individuals and effective control of inbreeding. However, capturing fish from the natural environment to form broodstocks can be challenging for certain species, such as piau três pintas (*Leporinus friderici*) (Souza et al. 2020) and piapara (*Leporinus elongatus*) (Pereira et al. 2017), due to difficulties in the capture, high costs, and the need to obtain complex authorizations from government agencies (Vidal and Batlouni 2023). Obtaining authorization can be particularly challenging when fishing is prohibited due to overfishing of the species.

Ensuring proper genealogy control and meticulous maintenance of breeders across generations can guarantee the integrity of stocks and prevent problems associated with breeding fish that exhibit a pure phenotype but are indeed hybrids (intraspecific or interspecific). The maintenance of varieties formed from captures in different locations necessitates preventing genetic contamination in the stock, which can occur due to errors in genealogy control or the improper use of breeders from other varieties during mating. Fish from different varieties, although belonging to the same species, can exhibit significant genetic divergence while appearing morphologically similar. Therefore, genetic contamination through intraspecific hybridization is difficult to detect and can easily occur in varieties maintained in both production and natural stocks that receive fish from restocking programs without adequate precautions.

Interspecific hybrids between genetically distant fish species can occur, and some hybrids may be fertile. For example, the hybrid resulting from the cross between cachara (*Pseudoplatystoma reticulatum*) and pintado (*Pseudoplatystoma corruscans*) (Hashimoto et al. 2012), which can be used in induced reproduction with certain species of the genus *Pseudoplatystoma* or with catfish from the Amazon basin (*Leiarius marmoratus*), resulting in a fish that combines traits from all three species. However, there is no consensus among scientific studies regarding the existence of heterosis in different traits when interspecific hybrids involving South American Neotropical species are used. Two main concerns arise in relation to interspecific hybridization: the uncertainty of the offspring produced (i.e., whether they are pure or hybrid *Pseudoplatystoma*) and the risk of these fish escaping into the natural environment and potentially contaminating wild stocks. In this context, genetic characterization of the broodstock is crucial and involves the use of microchip identification (Figure 1) and DNA analysis using molecular markers, as described in detail by Vaini et al. (2014) for fish of the genus *Pseudoplatystoma*, which allows for the determination of whether the fish is a pure species or a hybrid.

It is important to note that morphological differences, such as the pattern of spots, can easily confuse producers. Therefore, caution must be exercised to avoid a high incidence of hybrid fish in the formation of broodstocks, as demonstrated in the study by Vaini et al. (2014). Another common example of hybrid fish is documented by Hashimoto et al. (2014), who observed successful reproduction in the offspring resulting from the mating of female patinga hybrids (female pacu [*P. mesopotamicus*] x male pirapitinga [*Piaractus brachypomus*]) with male tambaqui, indicating that the female resulting from the cross between the two species of the genus *Piaractus* is fertile, at least under induced reproduction.

1.1.4 Gonadal Differentiation and First Gonadal Maturation: Definition of Concepts and Application in Broodstock Formation

Limited scientific studies in the literature explore the formation of new broodstocks of Neotropical South American migratory fish starting from the initial stage (larvae/juveniles). Consequently, batches produced under routine larviculture conditions may be used for breeding breeders without specific criteria. Therefore, it is important to properly understand the processes of sexual differentiation and first maturation in Neotropical South American migratory fish, as these areas remain poorly researched.

Knowledge regarding sex determination, sex differentiation, and sex chromosome identification in South American Neotropical fish is limited (Fernandino and Hattori 2019). Sexual differentiation in fish occurs at different ages, with undifferentiated gonads persisting for weeks to years. During the process of gonadal differentiation into ovaries, there is a substantial proliferation of germ cells; in general, ovarian differentiation precedes testicular differentiation (Martínez et al. 2014).

Gonadal differentiation is influenced by both genetic and environmental factors, although the determining factors for gonadal differentiation in South American migratory fish have been sparsely studied (Fernandino and Hattori 2019). Differentiation in some species occurs during the juvenile phase, such as in cachara, tambaqui, and lambari (*Astyanax* spp.), while in others, such as pacu and piracanjuba, it occurs later during the growth and growth-out phase (Table 1). Factors such as temperature (Baroiller et al. 2009) and stocking density (Ribas et al. 2017) can influence the process of gonadal differentiation, yet their effects on the reproductive process are understudied and not extensively considered in native fish aquaculture.

While sexual differentiation in South American migratory fish used in production remains poorly understood, there is an even greater knowledge gap regarding the process of first maturation. Following sexual differentiation, fish—in this case, future breeders—possess quiescent gonads that can remain in a resting state for months or years. These individuals have defined gonadal sex but are still young and pre-pubescent (Taranger et al. 2010). Although the gonads have differentiated into ovaries and testes, they do not produce gametes and instead harbor germ cells in the early stages of the first meiosis, similar to findings in model fish such as zebrafish (Leal et al. 2009). Consequently, a group of young fish with differentiated gonads exists that are sexually immature and cannot be used as breeders.

Table 1. Gonadal differentiation of some Neotropical South American species.

Species	Gonadal Differentiation	Culture System/Temperature/ Photoperiod	Density	References
Piaractus mesopotamicus	♀: 150 days after hatching 18.21 ± 1.22 g 9.55 ± 0.65 cm	Water recirculation tanks/24.29 ± 0.86°C/NA	400 fingerlings/m^3	Barbosa et al. 2022a
Pseudoplatystoma Fasciatum	♀: 39–45 days after hatching; 51.5 cm and 0.66 g	Water recirculation tanks/NA/NA	4–6 fish/L	Valentin et al. 2016
Pseudoplatystoma fasciatum	♂: 39–45 days after hatching; 45.1 mm	Raceway/25–30°C/NA	6 fish/L	Valentin et al. 2019
Astyanax altiparanae	♀: 58 days after hatching ♂: 73 days after hatching	NA/NA/NA	NA	Adolfi et al. 2015
Colossoma macropomum	♀: 4.0 cm	Polyethylene tanks/28°C/NA	NA	Lobo et al. 2020
Brycon orbignyanus	♀ and ♂: 323 days after fertilization	Earthen ponds/17.08°C–25.53°C/10–14 h	50 post-larval fish/m^2 (initial) – 1 fish/m^2 (final)	Zardo et al. 2021
B. orbignyanus	♀: 3 months – 15 cm ♂: 5 months – 20 cm	Water recirculation tanks/27 °C/natural photoperiod	NA	Quirino et al. 2022
Arapaima gigas	♀: 9.2 cm ♂: > 80 cm	Commercial farms/NA/NA	NA	Amaral et al. 2020

NA – not available

Physiologically, puberty onset in males is marked by the establishment of the blood-testis barrier, allowing the production of spermatozoa as the haploid cells are sequestered within the germinal epithelium through tight junctions formed between Sertoli cells (Leal et al. 2009; Batlouni et al. 2005). In females, in a very summarized explanation, the beginning of first maturation is characterized by the accumulation of cortical alveoli in the pre-vitellogenic oocytes, signaling the initiation of vitellogenesis (reviewed in Taranger et al. 2010). Consequently, young females exhibit cortical alveoli on the periphery of previtellogenic oocytes, indicating the start of the first maturation process and, for migratory native fish, the first reproductive cycle. Puberty onset is controlled by the activation of the brain-pituitary-gonadal axis and influenced by a range of internal and external factors that modulate and stimulate this activation, although these factors remain largely understudied (Taranger et al. 2010; Borella et al. 2020). Thus, the first maturation and the use of breeders in Neotropical South American species require a relatively long time (Table 2). However, the age and sexual maturity of fish may vary depending on factors such as (i) high stocking density, which hinders early attainment of the target body weight due to production nearing the tank's carrying capacity, and (ii) inadequate water quality throughout or in part of the breeder formation period.

In terms of production, it is known that after the first gonadal maturation, fish redirect energy toward gonadal development, potentially shifting energy allocation from growth to gonadal development, as observed in certain species (Taranger et al. 2010). Therefore, knowing the age at first maturation is important not only for forming new broodstock but also for assessing the need for monosex lots or sex reversal in species without differentiated growth rates between sexes up to the slaughter size.

Table 2. Information on the first sexual maturation of some Neotropical South American species.

Species	First sexual maturation	Reference
B. orbignyanus	♂: 730 days after fertilization	Zardo et al. 2021
P. mesopotamicus	♀: 39.90 cm and ♂: 40.45 cm	Barbosa et al. 2022a
P. corruscans	♀: 71.90 cm and ♂: 84.30 cm	Barbosa et al. 2022b
C. macropomum	♀: 58 cm; 4–5 years old	Val and De Oliveira 2021
A altiparanae	♀: 9–12 cm and ♂: 7–9 cm	Porto-Foresti et al. 2010
A. gigas	♀ and ♂: 157 cm; 3 years old	Amaral et al. 2019
Salminus brasiliensis	♀ and ♂: 55.5 cm	Barzotto and Mateus 2017

1.1.5 Production Systems and Regimes for Broodstock Formation

Few studies have evaluated breeders in various production systems, including earthen ponds, net ponds, elevated ponds, raceways, in-pond raceway systems, recirculating aquaculture systems (RAS), and biofloc technology (BFT), as well as in different production regimes (extensive, semi-intensive, and intensive). Regardless of the system or production regime, maintaining adequate water quality and density is crucial for successful broodstock formation.

The most commonly used method for breeding native migratory fish is maintaining breeders in earthen ponds. Keeping breeders at lower densities compared to the growth and fattening phase is recommended. Typically, a recommended range for medium to large fish is 3 m^2 to 8 m^2 per individual. Densities of 800 g fish/m^2 (Sato et al. 2021) and 0.25 fish/m^3 (Kuradomi et al. 2016) have been used for pacu breeders, while 50 fish/m^2 is recommended for lambari breeders (Roza de Abreu et al. 2021), and 1 fish/m^2 for curimbatás in earthen ponds. We recommend establishing nurseries in secluded areas, free from external sources of stress, with a minimum depth of 1.5 m and a water surface area of 200 m^2, such as in the case of cacharas (Leonardo et al. 2004).

Nurseries should have a proper water flow, ranging from 0.3 L/s (Batlouni et al. 2006) to values that indicate a water exchange similar to a raceway system (total volume exchange per hour). For the fattening phase, a flow rate of 10 to 15 L/s/ha has been used (Ribeiro et al. 2001), allowing the production of 1 kg of fish/m^2 without an aerator or 3 to 8 kg of fish/m^2 when an aerator is employed (1 hp/1,000 m^2). However, there is a lack of information regarding the appropriate flow rate for the formation and maintenance of breeders of South American Neotropical fish species, although the general recommendation is to maintain lower densities compared to production.

It is crucial to emphasize the significance of water flow for oxygenation and water quality, as insufficient water exchange can accumulate organic matter. In this regard, the utilization of semi-intensive production in earthen ponds can be an appealing alternative, provided there is minimal variation in water quality variables over 24 hours and throughout the year. Using aerators equipped with sensors that activate automatically based on dissolved oxygen levels in the water is recommended.

Another important consideration in fish production in earthen ponds is the regular removal of organic matter that accumulates at the pond's bottom. Ponds typically feature drainage systems, which often discard water from the bottom. Even so, periodic complete drying of the pond and removal of organic matter are necessary. Liming is recommended after this process to eliminate parasites and other microorganisms that can cause diseases affecting the fish, following the process outlined in Cecarelli et al. (2000). Purging, a practice involving rapid water removal from the pond to achieve a substantial water exchange in the shortest time possible in the case of large water volumes, can help control excessive levels of total ammoniacal nitrogen, thereby mitigating issues related to ammonia (NH_3) and nitrite (NO^{-2}) toxicity.

Using bacteria that consume organic matter deposited at the pond bottom is another described approach for water quality control. Although scientific knowledge in this area for South American Neotropical species is limited, using these probiotics as bioremediators in aquaculture has been reported (Soltani et al. 2019; Nayak 2021). In fish species, probiotics are often used as dietary supplements and are more closely associated with dietary supplementation, disease resistance, and immune response. The supplementation of native species, such as tambaqui (Costa et al. 2021) and lambari (Moraes et al. 2018), with probiotics, has been reported, although no significant differences in water quality were observed. However, in Nile tilapia, water supplementation with probiotics resulted in reduced concentrations of nitrogenous compounds in the water (Kord et al. 2022; Haraz et al. 2023).

Various products currently available on the market can be incorporated into feed or applied directly to the water. These probiotic additives can directly benefit the fish and/or enhance water quality. In addition to probiotics, several prebiotics and synbiotics (prebiotic + probiotic) are also commercially available. The use of these additives becomes more important under conditions of low water exchange in tanks, high production densities, and the use of high-protein levels in feeds.

Ponds should preferably be populated with a single species, particularly to avoid aggressive and invasive species [e.g., Nile tilapia, traíra (*Hoplias malabaricus*), and tucunaré (*Cichla* spp.)] causing stress to the breeders through territorial fights and consuming significant amounts of feed, compromising the proper nutrition of the breeders. In practice, invasive species tend to naturally occur in native fish breeding ponds, necessitating ongoing management for their removal. In order to mitigate this problem, screens should be employed in the water inlet pipes of the tanks. In many regions, natural threats such as otters, alligators, anacondas, and birds, especially cormorants, pose risks to breeders, particularly for small species like lambaris and the genus *Leporinus* spp. In such cases, covering the ponds with anti-bird screens and, in more severe cases, using wire mesh, preferably with a masonry base to prevent digging and entry by certain animals, is recommended.

Regarding the use of net cages, recent studies have demonstrated that this production system can be a suitable option for maintaining native fish breeders, such as jundiá (*Rhamdia quelen*) (Reidel et al. 2010) and cachara (Romagosa 2010). On the other hand, it may not be suitable for species that primarily feed at the bottom, such as curimbatá (*Prochilodus lineatus*), as reported by Hainfellner et al. (2012) and De Souza et al. (2015). These studies demonstrated that the net cage production system altered gametogenesis in males and females of curimbatá, even in the winter months, in the resting and early maturation reproductive classes. Therefore, maintaining good management conditions throughout the year—including during the months when vitellogenesis occurs and the initial stages of sperm formation can result in DNA damage despite reasonable motility rates—is essential (De Souza et al. 2015; Kuradomi et al. 2016). However, keeping broodstock in net cages in environments with a large water volume, such as rivers and dams, can be an interesting alternative since this production system typically exhibits less variation in water quality throughout the day compared to the earthen pond production system.

The adoption of more environmentally sustainable production technologies has gained increased attention. Biofloc technology (BFT) is a notable example, particularly considering that this approach minimizes or eliminates water exchange, while the microbial aggregates (bioflocs) formed in the system can be consumed by fish (Emerenciano et al. 2017) and act as immunostimulators (Figure 2). Some recent studies have indicated that maintaining broodstock in BFT systems has improved reproductive aspects and/or facilitated the production of fish juveniles, such as Nile tilapia (Ekasari et al. 2015) and African catfish (*Clarias gariepinus*) (Ekasari et al. 2016), as well as the reproduction of marine shrimp (Emerenciano et al. 2014; Cardona et al. 2016). In addition, positive outcomes have been observed in certain Neotropical

Figure 2. (A) 10-m^3 tanks employed for breeder preparation using biofloc technology (BFT); (B) volume measurement of settleable solids using an Imhoff cone; (C) 4-m^3 elevated tanks utilized for fish production with biofloc technology; and (D) microscopic visualization (10x magnification) emphasizing phytoplankton and zooplankton in the flakes.

South American fish species, such as lambari (Lima 2022; Valandro 2023). However, studies investigating the application of BFT specifically for reproduction in other South American Neotropical fish species are lacking. Therefore, it is important to emphasize that, in addition to its potential as a technology for fish reproduction, BFT aligns with sustainability goals. Given the importance of technology, increased research investment is necessary to evaluate the effects and mechanisms of BFT on the production and reproduction of different Neotropical South American fish species.

Some important considerations regarding the use of BFT are as follows: (i) maintaining continuous oxygenation in the system is crucial since even a brief period without oxygen can lead to system failure and fish mortality (a generator should be available in case of a power outage); (ii) meticulous attention to water quality, particularly in maintaining low levels (close to zero) of total ammoniacal nitrogen and nitrite, while observing high alkalinity levels (above 100 mg/L) and controlling settleable solids levels (Figure 2), as demonstrated by Emerenciano et al. (2017). The presence of elevated nitrate levels (~ 25–50 mg/L) indicates the proper functioning of the system mediated by bacteria that readily convert nitrogen. Notably, BFT has shown promising results in the reproductive performance of marine shrimp and Nile tilapia, yet its potential application in South American Neotropical fish cultivation requires further investigation.

1.1.6 General Aspects of Feeding and Control of Water Quality Variables

Regarding the nutritional aspect of fish breeders, a significant challenge lies in the lack of specific feeds tailored for Neotropical South American species. This inadequacy hampers both the formation of the broodstock and the preparation for reproduction. Unfortunately, this nutritional issue is far from settled for many South American Neotropical species. Despite the absence of species-specific diets or production phase feeds, high-quality foods are essential, as feed quality directly impacts reproduction (Bombardelli et al. 2017). Furthermore, excessive feeding, particularly before the reproductive phase, should be avoided to prevent compromising reproductive quality and water quality, especially when utilizing high-protein diets.

Water quality variables are fundamental for fish development. Among the main variables that require constant analysis are (i) temperature, (ii) dissolved oxygen, (iii) pH, (iv) total ammonia nitrogen, (v) nitrite, and (vi) water alkalinity/hardness. The impact of evaluating and adapting these variables is greater when the production regime is intensive or when production conditions are in closed systems, such as RAS and BFT. Tools providing better control, such as aerators connected to automatic or heating systems, are recommended to ensure optimal preparation of South American Neotropical fish breeders. Inadequate water quality variables, as advised by Boyd (1998) for traditional production and Emerenciano et al. (2017) for BFT, can negatively impact reproduction. Since water quality variables are interconnected and influenced by multiple factors, it is essential to address the collective suitability of these variables to determine the overall quality of the culture water. For instance, low dissolved oxygen becomes more critical when coupled with excess nitrite or toxic ammonia in the tank.

Various water characteristics can affect the quality of fish gametes produced by breeders. Classic studies have shown that factors such as dissolved oxygen concentrations in water (Wu et al. 2003; Shang et al. 2006; Landry et al. 2007; Thomas et al. 2007; Corriero et al. 2007) and high levels of stress (Buyukhatipoglu and Holtz 1984; Campbell et al. 1992; Campbell et al. 1994; Corriero et al. 2011) have a significant detrimental effect on fish breeding, leading to the formation of low-quality gametes (Bobe and Labé 2010). Additionally, other factors like the volume of water in the ponds (Buchet et al. 2008) and the concentration of toxic ammonia (Ram and Sathyanesan 1986) have a deleterious effect on gamete formation.

While limited research addresses these aspects in South American Neotropical fish, some studies highlight the detrimental effects of hypoxia on pirapitinga reproduction (Dabrowski et al. 2003), as well as the impact of hypoxia and other variables on gametogenesis and reproduction of curimbatá (Hainfellner et al. 2012; De Souza et al. 2015). Nevertheless, given the wealth of knowledge on other species, it is reasonable to assume that these variables impose similar adverse effects on South American Neotropical migratory species.

Finally, it is important to acknowledge that even with careful management and experienced producers, unexplained failures and widespread embryo mortality are often observed. Recently, a group of European researchers developed a study with yellow perch (*Perca flavescens*) where they classified spawnings from the same batch as bad, good, or excellent based on embryonic survival in identical incubation conditions. Proteomic analysis suggested that different spawning types may be associated with molecular traits of the eggs, independent of laboratory handling (Castets et al. 2012). Similarly, unexplained larval mortality in native fish may also be linked to underlying processes.

2. Broodstock Management

2.1 Preparation for Reproduction: Timing, Selection, and Transport of Breeders to the Laboratory

Many Neotropical South American migratory species typically undergo reproduction between October and March (Mechaly et al. 2023). However, being in the reproductive period does not guarantee that the female or male can spawn at the time of capture. This aspect is crucial as inducing breeding outside the appropriate reproductive period may result in reproductive failure and, in some cases, lead to death, particularly in females. In order to mitigate this situation in females, one practice involves analyzing the germinal vesicle using Serra's fluid (60 mL of 90% alcohol, 30 mL of formalin, and 10 mL of glacial acetic acid) (Lemanova and Sakun 1975) (Figure 3A; 3B). Although the effectiveness of this approach in different species has not been confirmed through studies, it is anticipated that females with

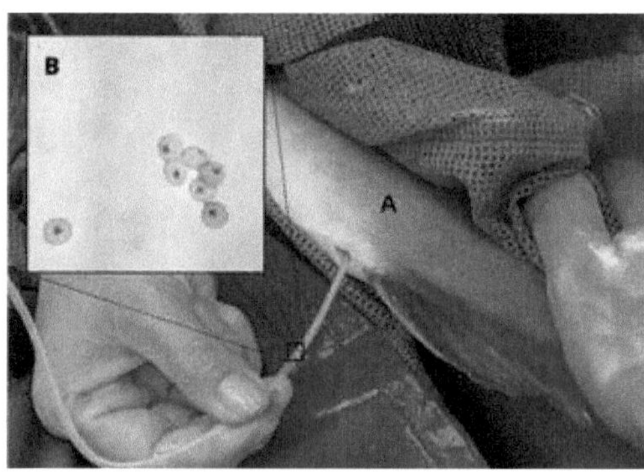

Figure 3A. (A) Cannulation procedure performed on a female pacu (*Piaractus mesopotamicus*) at reproductive age, and (B) analysis of the positioning of the germinal vesicle in the oocyte by clarifying the sample (~ 2 minutes) in Serra's fluid for examination under a magnifying glass.

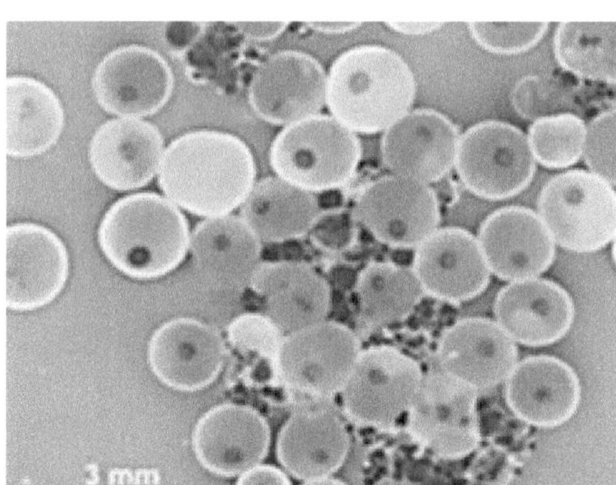

Figure 3B. Oocytes of *Astyanax altiparanae* obtained by biopsy, counterstained with Trypan blue, and cleared with Serra's fluid, displaying central, eccentric germinal vesicles and GVBD (germinal vesicle breakdown) oocytes with ruptured vesicles (photograph by Cristiane Benevente; doctoral thesis).

partially migrated germinal vesicles might respond more effectively to hormonal induction.

Despite being a scientifically based concept, the application of the cannulation technique faces some dissent among researchers and producers, attributed to various reasons, including (i) concerns about stressing the breeders, (ii) lack of familiarity with the procedure, and (iii) the requirement for specialized equipment and trained personnel. One of the first works to report the different types of response to hormonal treatments was developed in carp by Levavi-Zermonski and Yaron (1986), who demonstrated that the change in circulating steroid levels depends on the maturational stage of the oocytes. Fish with central germinal vesicles responded to increased gonadotropin with elevated circulating levels of 17β-estradiol (vitellogenesis-inducing substance) but not of 17α,20β-dihydroxy-4-pregnen-3-one (17,20-P; substance inducing final maturation and ovulation). However, in fish with ovaries containing peripheral vesicles, a positive response to hormone induction was observed, with an increase in 17,20-P levels and a reduction in estradiol.

Subsequently, Patiño et al. (2001) introduced the concept of "oocyte maturational competence," proposing that the production of Maturation Inducing Substance (MIH) occurs only at advanced stages of vesicle migration. They defined oocyte maturational competence as the ability of oocytes to respond to hormonal stimulation, which is generally correlated with the germinal vesicle migration process. Based on this concept, it is hypothesized that oocytes with eccentric germinal vesicles or ovaries with a higher percentage of oocytes displaying eccentric germinal vesicles are theoretically more capable of responding to hormonal treatments. While numerous studies have explored these concepts in species mainly produced in the northern hemisphere through *in vitro* studies (Patiño et al. 2001), this concept remains largely unexplored in migratory tropical species.

When cannulation is employed for South American Neotropical migratory fish, researchers often adopt a minimum percentage of eccentric oocytes for the selection of breeders, such as 30% eccentric germinal vesicle for pacu (Kuradomi and Batlouni 2018; Schorer et al. 2016; Table 3). When evaluated in pacu, the method was much more accurate to compare with the ovarian maturation scale than other parameters like analyzing the papillary hemorrhage or tenderness of the abdomen, which are much more subjective for the choice of breeders than biopsy (Schorer et al. 2016).

In practice, the process used to select males is simpler and usually involves gentle compression of the reproductive region (ventral in *Siluriformes* fish; lateral in other fish), with males capable of releasing semen considered fit for reproduction (Figure 4). However, such a practice, without further analysis, may lead to the use of poor-quality semen. Gamete quality, defined as the ability to fertilize the opposite gamete and generate viable embryos (Bobe and Labé 2010), is compromised when males are selected solely based on the quantity of semen released, as observed in curimbatá kept under inadequate handling conditions (De Souza et al. 2015). In such cases, although broodstock in cages may produce a reasonable amount of semen, the quality of the generated embryos is significantly reduced, attributed to damage to sperm DNA integrity. Additionally, it is important to consider that fish spermatozoa

Table 3. Minimum percentage of oocytes with eccentric nuclei for selecting South American Neotropical females suitable for spawning.

Species	**Percentage**	**Reference**
P. fasciatum	82.5% with central nuclei and 17.5% with peripheral nuclei	Leonardo et al. 2004
P. mesopotamicus	≥ 30% of oocytes with nuclei shifted toward the periphery	Kuradomi and Batlouni, 2018; Sato et al. 2021
P. mesopotamicus	> 50% oocytes with peripheral nuclei	Bittencourt et al. 2012
C. macropomum	> 50% oocytes with peripheral nuclei	Pires et al. 2018
Brycon opalinus	35% with central nuclei and 65% with peripheral nuclei	Narahara et al. 2002
Brycon insignis	> 60% oocytes with peripheral nuclei	Andrade-Talmelli et al. 2002
Brycon cephalus	80% oocytes with excentric nuclei	Romagosa 1998

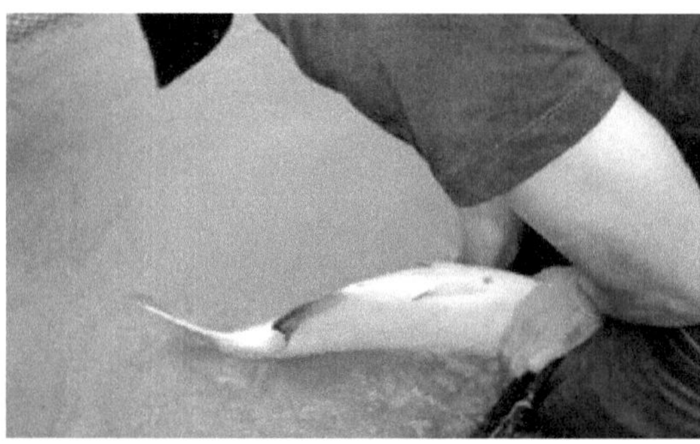

Figure 4. Tambaqui male selection through gentle pressure applied in the encephalocaudal direction.

are short-lived cells and may lose their quality when stored in the testes for extended periods (Kuradomi et al. 2016). Therefore, whenever possible, more comprehensive techniques, such as sperm concentration and assessment of viable spermatozoa, should be utilized to increase the probability of using high-quality semen, always coupled with proper breeder management.

Although little applied in breeding laboratories, alternatives for analyzing semen quality could also prove useful in male selection. Established insemination dose ratios exist for some species, such as tambaqui, which has a recommended dose of 100,000 spermatozoa/oocyte (Leite et al. 2013). Consequently, the number of males selected for reproduction will depend on the number of females, semen production, and sperm concentration obtained following the standard breeding protocol for the species. Other variables further explored in subsequent chapters of this book, such as the amount and volume of water used in dry fertilization, are also essential for successful fish reproduction.

Transporting breeders from maintenance tanks to the breeding laboratory should be conducted rapidly and with minimal stress to avoid detrimental effects on reproduction. The general steps involved in the process include:

i) Fasting the breeders for approximately 24 hours before management;
ii) Moving the trawl net during cooler times of the day (early morning or late afternoon, depending on the reproductive protocol), closing the trawl net near the entry of water;
iii) Ensuring the trawl is not excessively tight to prevent fish from struggling;
iv) Calmly selecting breeders without causing unnecessary stress in the trawl;
v) After selecting the males, preferably transporting them in non-abrasive bags (e.g., plastic bags) filled with water from the capture tank and oxygenated;
vi) Upon arrival at the laboratory, acclimatize the fish with the water in the containment box or tank;
vii) Subsequently, marking the breeders for easy identification, ideally using a microchip or earrings or applying a colored thread to the dorsal fin;
viii) Ensure to weigh the breeders. The fish containment tank or box must be adequately sized, and constant water inflow and oxygenation from a blower are preferred.

2.2 Definition of Reproductive System

Fish reproduction can be categorized into three distinct reproductive systems: (i) natural, (ii) semi-natural, and (iii) by extrusion. The natural reproduction system is applicable to fish that spawn spontaneously in ponds, requiring no hormonal intervention for reproduction. In this system, fish are kept in tanks or hapas within a tank, and eggs, larvae, or juveniles are collected, depending on the species. Fish that do not reproduce naturally require hormonal induction. When this induction is followed by the release of fish into a water tank or designated reproduction tank, it is characterized as a semi-natural reproduction system. The process involving gamete extrusion is known as a gamete extrusion system (Figure 5).

The definition of the appropriate reproduction system depends on several factors, primarily the species, ease of gamete extrusion, and the available infrastructure for reproduction. The semi-natural system necessitates a generally circular tank with constant water intake to ensure proper circulation. Once the species reaches the required degree-hours, a valve connected to the central part of the tank is opened, facilitating the transfer of eggs from the breeders' tank (usually from the central region) to a collection incubator (Figure 6). The dimensions of tanks need to be adjusted according to the size of the species.

The semi-natural reproduction system has shown promising results, particularly toward the end of the reproductive period. During this time, breeders may experience increased stress due to management practices and aging of gametes (Kuradomi et al. 2016), which tend to influence reproductive efficiency. Few studies comparing different breeding systems have been reported. For instance, Zanoni et al. (2016)

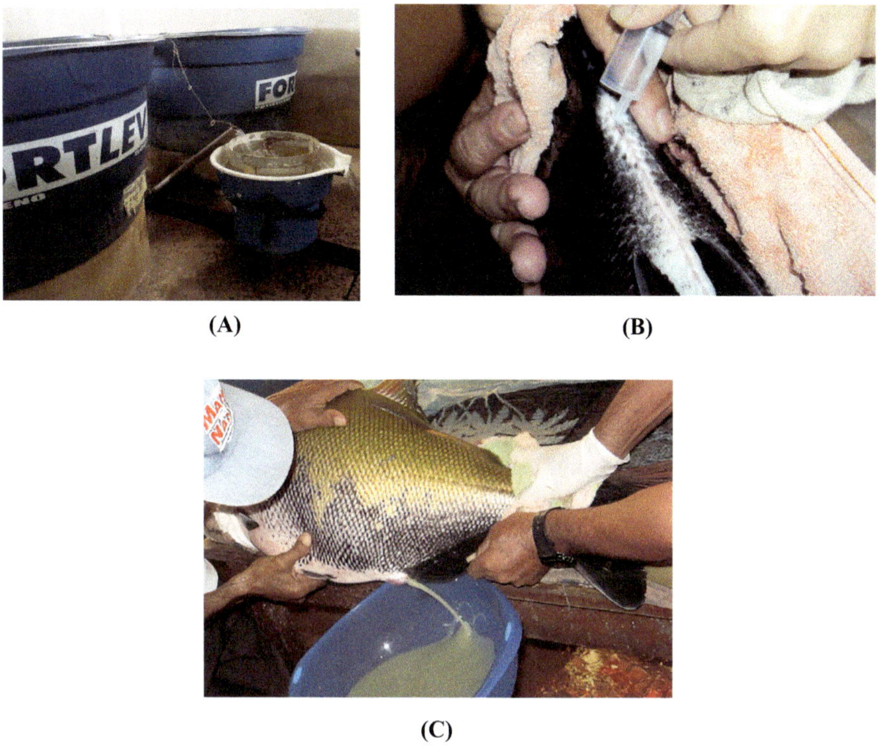

Figure 5. Breeding systems: (A) semi-natural, with box/tank size adjusted according to breeder size; (B) extrusion, demonstrating the collection of tambaqui semen into a syringe for subsequent fertilization [the collected semen can be stored for a few hours under appropriate conditions, as demonstrated by Pires et al. (2019)]; (C) extrusion, illustrating the collection of tambaqui oocytes.

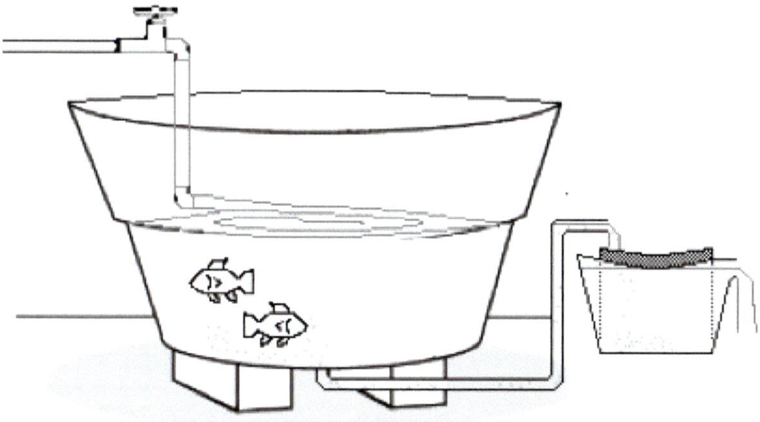

Figure 6. Details of the semi-natural reproduction system.

found that the semi-natural reproduction system was more efficient for piracanjuba than the extrusion reproduction system. Among the 16 females evaluated in the semi-natural system, 62.5% spawned, with an 87.9% fertilization rate, while in the extrusion system, all five females spawned, but the fertilization rate was only 8.2%.

For lambari reproduction, the semi-natural system is widely preferred as gamete extrusion from breeders to produce large quantities of juveniles is challenging. However, there are reports of using one or two hormone doses during the induction of lambari reproduction, and the sexual proportions, volume of water in the boxes, and induction methods vary significantly between experiments with this species. Some producers use the natural system for lambari reproduction, placing males and females in nurseries to obtain young forms. Reports indicate that, under specific management conditions or after heavy rain, lambari reproduction can occur naturally.

In response to the high demand for lambari as bait and the need to increase reproductive efficiency, scientific studies have demonstrated successful results with hormone induction using two doses (Roza de Abreu et al. 2021; Roza de Abreu et al. 2022). Maintaining females of this species at 146 to 262 degree-hours after induced spawning has been shown to enable reproduction again seven days after spawning (Lira et al. 2018). Additionally, a recent patent has highlighted lambari as an excellent fish for providing pituitary extract for hormone induction in larger species, such as curimbatá, piau-três-pintas, pacu, and tambaqui (Patent Deposit: BR 102021 0183306). This protocol includes specific management conditions and the period of the year for lambari pituitary extract collection, ensuring its efficiency in inducing ovulation.

The semi-natural reproduction system can potentially maximize genetic variability, as reported for pacu species (Povh et al. 2010). However, for this situation to occur, the species should not exhibit dominant behavior, where only one or a few males fertilize the oocytes, as noted by Porta et al. (2006). Furthermore, for certain species, the typical reproductive induction protocols are ineffective in obtaining semen by extrusion. For example, this is the case for species of the genus *Leporinus*, as well as for some species sensitive to extrusion handling, where it may cause mortality in the breeders, such as species of the genus *Brycon* (Lopera-Barrero et al. 2014). In such instances, using semi-natural reproduction management is a favorable alternative for reproduction. Moreover, when dealing with species sensitive to management, using an inducer in a single dose may offer satisfactory results. For instance, studies have demonstrated successful reproduction in tambaqui with a single dose of Ovopel for both males (Martins et al. 2017) and females (Souza et al. 2018). However, it is essential to evaluate these aspects in different South American Neotropical species, particularly in Siluriformes species (e.g., species of the genus *Pseudoplatystoma*), considering that observations of reproductive success in a semi-natural system were only observed in scaled fish. Another critical factor in the semi-natural reproduction system is maintaining an appropriate proportion of males and females. The inseminating dose (number of spermatozoa:oocyte) in the extrusion reproduction system is generally high, and in the semi-natural reproduction system, the ratio of spermatozoa:oocytes might need to be higher due to the greater dilution of gametes in water during fertilization compared to the extrusion reproduction system.

An improper ratio of males and females can significantly impact the fertilization rate and, consequently, the hatching rate.

2.3 Hormone Induction Protocols for Rheophilic Fish

Various hormone induction protocols have been explored as potential replacements or complement to the traditional carp pituitary extract protocol in fish. These alternative inducer tests have been applied to different South American Neotropical migratory fish species. Notable results have been observed, including the use of prostaglandin in combination with the traditional protocol and the application of synthetic hormones like Ovopel. Table 4 below summarizes selected studies involving different inducers in South American Neotropical fish, providing insights into the possibility of complementing or substituting carp pituitary extract in specific species' reproduction protocols.

2.4 Reuse of Rheophilic Fish

The reuse of male and female fish within the same reproductive period is a practice that has received relatively little attention in fish farms, mainly due to the lack of scientific information. However, it holds promising potential, particularly when harnessing fish of greater genetic value for selective breeding programs. Additionally, reusing fish in the same reproductive period can lead to cost savings for producers, as maintaining a smaller number of breeders becomes feasible.

The reuse of male rheophilic fish has been primarily investigated, as evidenced by studies in tambaqui by Pires et al. (2017), catfish from the Amazon basin (*L. marmoratus*) by Spica et al. (2021), and pacu by Kuradomi et al. (2016). Reusing males can offer benefits, particularly in species that release a small volume of semen, as demonstrated by Spica et al. (2021), who successfully maintained the quantitative and qualitative traits of semen through this practice. It is worth noting that the insemination dose in fish is generally high, underscoring the importance of maximizing the use of males. Furthermore, in the case of pacu, re-spermiation has proven to be an excellent method for eliminating aged spermatozoa and obtaining semen with reasonable volume and quality (Kuradomi et al. 2016).

The reuse of female rheophilic fish is less common, but some studies have shown the possibility of obtaining a reproductive return from females within the same reproductive period. For instance, Pires et al. (2018) observed successful reproduction of certain tambaqui females in the same reproductive season. However, further investigations involving additional analyses, such as histological examinations of gonads, are essential to clarify whether this represents a new recruitment of oocytes or a residual batch from the initial spawning. Regardless of this aspect, achieving a second spawning, particularly from fish of high genetic value, can significantly enhance the utilization of females in breeding programs.

Table 4. Evaluation of reproductive inducers as alternatives to the traditional carp pituitary extract protocol (in females: between 5 and 6 mg/kg of live weight in two applications – 10 and 90% – with an interval between 12 and 24 h between applications; in males: 2.5 mg/kg in a single dose).

Species	Hormone	Application Protocol	Results	Author
B. orbignyanus	Buserelin	0.125 mL (preliminary dose) administered only in females; 0.25 mL (first dose) and 0.75 mL/kg of fish (second dose)	Spawning was observed, but there was no fertilization rate	Paulino et al. 2011
P. mesopotamicus	Prostaglandin	0.6 mg/kg (first dose) and 5.4 mg carp pituitary extract/kg of fish weight (second dose) associated with 2 or 5 mL prostaglandin	All females spawned with the treatment involving carp pituitary extract associated with prostaglandin. The fertilization and hatching rate was above 50%	Criscuolo-Urbinati et al. 2012
L. marmoratus	Ovopel (mGnRHa + MET)	0.2, 0.4, and 0.6 pellet/kg of fish (single dose)	The use of Ovopel applied to males showed results similar to carp pituitary extract	Araújo et al. 2014
L. obtusidens	Gonadorelin	20, 40, 60, 80, and 100 μg/kg of fish (single dose)	The dose of 60 μg/kg increased semen volume compared to the negative control; however, it was similar to the positive control	Bernardes Júnior et al. 2017
C. macropomum	Ovopel (mGnRHa + MET)	0.1, 0.2, 0.3, 0.4, 0.5, 0.6, and 0.7 pellet/kg of fish	The doses of 0.2, 0.3, and 0.7 pellet/kg were similar to carp pituitary extract when compared in terms of sperm motility	Martins et al. 2017
B. cephalus	Ovopel (D-ala6, pro9net-mGnRH + Metoclopramide)	⅓ pellet/kg (T1), ⅔ pellet/kg (T2) and 1⅓ pellet/kg	An Ovopel dose of ⅔ pellet/kg was more suitable for hormonal induction of males compared to the traditional protocol with carp pituitary extract	Bashiyo-Silva et al. 2015
L macrocephalus	Conceptal (mGnRHa + MET)	7 μg/kg mGnRHa + 10 mg metoclopramide (single dose)	All females spawned, but there was no hatching rate	Pereira et al. 2017
L. elongatus	Conceptal (mGnRHa + MET)	3.5 μg mGnRHa + 10 mg metoclopramide/kg of fish (single dose) 5 μg mGnRHa + 10 mg metoclopramide/kg of fish (single dose)	All females spawned at both doses, but there was no hatching rate	Pereira et al. 2018

C. macropomum	Ovopel (mGnRHa + MET)	0.2 and 0.4 pellet/kg of fish (single dose)	All females spawned with the lowest dose, fertilization and hatching rates above 45%. At the highest dose, however, 62.5% of females spawned and fertility and hatching rates were above 19%	Souza et al. 2018
C. macropomum	Buserelin	0.5 mL/kg of fish (single dose)	Spawning rate of 40%, fertility and hatching rate above 38%	Kozen-Freitas et al. 2020
L. fredereci	mGnRHa + MET	40 µg mGnRHa + 20 mg metoclopramide/kg of fish (single dose) 4 µg/kg mGnRHa + 2 mg metoclopramide/kg of fish (first dose)/8 µg mGnRHa + 4 mg MET/kg of fish (second dose)	Spawning rate above 50% and 70% with the single and split doses, respectively, yet no fertility or hatching	Souza et al. 2020
A. altiparanae	sGnRHa	1, 10, 20, 40, 80, and 160 µg/kg of fish (single dose) (experiment 1) 10, 100, and 1,000 µg sGnRHa/kg of fish (two doses) (experiment 2)	Spawning observed in all groups except 1 µg/kg of fish, fertilization and hatching rate above 60% (experiment 1); Spawning observed in all groups, fertilization and hatching rate above 70% (experiment 2)	Roza de Abreu et al. 2021
A. altiparanae	LHRHa LHRa + domperidone	20, 60, 180, 540, and 1,620 µg/kg LHRHa of fish (experiment 1) 180 µg/kg LHRHa; 180 µg/kg LHRHa+ 20 mg domperidone/kg of fish (experiment 2)	Spawning observed in all groups except 20 µg/kg of fish, fertilization and hatching rate above 80% at lower doses (experiment 1); Spawning observed in all groups, fertilization rate above 45% (experiment 2)	Roza de Abreu et al. 2022

mGnRHa: mammalian gonadotropin-releasing hormone analog; sGnRHa: salmon gonadotropin-releasing hormone analog; MET: metoclopramide; LHRHa: luteinizing hormone-releasing hormone analog.

3. Final Remarks

Increasing the production of South American Neotropical species holds paramount importance in bolstering fish production, particularly in areas where the cultivation of exotic species is prohibited. Furthermore, significant progress in knowledge acquisition and application is essential to ensure the successful cultivation of numerous species with superior quality, meeting the domestic market demand, especially in diverse countries such as Brazil. It is also important to emphasize that the development and refinement of production techniques for native species serve to reinforce the connection between aquaculture, environmental preservation, and fishing. A pertinent example is restocking programs that release thousands of native fish fingerlings into Brazilian basins annually, relying on improved production and reproduction techniques for these species. If properly implemented, such programs have the potential to increase fish stocks significantly. It is, therefore, essential to develop technology packages tailored to South American Neotropical species. In order to achieve this, it is imperative to comprehend the aspects pertaining to the maintenance and management of breeders of these species, considering the little existing knowledge about many of the variables that influence reproductive success. Ultimately, advancing knowledge will pave the way for implementing various biotechnological techniques, such as gamete cryopreservation, which holds promise in enhancing the captive reproduction of these species.

Acknowledgments

This study was financed in part by the Coordenação de Aperfeiçoamento de Pessoal de Nível Superior, Brazil (CAPES), Finance Code 001, Universidade Estadual Paulista Jaboticabal/UNESP, Federal University of Mato Grosso do Sul (UFMS), and Conselho Nacional de Desenvolvimento Científico e Tecnológico (CNPq).

References

Adolfi, M.C., A.C.O. Carreira, L.W.O. Jesus, J. Bogerd, R.M. Funes, M. Schartl et al. 2015. Molecular cloning and expression analysis of dmrt1 and sox9 during gonad development and male reproductive cycle in the lambari fish, *Astyanax altiparanae*. Reprod. Biol. Endocrinol. 13: 1–15.

Alexandru, V.B., C. Gabriel, I. Lucian and B.I. Alin. 2014. Morphological abnormalities as a result of inbreeding in controlled reproduction in *Oreochromis niloticus*, Linnaeus, 1757. J. Biotechnol. 185: S19.

Amaral, A.C., A.F. Lima, L.N. Ganeco-Kirschnik and F.L. de Almeida. 2020. Morphological characterization of pirarucu *Arapaima gigas* (Schinz, 1822) gonadal differentiation. J. Morphol. 281: 491–499.

Amaral, J.S., R.L. Venturieri and R.G. Moreira. 2019. Gonadal steroids and energy availability during ovarian maturation stages of the Amazonian pirarucu *Arapaima gigas* (Teleostei: Osteoglossidae) in the wild. Comp. Biochem. Physiol. - Part A Mol. Integr. Physiol. 230: 106–114.

Andrade-Talmelli, E.F., E.T. De Kavamoto, M.Y. Narahara and N. Fenerich-Verani. 2002. Reprodução induzida da piabanha, *Brycon insignis* (Steindachner, 1876), mantida em cativeiro. Rev. Bras. Zootec 31: 803–811.

Araújo, F.C.T., R.P. Ribeiro, E.C. Campos, H. Todesco, K.M. Tsujii, L.S.C. Mantovani et al. 2022. Could serum glucose be a selection criterion in Nile tilapia breeding programs? Aquac. 548: 737573.

Araujo, J.E.X.S., D.P. Streit Jr., J.S. Abreu, E.F.F. Martins, F.N. Souza, C.A.L. Oliveira et al. 2014. Ovopel and carp pituitary extract as spawning inducers in males of the Amazon catfish *Leiarius marmoratus* (Gill, 1970). Braz Arch. Biol. Technol. 57: 882–886.

Barbosa, R.P., R.Y. Kuradomi, R.T. Sato and S.R. Batlouni. 2022a. *Piaractus mesopotamicus* gonad differentiation. Aquac. Res. 53: 4106–4115.

Barbosa, L.M., R.M.R. Costa, C.C. Muniz and L.A.F. Mateus. 2022b. Tamanho da primeira maturação de duas espécies de interesse pesqueiro na bacia do alto Rio Paraguai – Mato Grosso: subsídios para o estabelecimento do tamanho mínimo de captura. Rev. Biodiversidade 21: 46–57.

Barzotto, E. and L. Mateus. 2017. Reproductive biology of the migratory freshwater fish *Salminus brasiliensis* (Cuvier, 1816) in the Cuiabá River basin, Brazil. J. Appl. Ichthyol. 33: 415–422.

Bashiyo-Silva, C., R.S. Costa, D.C. Ribeiro, J.A. Senhorini, R. Veríssimo-Silveira and A. Ninhaus-Silveira. 2016. Hormonal induction of *Brycon cephalus* (Characiformes, Characidae) to spermiation using D-ala6, pro9net-mGnRH þ metoclopramide. Zygote 24: 319e21.

Baroiller, J.F., H. D'Cotta and E. Saillant. 2009. Environmental effects on fish sex determination and differentiation. Sex Dev. 3: 118–135.

Batlouni, S.R., F.R. Carreno, E. Romagosa and M.I. Borella. 2005. Cell junctions in the germinal epithelium may play an important role in spermatogenesis of the catfish *P. fasciatum* (Pisces, Siluriformes). J. Mol. Histol. 36: 97–110.

Batlouni, S.R., E. Romagosa and M.I Borella. 2006. The reproductive cycle of male catfish *Pseudoplatystoma fasciatum* (Teleostei, Pimelodidae) revealed by changes of the germinal epithelium: An approach addressed to aquaculture. Anim. Reprod. Sci. 96: 116–132.

Batlouni, S.R., L.M.J. Silva and M.R. Abreu. 2021. Obtenção de extrato bruto de hipófise de *Astyanax altiparanae* para uso na reprodução induzida de peixes. BR. Patent # 10202101833-6.

Bernardes Jr., J.J., R.A. Bombardelli and A.P. de Oliveira Nuñer. 2017. Gonadorelin increases semen production and does not affect its quality in *Leporinus obtusidens*. Anim. Reprod. Sci. 185: 154–160.

Bittencourt, F., B.E. de Souza, T.A. Lui, M.I. Borella, W.R. Boscolo, A. Feiden et al. 2012. Protein diets promote the maturation of oocytes and spawning of *Piaractus mesopotamicus* kept in cages. J. Appl. Ichthyol. 28: 886–893.

Bobe, J. and C. Labbé. 2010. Egg and sperm quality in fish. Gen. Comp. Endocrinol. 165: 535–548.

Bombardelli, R.A., E.S.R Goes, S.M.N. Souza, M.A. Syperreck, M.D. Goes, A.C.O. Pedreira et al. 2017. Growth and reproduction of female Nile tilapia fed diets containing different levels of protein and energy. Aquac. 479: 817–823.

Borella, M.I., C. Chehade, F.G. Costa, L.W.O. De Jesus, M. Cassel and S.R. Batlouni. 2020. The brain-pituitary-gonad axis and the gametogenesis. pp. 315–341. *In*: Baldisserotto, B., E.C. Urbinati and J.E.P. Cyrino [eds.]. Biology and Physiology of Freshwater Neotropical Fish. Academic Press, Amsterdam, NL.

BOYD, C.E. 1998. Water Quality for Pond Aquaculture. Auburn University, Alabama.

Brambila-Souza, G., C.C. Mylonas, P.H. Mello, R.Y. Kuradomi, S.R. Batlouni, C.E. Tolussi et al. 2019. Thermal manipulation and GnRHa therapy applied to the reproduction of lambari-do-rabo-amarelo, *Astyanax altiparanae* females (Characiformes: Characidae) during the non-breeding season. Gen Comp. Endocrinol. 279: 120–128.

Buchet, V., E. Coquard, A. Sévère and H. Barone. 2008. Influence of tank volume on vitellogenesis and spawning performances in sea bass (*Dicentrarchus labrax* L.) Aquac. Res. 9: 420–426.

Büyükhatipoglu, S. and W. Holtz. 1984. Sperm output in rainbow trout (*Salmo gairdneri*) effect of age, timing and frequency of stripping and presence of females. Aquac 37: 63–71.

Campbell, P.M., T.G. Pottinger and J.P. Sumpter. 1992. Stress reduces the quality of gametes produced by rainbow trout. Biol. Reprod. 47: 1140–1150.

Campbell, P.M., T.G. Pottinger and J.P. Sumpter. 1994. Preliminary evidence that chronic confinement stress reduces the quality of gametes produced by brown and rainbow trout. Aquac. 120: 151–169.

Cardona, E., B. Lorgeoux, L. Chim, J. Goguenheim, H.L. Delliou and C. Cahu. 2016. Biofloc contribution to antioxidant defence status, lipid nutrition and reproductive performance of broodstock of the shrimp *Litopenaeus stylirostris*: Consequences for the quality of eggs and larvae. 2016. Aquac. 452: 252–262.

Castets, M.D., B. Schaerlinger, F. Silvestre, J.N. Gardeur, M. Dieu, C. Corbier et al. 2012. Combined analysis of *Perca fluviatilis* reproductive performance and oocyte proteomic profile. Theriogenology 78: 432–442.

Ceccarelli, P.S., J.A. Senhorini and G. Volpato. 2000. Dicas em piscicultura; perguntas e respostas. Santana Gráfica Editora, Maceió.

Corriero, A., R. Zupa, G. Bello, C.C. Mylonas, M. Deflorio, S. Genovese et al. 2011. Evidence that severe acute stress and starvation induce rapid atresia of ovarian vitellogenic follicles in Atlantic bluefin tuna, *Thunnus thynnus* (L.) (Osteichthyes: Scombridae). J. Fish Dis. 34: 853–860.

Costa, L.F.A., J.S. Lima, A.M.A. Siqueira, E.C.C. da Silva, V.C.F Lima, L.A. da, et al. 2021. Effect of multi-species probiotic administration in *Colossoma macropomum* juvenile rearing: supplementation and bioremediation. Aquac. Nutrition 27: 1721–1729.

Criscuolo-Urbinati, E., R.Y. Kuradomi, E.C. Urbinati and S.R. Batlouni. 2012. The administration of exogenous prostaglandin may improve ovulation in pacu (*Piaractus mesopotamicus*). Theriogenology 78: 2087–2094.

Dabrowski, K., J. Rinchard, J.S. Ottobre, F. Alcantara, P. Padilla, A. Ciereszko et al. 2003. Effect of oxygen saturation in water on reproductive performances of pacu *Piaractus brachypomus*. JWAS 34: 441–449.

De Souza, T.G., P. Hainfellner, R.Y. Kuradomi, M.E. Muñoz, R.M. Honji, R.G. Moreira et al. 2015. Inappropriate management conditions, especially for the regressed class, are related to sperm quality in *Prochilodus lineatus*. Theriogenology 83: 797–807.

Ekasari, J., M. Zairin Jr., D.U. Putri, N.P. Sari, E.H. Surawidjaja and P. Bossier. 2015. Biofloc-based reproductive performance of Nile tilápia *Oreochromis niloticus* L. broodstock. Aquac. Res. 46: 509–512.

Ekasari, J., M.A. Suprayudi, W. Wiyoto, R.F. Hazanah, G.S. Lenggara, R. Sulistiani et al. 2016. Biofloc technology application in African catfish fingerling production: The effects on the reproductive performance of broodstock and the quality of eggs and larvae. Aquac. 464: 349–356.

Emerenciano, M., G. Cuzon, M. Arévalo and G. Gaxiola. 2014. Biofloc technology in intensive broodstock farming of the pink shrimp *Farfantepenaeus duorarum*: spawning performance, biochemical composition and fatty acid profile of eggs. Aquac. Res. 45: 1713–1726.

Emerenciano, M.G.C., L.R. Martínez-Córdova, M. Martínez-Porchas and A. Miranda-Baeza. 2017. Biofloc technology (BFT): a tool for water quality management in aquaculture. pp. 91–109. *In*: Hlanganani, T. [ed.]. Water Quality. InTech, London, UK.

FAO. 1999. Food and Agriculture Organization of the United States. Inbreeding and Broodstock Management, FAO, Rome, Italy.

FAO. 2022. Food and Agriculture Organization of the United Nations. The State of World Fisheries and Aquaculture, FAO, Rome, Italy.

Fernandino, J.I. and R.S. Hattori. 2019. Sex determination in Neotropical fish: Implications ranging from aquaculture technology to ecological assessment. Gen. Comp. Endocrinol. 273: 172–183.

Garcia, A.L.S., C.A.L. Oliveira., H.M. Karim, C. Sary, H. Todesco and R.P. Ribeiro. 2017. Genetic parameters for growth performance, fillet traits, and fat percentage of male Nile tilapia (*Oreochromis niloticus*). J. Appl. Genet. 58: 527–533.

Gjedrem, T. 2005. Selection and Breeding Programs in Aquaculture. Springer Dordrecht, Berlin.

Haraz, Y.G., R.M. Shourbela, W.N. El-Hawarry, A.M. Mansour and S.S. Elblehi. 2023. Performance of juvenile *Oreochromis niloticus* (Nile tilapia) raised in conventional and biofloc technology systems as influenced by probiotic water supplementation. Aquac. 566: 739180.

Hainfellner, P., T.G.D. Souza, R.G. Moreira, L.S.O. Nakaghi and S.R. Batlouni. 2012. Gonadal steroids levels and vitellogenesis in the formation of oocytes in *Prochilodus lineatus* (Valenciennes) (Teleostei: Characiformes). Neotrop. Ichthyol. 10: 601–612.

Hainfellner, P., R.Y. Kuradomi, T.G. de Souza, R.T. Sato, D.G. Figueiredo Ariki, G.A. de Freitas et al. 2019. Reproductive cycle of the Amazonian planktivorous catfish *Hypophthalmus marginatus* (Siluriformes, Pimelodidae). Aquac. Res. 50: 3382–3391.

Hashimoto, D.T., F.D. Prado., J.A. Senhorini, F. Foresti and F. Porto-Foresti. 2012. Detection of post-F1 fish hybrids in broodstock using molecular markers: approaches for genetic management in aquaculture. Aquac. Res. 44: 876–884.

Hashimoto, D.T., J.A. Senhorini, F. Foresti, P. Martínez and F. Porto-Foresti. 2014. Genetic identification of F1 and post-F1 Serrasalmid juvenile hybrids in Brazilian Aquaculture. Plos One 9: e89902.

Iwersen, L.H.L., C. Lazoski, E. Zaniboni-filho and J. Ribolli. 2019. Genetic implications of restocking programs on wild populations of streaked prochilod *Prochilodus lineat*us. Bol. Inst. Pesca 45.

Kincaid, H.L. 1983. Inbreeding in fish populations used for aquaculture. Aquac. 33: 215–227.

Konzen-Freitas, A.R., J.G. de Abreu, J.S. de Abreu, V.L. de Queiroz Dantas, R.A.C. Corrêa Filho and J.A. Povh. 2020. Tambaqui females (*Colossoma macropomum*) spawn after hormonal induction with buserelin acetate. Anim. Reprod. Sci. 221: 106594.

Kord, M.I., S. Maulu, T.M. Srour, E.A. Omar, A.A. Farag, A.A.M. Nour et al. 2022. Impacts of water additives on water quality, production efficiency, intestinal morphology, gut microbiota, and immunological responses of Nile tilapia fingerlings under a zero-water-exchange system. Aquac. 547: 737503.

Kuradomi, R.Y., T.G. De Souza, F. Foresti, R.W. Schulz, J. Bogerd, R.G. Moreira et al. 2016. Effects of re-strippingon the seminal characteristics of pacu (*Piaractus mesopotamicus*) during the breeding season. Gen. Comp. Endocrinol. 225: 162–173.

Kuradomi, R.Y. and S.R. Batlouni. 2018. PGF2α and gonadal steroid plasma levels of successful and unsuccessful spawning *Piaractus mesopotamicus* (Teleostei, Characiformes) females. Aquac. Int. 26: 1083–1094.

Landry, C.A., S.L. Steele, S. Manning and A.O. Cheek. 2007. Long term hypoxia suppresses reproductive capacity in the estuarine fish, *Fundulus grandis*. Comp. Biochem. Physiol. Mol. Amp; Integr. Physiol. 148: 317–323.

Leal, M.C., E.R. Cardoso, R.H. Nóbrega, S.R. Batlouni, J. Bogerd, L.R. França et al. 2009. Histological and stereological evaluation of zebrafish (*Danio rerio*) spermatogenesis with an emphasis on spermatogonial generations. Biol. Reprod. 81: 177–187.

Leite, L.V., M.A.P. Melo, F.C.E. Oliveira, J.P.S. Pinheiro, C.C. Campello, J.F. Nunes et al. 2013. Determination of insemination dose and embryonic development in the artificial fertilization of tambaqui (*Colossoma macropomum*). Arq Bras Med. Vet. Zootec 65: 421–429.

Lemanova, N.A. and O.F. Sakum. 1975. Metodisceskoe posibilie pogor monalnoistimulacji proizvoditielei Karpa pré rannom polucenii licinok. Izvestiya Gosudartvennogo Nauchno – Isseledivatel Skogo Instituta 88: 3–23.

Leonardo, A.F.G., E. Romagosa, M.I. Borella and S.R. Batlouni. 2004. Induced spawning of hatchery-raised Brazilian catfish, cachara *Pseudoplatystoma fasciatum* (Linnaeus, 1766). Aquac. 240: 451–461.

Levavi-Zermonsky, B. and Z. Yaron. 1986. Changes in gonadotropin and ovarian steroids associated with oocytes maturation during spawning induction in the carp. Gen. Comp. Endocrinol. 62: 89–98.

Lima, J.J.F. 2022. Desenvolvimento e desempenho reprodutivo de lambaris (*Astyanax lacustris*) cultivados em sistema de bioflocos e em sistema de recirculação com água clara. MSc. Dissertation, University of State Paulista, São Paulo, BR.

Lira, L.V.G., R.Y. Kuradomi, T.G. de Souza, P. Hainfellner and S.R. Batlouni. 2018. *Astyanax altiparanae* Ovarian maturation after spawning in water recycling systems. Bol. Inst. Pesca 44: 1–10.

Lobo, I.K.C., A.R. do Nascimento, M.E.B. Yamagishi, Y. Guiguen, G.F. da Silva, D. Severac et al. 2014. Diversidade genética e paternidade de progênies de *Brycon orbignyanus* obtidas por diferentes sistemas reprodutivos. Semina: Ciências Agrárias 35: 541–554.

Martínez, P., A.M. Viñas, L. Sánchez, N. Díaz, L. Ribas and F. Piferrer. 2014. Genetic architecture of sex determination in fish: applications to sex ratio control in aquaculture. Front Genet. 5: 340.

Martins, E.D.F.F., D.P. Streit Jr, J.S. de Abreu, R.A.C. Corrêa-Filho, C.A.L. de Oliveira et al. 2017. Ovopel and carp pituitary extract for the reproductive induction of *Colossoma macropomum* males. Theriogenology 98: 57–61.

Mechaly, A.S., S.R. Batlouni, M. Elisio, E.A. Sanches, J.C. Gusman, M.M. García et al. 2023. Spawning induction for Latin American fishes. Rev. Aquac. 1–20

Moraes, A.V.D., M.D.O. Pereira, K.N. Moraes, J.P. Rodrigues-Soares, G.F.A. Jesus and A. Jatobá. 2018. Autochthonous probiotic as growth promoter and immunomodulator for *Astyanax bimaculatus* cultured in water recirculation system. Aquac. Res. 49: 2808–2814.

Narahara, M.Y., E.F. De Andrade-Talmelli, E.T. Kavamoto and H.M. Godinho. 2002. Reprodução induzida da Pirapitinga-do-Sul, *Brycon opalinus* (Cuvier, 1819), mantida em condições de Confinamento. Rev. Bras Zootec 31: 1070–1075.

Nayak, S.K. 2021. Multifaceted applications of probiotic *Bacillus* species in aquaculture with special reference to *Bacillus subtilis*. Rev. Aquac. 13: 862–906.

Nunes, A. 2019. Qualidade de carne, desempenho produtivo e curvas de crescimento de três grupos genéticos de tilápia-do-Nilo (*Oreochromis niloticus*). 2019. MSc dissertation. Universidade Federal do Mato Grosso do Sul, Mato Grosso do Sul, Brazil.

Patiño, R., G. Yoshizaki, P. Thomas and H. Kagawa. 2001. Gonadotropic control of ovarian follicle maturation: the two-stage concept and its mechanisms. Comp. Biochem. Physiol. B Biochem. Mol. Biol. 129: 427–439.

Paulino, M.S., A.B. Miliorini, L.D.S. Murgas, F.S.M. Lima and V.O. Felizardo. 2011. Reproductive performance of pacu, piracanjuba and curimba induced by buserelinc extract. Bol. Inst. Pesca 37: 39–45.

PEIXE BR. 2023. Anuário Brasileiro da Piscicultura, PEIXE BR, São Paulo, Brazil.

Pereira, T.S.B., C.N.P. Boscolo, R.G. Moreira and S.R. Batlouni. 2017. The use of mGnRHa provokes ovulation but not viable embryos in *Leporinus macrocephalus*. Aquac. Int. 25: 515–529.

Pereira, T.S.B., C.N.P. Boscolo, R.G. Moreira and S.R. Batlouni. 2018. *Leporinus elongatus* induced spawning using carp pituitary extract or mammalian GnRH analogue combined with dopamine receptor antagonists. Anim. Reprod. 15: 64.

Pires, L.B., E.A. Sanches, E. Romagosa, R.A.C. Corrêa Filho, D.P. Streit Jr., R.A.R. Nass et al. 2017. Semen characteristics of *Colossoma macropomum* from three successive sample collections in the same reproductive cycle. Aquac. Res. 48: 5104–5110.

Pires, L.B., R.A.C. Corrêa Filho, E.A. Sanches, E. Romagosa, T.G. da Silva, S. Rech et al. 2018. *Colossoma macropomum* females can reproduce more than once in the same reproductive period. Anim. Reprod. Sci. 196: 138–142.

Pires, L.B., E.A. Sanches, E. Romagosa, R.A.C. Corrêa Filho, R.A.R. Araujo, N.M. Lopera-Barrero et al. 2019. Sperm quality of *Colossoma macropomum* after room-temperature and cold storage. J. Appl. Ichthyol. 35: 747–753.

Porta, J., J.M. Porta, G. Matínez-Rodríguez and M.C. Alvarez. 2006. Development of a microsatellite multiplex PCR for Senegalese sole (*Solea senegalensis*) and its application to broodstock management. Aquac 256: 159–166.

Porto-Foresti, F., R.B. Castilho-Almeida, J.A. Senhorini and F. Foresti. 2010. Biologia e criação do lambari do rabo amarelo (*Astyanax altiparanae*). pp. 101–115. *In*: Baldisserotto, B. and L.C. Gomes [eds.]. Espécies nativas para piscicultura no Brasil. Editora UFSM, Mato Grosso do Sul, MS, Brazil.

Povh, J.A., N.M. Lopera-barrero, R.P. Ribeiro, E. Lupchinski Jr., P.C. Gomes and T.S. Lopes. 2008. Monitoreo genético enprogramas de repoblamiento de peces mediante marcadores moleculares. Cien Inv. Agr. 35: 5–15.

Povh, J.A., R.P. Ribeiro, R.N. Sirol, D.P. Streit Jr., H.M. Moreira, F. Siewerdt et al. 2010. Microsatellite analysis of the parental contribution of *Piaractus mesopotamicus* to the production of offspring in the semi-natural system of reproduction. Braz Arch. Biol. 53: 389–396.

Povh, J.A., C.A.L. Oliveira, R.P. Ribeiro, L. Shiotsuki, S.A. Texeira, D.C. Fornari et al. 2022. Perspectivas para o melhoramento genético de peixes nativos. Anais do 31° congresso brasileiro de zootecnia: produtividade e conservação: o futuro da zootecnia. São Carlos, SP: Aptor Software.

Pozzer, R.F., C.F. Kuhn, R.A. Chedid, R.M. Hideo, S.R. Batlouni, I.B.V. Franceschini et al. 2023. Morphological analysis of the digestive tract of *Hypophthalmus marginatus* (Valenciennes, 1840). Anat. Histol. Embryol. 52: 603–610.

Quirino, P.P., M.L.R. Delgado, L. Gomes-Silva, C.F. Benevente, M.L. Grigoli-Olivio, B.C. Bianchini et al. 2022. Female sex inversion as a reason for an unbalanced sex ratio in the neotropical species *Brycon orbignyanus*. Aquac. Res. 53: 1706–1726.

Ram, R.N. and A.G. Sathyanesan. 1986. Ammonium sulfate induced nuclear changes in the oocyte of the fish, *Channa punctatus* (Bl.). Bull Environ. Contam. Toxicol. 36: 6.

Reidel, A., W.R. Boscolo, A. Feiden and E. Romagosa. 2010. The effect of diets with different levels of protein and energy on the process of final maturation of the gametes of *Rhamdia quelen* stocked in cages. Aquac. 298: 354–359.

Ribas, L., A. Valdivieso, N. Díaz and F. Piferrer. 2017. Appropriate rearing density in domesticated zebrafish to avoid masculinization: links with the stress response. J. Exp. Biol. 220: 1056–1064.

Ribeiro, R.P. 2001. Construção de tanques. pp. 45–52. *In*: Moreira, H.L.M., L. Vargas, R.P. Ribeiro and S. Zimmermann [eds.]. Fundamentos da moderna Aquicultura. Editora ULBRA, Canoas, RS, Brazil.

Romagosa, E. 1998. Desenvolvimento gonadal (morfologia; ultraestrutura) e indução da reprodução do matrinchã, *Brycon cephalus* (Günther, 1869) em cativeiro Vale do Ribeira. Ph.D. Thesis, Federal University of São Carlos, SP, Brazil.

Romagosa, E. 2010. Reproductive status in females of the Brazilian catfish, *Pseudoplatystoma fasciatum* reared in cages. J. Appl. Ichthyol. 26: 806–811.

Roza de Abreu, M., L.M.J. Silva, D.G. Figueiredo-Ariki, R.T. Sato, R.Y. Kuradomi and S.R. Batlouni. 2021. Reproductive performance of lambari (*Astyanax altiparanae*) in a seminatural system using different protocols. Aquac. Res. 52: 471–483.

Roza de Abreu, M., L.M. de Jesus Silva, D.G. Figueiredo-Ariki, R.T. Sato, R.T. Kuradomi and S.R. Batlouni. 2022. The effect of LHRHa with and without dopamine antagonist on reproductive performance in lambari *Astyanax altiparanae*. Aquac. 550: 737883.

Sarder, M.R.I., D.J. Penman, J.M. Myers and B.J. McAndrew. 1999. Production and propagation of fully inbred clonal lines in the Nile tilápia (*Oreochromis niloticus* L.). J. Exp. Zool. 284: 675–685.

Sato, R.T., R.Y. Kuradomi, M.C. Calil, L.M. de Jesus-Silva, M.R. de Abreu, D.G. Figueiredo-Ariki et al. 2021. Resumption and progression of meiosis and circulating levels of steroids and prostaglandin F2α of *Piaractus mesopotamicus* induced by hypophysation with prostaglandin F2α. Aquac. Res. 52: 1026–1037.

Schorer, M., R.G. Moreira and S.R. Batlouni. 2016. Selection of pacu females to hormonal induction: Effect of age and of evaluation methods. Bol. Inst. Pesca 42: 901–913.

Shang, E.H., R.M. Yu and R.S. Wu. 2006. Hypoxia affects sex differentiation and development, leading to a male-dominated population in zebrafish (*Danio rerio*). Environ. Sci. Technol. 40: 3118–3122.

Shiotsuki, L., L.C.V. Villela, L.N.G. Kirschnik, L.E.L. Freitas, F.P. Rezende et al. 2019. Fundação das bases genéticas para um futuro Programa de Melhoramento de Tambaqui (*Colossoma macropomum*). Aquac. Brasil 17: 10.

Silva, A.C.F., R.A.C. Corrêa-Filho, A.S. Ventura, A.L. Nunes, L.M. Laice, R.P. Ribeiro et al. 2020. Reproductive traits in different Nile tilapia genetic groups. Arq Bras Med. Vet. Zootec 72: 1797–1804.

Soltani, M., K. Ghosh, S. Hoseinifar, V. Kumar, A.J. Lymbery and E. Ringø. 2019. Genus Bacillus, promising probiotics in aquaculture: aquatic animal origin, bio-active components, bioremediation and efficacy in fish and shellfish. Rev. Fish Sci. Aquac. 27: 331–379.

Souza, F.N., E.D.F.F. Martins, R.A.C. Corrêa Filho, J.S. De Abreu, L.B. Pires, D.P. Streit Jr. et al. 2018. Ovopel® and carp pituitary extract for induction of reproduction in *Colossoma macropomum* females. Anim. Reprod. Sci. 195: 53–57.

Souza, T.G.D., R.Y. Kuradomi, S.M. Rodrigues and S.R. Batlouni. 2020. Wild *Leporinus friderici* induced spawning with different dose of mGnRHa and metoclopramide or carp pituitary extract. Anim. Reprod. 17.

Spica, L.N., E.A. Sanches, D.P. Streit Jr., R.A.C. Corrêa-Filho, L.O. Brasileiro, R.B. Rodrigues et al. 2021. Successive seminal collections from *Leiarius marmoratus* during the reproductive period. Anim. Reprod. Sci. 234: 06852.

Streit Jr., D.P., J.A. Povh, D.C. Fornari, J.M. Galo, L.R.J. Guerreiro, D. Oliveira et al. 2012. Recomendações técnicas para a reprodução do tambaqui. Documentos/Embrapa Meio-Norte, Teresina, PI, Brazil.

Taranger, G.L., M. Carrillo, M. Schulz, R.W. Fontaine, P. Zanuy, S. Felip et al. 2010. Control of puberty in farmed fish. Gen. Comp. Endocrinol. 165: 483–515.

Thomas, P., M.S. Rahman, I.A. Khan and J.A. Kummer. 2007. Widespread endocrine disruption and reproductive impairment in an estuarine fish population exposed to seasonal hypoxia. Proc. R Soc. B: Biol. Sci. 274: 2693–701.

Vaini, J.O., A.B. Grisolia, F.D. Prado and F. Porto-Foresti. 2014. Genetic identification of interspecific hybrid of Neotropical catfish species (*Pseudoplatystoma corruscans* vs. *Pseudoplatystoma reticulatum*) in rivers of Mato Grosso do Sul State, Brazil. Neotrop. Ichthyol. 12: 635–641.

Val, A.L. and A.M. de Oliveira. 2021. *Colossoma macropomum*—A tropical fish model for biology and aquaculture. J. Exp. Zool Part A Ecol. Integr. Physiol. 335: 761–770.

Valandro, J.S.I. 2023. Desempenho reprodutivo do lambari-do-rabo-amarelo (*Astyanax lacustris*) criados em diferentes sistemas fechados de produção. MSc. Dissertation, University of State Paulista, São Paulo, BR.

Valentin, F.N., S.R. Batlouni, N.F. Nascimento, R.C. Silva, B. Manzini, C.C. Hilbig et al. 2016. Ovarian differentiation and development in cachara *Pseudoplatystoma fasciatum*. J. Fish Biol. 89: 445–466.

Valentin, F.N., L.S.O. Nakaghi, S.R. Batlouni, N.F. do Nascimento, M.C.F. Paes, M.C. Silva et al. 2019. Testicular differentiation and development in South American catfish, surubim, *Pseudoplatystoma fasciatum*. J. Appl. Ichthyol. 35: 529–540.

Vianna, N.C., M.C. Filho, R.M. Antônio, M.S. Jardim, S.R. Batlouni, S.G.C. Britto et al. 2013. Projeto Reintrodução Adequada de Peixes em Represas. UNESP, São Paulo.

Vidal, M.V. and S.R. Batlouni. 2023. The environmental licensing of hydroelectrics and the interface with migratory fish and aquaculture in Brazil. Bol. Inst. Pesca 48.

Wu, R.S., B.S. Zhou, D.J. Randall, N.Y. Woo and P.K. Lam. 2003. Aquatic hypoxia is an endocrine disruptor and impairs fish reproduction. Environ. Sci. Technol. 37: 1137–1141.

Yoshida, G.M., C.A.L. Oliveira, N.M. Kunita, G.S. Rizzato and R.P. Ribeiro. 2015. Reproduction performance of female Nile tilapia under different environments and age classes. Acta Sci. Anim. Sci. 37: 221–226.

Yoshida, G.M., C.A.L. Oliveira, E.C. Campos, H. Todesco, F.C.T. Araújo, H.M. Karin et al. 2021. A breeding program for Nile tilapia in Brazil: results from nine generation of selection to increase growth rate in cages. J. Anim. Breed Genet. 139: 127–135.

Zanoni, M.A., F.G. Costa, S. Carvalho and F.R.F. Seiva. 2016. Physiological and biochemical changes of females of Piracanjuba, subjected to induced reproduction. J. Anim. Physiol. Anim. Nutr. 100: 673–679.

Zardo, E.L., D.C. Fornari, J. Gioria, D.A. Rotili, I.C. Gomes, J.R. Esquivel-Muelbert et al. 2021. Gonadal development period and sexual differentiation through histological analysis in *Brycon orbignyanus* (Valenciennes, 1850) (Characiformes: Bryconidae). Aquac. 539: 736636.

CHAPTER 8

Embryos and Ovarian Tissues of South American Neotropical Fish

Current State of Development

Leandro Andrés Miranda,[1,*] *Thaíza Rodrigues de Freitas,*[2] *Monica Rodrigues Ferreira Machado*[3] and *Lis Santos Marques*[4]

INTRODUCTION

Cryopreservation is a biotechnology used to conserve living cells, gametes, and embryos during a short or prolonged storage time, using liquid nitrogen as a refrigerant medium. Its mechanism is based on achieving cell dehydration in a controlled manner, allowing intracellular and extracellular fluid exchange, thus mitigating the formation of ice crystals in the cytosol (Chao and Liao 2001). Cell deterioration is stopped in biological tissues stored at sufficiently low temperatures, such as below –150°C (Leibo and Mazur 1971). Thus, this cessation of biological time allows cells to be preserved indefinitely. Cryopreservation of maternal and paternal gametes is an essential technique for establishing biological material banks, as it preserves

[1] Laboratorio de Ictiofisiología y Acuicultura, Instituto Tecnológico de Chascomús (CONICET-UNSAM). Escuela de Bio y Nanotecnologías (UNSAM). Av. Intendente Marino Km. 8.2 (B7130IWA) Chascomús, Buenos Aires, Argentina.
[2] Animal Science Research Program, Av. Bento Gonçalves, 7712, Federal University of Rio Grande do Sul (UFRGS) – CEP: 91540-000, Porto Alegre, RS, Brazil.
[3] Department of Biological Sciences, Cidade Universitária BR 364, km 195, 3800 Federal University of Jataí (UFJ) – CEP: 75801-615, Jataí, GO), Brazil.
[4] Centro de Modelos Biológicos Experimentais – CeMBE da PUCRS, Porto Alegre, RS, Brasil.
Emails: thaiza.rfreitas@gmail.com, monica_rodrigues@ufj.edu.br, lismarx@gmail.com
* Corresponding author: lmiranda@intech.gov.ar

the genetic variability of the population and maintains the characteristics of the species. Cryopreservation of fish sperm is widely used, but some issues concern the female genome (embryo and oocyte). Among them is the high sensitivity to cold, the presence of compartments, the low permeability to the membrane, and the high amount of intracytoplasmic lipids (Tsai and Lin 2012; Diwan et al. 2020). These structural differences happen mainly because of external fertilization. Thus, over the years, several studies have been conducted to overcome these problems and preserve the maternal genetic material of fish. Therefore, this chapter will present the current state of development of embryo and ovarian tissue cryopreservation techniques for South American Neotropical fish species and their prospects.

1. Embryo Cryopreservation

Many studies have been carried out about the cryopreservation of fish embryos on different species, but practically all have negative results. This is due to their large size (several times larger than that of humans), the presence of yolk, and the low and different permeability of their membranes that make it difficult for cryoprotectants to enter (Hagedorn et al. 1997; Valdez et al. 2005; Lahnsteiner 2008; Diwan et al. 2020). In order to overcome these obstacles, it is necessary to evaluate the toxicity of cryoprotectants, the method of their incorporation, and the application of different freezing protocols (Zhang et al. 2005). Different combinations of external and internal cryoprotectants have been used, including glucose, sucrose, trehalose, methanol, dimethyl sulfoxide, ethylene glycol, and glycerol, among others (Diwan et al. 2020). The correct mixture of these cryoprotectants will allow adequate cell protection; however, all cryoprotectants present moderate toxicity, so it is necessary to establish the most resistant embryonic stage (Lahnsteiner 2008). Also, other types of cryoprotectants have been used, such as antifreeze proteins (AFPs) that have been isolated and characterized in different subarctic marine fishes and that would prevent the formation of intracellular crystals (Beirão et al. 2006; Robles et al. 2004, 2006, 2007; Robles and Valcarce 2019). Regarding the methods of incorporating cryoprotectants, electroporation, treatment with aquaporins, pressurization, dechorionation, dehydration, sonication, and microinjection have been tested (Janik et al. 2000; Zhang et al. 2005; Alam et al. 2018). Recently, the ability to rewarm cryopreserved zebrafish embryos that grow into adult fish capable of breeding normally was reported for the first time. This protocol employed a single injection of cryoprotective agents and gold nanorods into the yolk and immersion in a precooling bath to dehydrate the perivitelline space. Afterward, they were plunged into liquid nitrogen for a few seconds and immediately exposed to laser irradiation to heat the nanoparticles (Khosla et al. 2020).

Chilling and freezing at non-cryogenic subzero temperatures is an alternative to cryopreservation of fish embryos. This technique consists of exposing the embryos to a cryoprotective solution and storing them at a temperature below 0°C for a short period, leading to a decrease in enzymatic and cellular activity (Cloud et al. 2000). Chilling protocols have been evaluated in embryos of different fish species such as *Labeo rohita, Catla catla, Cirrhinus mrigala,* and *Cyprinus carpio* (Ahammad et al. 1998, 2003a, b). In the case of South American Neotropical fishes, this technique was used with some success, using different cryoprotective solutions, temperatures, and storage times. In this sense, it has been possible to cool *Piaractus mesopotamicus* embryos by combining different concentrations of methanol and sucrose, obtaining 69.2 percent hatching after keeping them at –8°C for 6 hours. (Streit et al. 2007; Lopes et al. 2011; Fornari et al. 2012). Using similar protocols, it was also possible to freeze embryos of *Rhinelepsis aspera* (Fornari et al. 2011), *Salminus brasiliensis* (Fornari et al. 2013), *Piaractus brachypomus* (Pessoa et al. 2014) and *Steindachneridion parahybae* (Lopes et al. 2015). Different temperatures and storage times have also been tested in *Brycon orbignyanus* embryos, observing that after 24 hours, all embryos cooled to 0°C or 8°C died (Faria Paes et al. 2014).

In Argentina, the pejerrey *Odontesthes bonariensis* is considered one of the most promising fish for aquaculture due to the interest in its sport fishing and the quality of its meat (Somoza et al. 2008). Given the interest in the species' breeding, its reproductive physiology (Elisio et al. 2014, 2015), the management of reproduction in captivity (Miranda et al. 2005; Miranda and Somoza 2009; Chalde et al. 2016; Chalde and Miranda 2020), and sperm cryopreservation (Lichtenstein et al. 2010) have been studied.

In this species, the first tests of toxicity and permeability of a cryoprotectant (dimethyl sulfoxide) to its embryos were reported by Suzuki et al. (1995). Subsequently, given the difficulty of cryopreserving complete embryos, isolated blastomeres were successfully frozen for 90 min at –75°C, but with survival rates close to 20 percent (Strüssmann et al. 1999). Then, the convenience of using trehalose as an embryonic cryoprotectant was demonstrated (Routray et al. 2002). In 2013, the first preliminary studies were performed using combinations of dimethyl sulfoxide (DMSO), ethylene glycol, methanol, sucrose, and trehalose and freezing mid-stage embryos at –14°C for 1 h. The results were very promising, obtaining embryonic survival rates of 100 percent and a hatching rate of 40 percent with the combination of 20 percent methanol and sucrose (del Fresno and Miranda 2013). All these studies served as the basis for freezing pejerrey embryos at temperatures close to –20°C. In this regard, pejerrey embryos have recently been successfully frozen for 1 h at –14 and –20°C. In that work, two cooling curves (fast and slow) were tested, and cryoprotective solutions based on methanol, DMSO, and sucrose were applied by exposure or for the first time in pejerrey by microinjection. The results obtained showed high rates of embryonic survival (80–100%) and hatching (68–87%) for all trials. In particular, the most appropriate protocol for freezing *O. bonariensis* embryos at –14°C for 1 h was the slight decrease in temperature in combination with a cryoprotective solution of 2.5 M methanol, 1.4 M DMSO, 0.3 M sucrose, and 0.08 M NaCl, obtaining 87% hatching, with up to 30% normal-looking larvae. The hatching percentage decreased to 60% when cooling was carried out to –20°C, and only 4.5% of normal-looking larvae were obtained when the embryos were microinjected into the perivitelline space (Macoretta and Miranda 2020).

In conclusion, we can say that chilling protocols of embryos to sub-zero temperatures in South American fish species from the Neotropics is possible, but it is necessary to continue refining adequate protocols for each species.

2. Cryopreservation of Fish Ovarian Tissue

Cryopreservation of the maternal genome is particularly important, as it preserves mitochondrial DNA and mRNAs that control the early stages of embryonic development (Tsai et al. 2010). However, cryopreservation of mature fish oocytes remains a challenge.

Follicular development begins with the formation of primordial germ cells. These germ cells then transform into oogonia and later into primary growing oocytes. At this moment, the first meiotic division begins, in which structural changes occur, and each oogonia multiplies by mitosis. At this stage, the oogonia is surrounded by a single layer of granulosa cells secreting the basal lamina, which separates the ovarian stroma (Lubzens et al. 2010). Around the basal lamina is the thecal layer associated with blood vessels (Cao et al. 2019; Lubzens et al. 2010). Ovarian follicles can be classified according to the oocyte development present in them and are classified as:

I) *Follicles in Primary Growth*: These follicles show many nucleoli in the nucleus of the oocytes. These follicles have two phases: the pre-follicular phase, in which the oocyte is inside a nest of oocytes, and the follicular phase, which is determined when the oocyte leaves the nest and lives in a definitive follicle (Selman et al. 1993; Yön et al. 2008).

II) *Cortical Alveolar Follicle*: At this stage, the nucleus doubles in size, the nucleoli are pushed to the periphery of the nucleus, and the cortical alveoli, which are vesicles rich in glycoproteins and hormones, are visualized. As the follicles increase in size, cortical alveoli multiply; consequently, the follicle diameter also increases (Yön et al. 2008; Lubzens et al. 2010).

III) *Vitellogenic Follicle*: In this phase, there is an increase in the thickness of the vitelline envelope and the absorption via endocytosis of vitellogenin, which is produced by the liver and will later be responsible for the nutrition of the post-larvae. The oocyte becomes competent for fertilization at the end of this process (Lubzens et al. 2010).

IV) *Mature Follicle*: In this phase, the particles of lipids and proteins merge, giving a homogeneous appearance, with the vitelline envelope evident.

The high amount of water and lipids inside mature follicles makes dehydration difficult, and the high sensitivity to cold and the low permeability to cryoprotectants make it difficult to obtain success in cryopreservation of this phase of the follicles. Therefore, protocols for cryopreservation of isolated immature oocytes in stage III began to be developed. At this stage, oocytes have the advantage of being smaller, having fewer lipids and greater permeability since they do not yet have a chorionic membrane. The studies by Zhang et al. (2008) and Guan et al. (2008) pioneered the cryopreservation of fish oocytes. The authors carried out slow freezing of vitellogenic follicles, also known as stage III, isolated from zebrafish (*Danio rerio*). The first study used 2 M dimethylsulfoxide (DMSO) as a permeable cryoprotectant and obtained 14.9% membrane integrity (Zhang et al. 2008). The second used 4 M methanol as a cryoprotectant and obtained greater membrane integrity (29.9%) (Guan et al. 2008). From these studies, new slow-freezing protocols of isolated immature oocytes in stages I (primary growth stage), II (cortical alveolus stage), and III (vitellogenesis) were developed, where membrane integrity between 12% and 50.7% was observed (Tsai et al. 2009; Tsai et al. 2010; Zampolla et al. 2012).

Another methodology tested for the cryopreservation of isolated zebrafish stage III oocytes was vitrification using 3 M DMSO, 2 M methanol, and 2 M propylene glycol diluted in KCl buffer as a cryoprotectant (Guan et al. 2010). Vitrification is the rapid cooling of the liquid medium in which ice crystals are not formed (Vajta 2000). However, studies have reported a high loss of follicular viability after warming, in which all oocytes became translucent after some minutes (Guan et al. 2010).

Zampolla et al. (2011) were the first to propose the cryopreservation of zebrafish ovarian tissue fragments using slow freezing and 1 M methanol as a cryoprotectant. The authors observed that the stages I and II oocytes were less sensitive to cryopreservation than stage III oocytes (Zampolla et al. 2011). This suggests that oocytes in the early stages of development are less vulnerable to the negative effects of cryopreservation, as they have a smaller volume and lower amount of lipids. From this, studies were also carried out using the vitrification of zebrafish ovarian tissue. Godoy et al. (2013), when vitrifying in fibreplug ovarian tissue containing oocytes in stage III, using 1.5 M methanol and 4.5 M propylene glycol diluted in 90% Leibovitz-15 medium, observed 59.9% of membrane integrity. However, it showed a decrease in ATP levels after 2 hours (Godoy et al. 2013). Furthermore, Marques et al. (2015), using the same concentration of cryoprotectants, after vitrifying in cryovial ovarian tissue containing oocytes at all stages of development, observed a high percentage of viability of immature oocytes (76% in stage I and 43% in stage II).

When comparing the most common cryopreservation methodologies of zebrafish ovarian tissue, it was observed that vitrification had greater preservation of mitochondrial activity and lower levels of oxidative stress than slow freezing (Marques et al. 2019). These studies demonstrate that cryopreservation of ovarian tissue containing follicles in primary growth can be a promising alternative for the cryopreservation of oocytes in fish. Among the advantages of cryopreservation of ovarian tissue is the protection of follicles against mechanical stress during the cryopreservation process, which can increase the viability of cryopreserved oocytes. Furthermore, the presence of neighboring cells and tissue components around the follicles may provide metabolic support, which protects the follicles during the cryopreservation process, ensuring the integrity and functionality of the oocyte. Thus, using ovarian tissue is an efficient and effective strategy for the cryopreservation of immature fish oocytes, minimizing the adverse effects of cryopreservation that often occur in isolated follicles.

It is possible to observe that the basis of the studies on the cryopreservation of fish oocytes was established using the zebrafish model. However, studies on the cryopreservation of ovarian tissue from South American Neotropical fish are still scarce and so far involving only two species: Pacu (*Piaractus mesopotamicus*) and Piracanjuba (*Brycon orbignyanus*) (Marques et al. 2018; Marques et al. 2021). The protocol for vitrification of zebrafish ovarian tissue developed by Marques et al. (2015) was applied to these cited species, and Pacu observed 70% of membrane integrity in immature follicles (Marques et al. 2018). In Piracanjuba, the results were not different, and even the membrane integrity of oocytes in primary growth in vitrified tissues did not differ from the integrity of oocytes in fresh tissue (not cryopreserved) (Marques et al. 2021). Therefore, these results show that the zebrafish is a potentially suitable model for studies of ovarian tissue cryopreservation from South American fish species.

The lack of cryopreservation protocols for ovarian tissue of South American Neotropical fish species is a challenge to the preservation of the genetic diversity of these species. Although the results obtained with Pacu and Piracanjuba are promising, it is important to emphasize that the diversity of fish species in South America is vast, and there is still much to be done to develop protocols for other species. Expanding studies to include more fish species is crucial to ensure the preservation of the genetic diversity of these species.

2.1 In Vitro Maturation of Fish Oocytes

In order to make it possible to generate new individuals from cryopreserved immature oocytes, these oocytes must be matured *in vitro* for subsequent fertilization. *In vitro* maturation of fish oocytes has been described in some fish species since the 1970s, such as rainbow trout (*Oncorhynchus mykiss*) (Jalabert 1976; Nagahama et al. 1980), goldfish (*Carassius auratus*) (Jalabert 1976), yellow perch (*Perca flavescens*) (Goetz and Theofan 1979), amago salmon (*Oncorhynchus rhodurus*) (Nagahama et al. 1980), medaka (*Oryzias latipes*) (Iwamatsu et al. 1987), and zebrafish (*Danio rerio*) (Li et al. 1993; Selman et al. 1993, 1994; Seki et al. 2008; Silva et al. 2018). These studies used different culture media, growth factors, hormones, and other substances to stimulate oocyte maturation.

The first studies were carried out in cyprinids and salmonids in which the hormone 17α,20β-Dihydroxy-4-pregnen-3-one (DHP) was demonstrated to be efficient in the *in vitro* maturation of oocytes in stage III until complete maturation (Jalabert 1976; Nagahama et al. 1980; Goetz and Theofan 1979; Iwamatsu et al. 1987; Selman et al. 1994). Later studies showed that the growth factors cyclin B, activin, and transforming growth factor β1 play an important role in the *in vitro* maturation of zebrafish oocytes (Kondo et al. 1997; Pang and Ge 2002; Kohli et al. 2005).

One of the first reports of zebrafish stage III oocyte maturation followed by fertilization found cleavage and hatching rates of 70% and 63%, respectively (Seki et al. 2008). In this work, the hormone DHP and Leibovitz-15 90% medium (pH 9.0) supplemented with bovine serum albumin (BSA) were used. Other authors report obtaining live larvae using luteinizing hormone (LH) and follicle-stimulating hormone (FSH) in the maturation of stage III zebrafish oocytes (Silva et al. 2018). In contrast, complete *in vitro* maturation of oocytes in early stages (I and II) has not yet been achieved. An *in vitro* maturation study of zebrafish oocytes reported that the stage I oocytes were differentiated into stage II and stage II oocytes into stage III using human chorionic gonadotropin (hCG) (Tsai et al. 2010). Another study carried out the *in vitro* maturation of follicles included in fragments of ovarian tissue (Anil et al. 2018). These authors used 100 mIU/ml of FSH and 20% fetal bovine serum (FBS) and observed the maturation of oocytes from stage I to stage III. However, the authors did not achieve the same results when performing the *in vitro* maturation of cryopreserved ovarian tissue fragments.

In this context, the need to improve maturation protocols for immature fish oocytes after cryopreservation is evident, especially for those in the early stages of development, as they are less sensitive to the cryopreservation process. The lack of an efficient protocol for the *in vitro* maturation of cryopreserved oocytes is a limiting factor for the practical application of this biotechnology in fish farming and the conservation of aquatic species. Therefore, additional studies are needed to develop reliable protocols that allow efficient and consistent maturation of cryopreserved fish oocytes.

2.2 Ovarian Tissue Cryopreservation Aiming to Preserve Oogonia

An additional alternative to the use of ovarian tissue cryopreservation in fish is the preservation of oogonia, which are diploid germ cells containing the complete genetic information of an individual. Depending on the somatic environment, these cells can develop into oocytes or spermatozoa (Yoshizaki et al. 2010). Several fish species have already demonstrated success in cryopreserving oogonia in ovarian tissue, with membrane integrity levels ranging from 40% to 93%, such as salmonids (Lee et al. 2016; Lujic et al. 2017; Lee et al. 2021), Siberian sturgeon (Pšenička et al. 2016), common carp (Franěk et al. 2019), eel (Šćekić et al. 2020), blue catfish (Abualereesh et al. 2021; Hettiarachchi et al. 2022) and American paddlefish (Ye et al. 2021). Most of the aforementioned studies used the slow-freezing methodology, mainly using 1.5 to 1 M DMSO as a permeable cryoprotectant associated with sugars such as trehalose, glucose, and lactose, as well as other nutrient sources such as BSA, fetal bovine serum FBS, and egg yolk. Only two studies addressed the vitrification methodology, where both used 3 M propylene glycol and 3 M DMSO as cryoprotectants and vitrified by immersion in a needle (Lujic et al. 2017; Šćekić et al. 2020).

When comparing the cryopreservation of sturgeon oogonia in ovarian tissue with dissociated cells, Pšenička et al. (2016) observed higher membrane integrity in cells cryopreserved in tissue (93% and 56%, respectively), which demonstrates that cryopreservation in tissue can be an advantageous alternative to dissociated oogonia. Cryopreservation in tissue facilitates collection out of labs and cryopreservation, as enzymatic dissociation is a process that takes hours to complete and requires more equipment (centrifuge and shaker). In addition, whole-tissue cryopreservation will provide essentially non-contaminated live cells (Pšenička et al. 2016).

A maturation process is required for oogonia to become viable gametes for fertilization, as well as for immature oocytes. However, no established *in vitro* protocols for oogonia maturation have motivated researchers to invest *in vivo* maturation methods, such as transplantation to sterile individuals. Several studies have demonstrated successful transplantation and cell proliferation of cryopreserved oogonia in different fish species (Pšenička et al. 2016; Lee et al. 2016; Franěk et al. 2019; Ye et al. 2021; Lee et al. 2021; Hettiarachchi et al. 2022). Lee et al. (2016) observed that cryopreserved ovarian germ cells from Orange-colored transgenic rainbow trout transplanted into tripled-diploid rainbow trout larvae resumed spermatogenesis and oogenesis in male and female recipients. In addition, these differentiated sperm and oocytes, when fertilized, generated rainbow trout with the donor's orange body color characteristics. These results demonstrate that a practical application of oogonia cryopreservation is already possible; however, studies reporting this success are still scarce, and more efforts must be taken to develop protocols for different species of fish.

This technique has not yet been applied to South American fish species and needs to be further explored. Establishing a genetic bank through oogonia preservation and subsequent reconstitution from transplantation to sterile recipients would be an important strategy to conserve the maternal genome and reconstitute endangered and commercially important South American Neotropical fish species.

Acknowledgments

Part of this study was supported by grants from ANPCyT to L. A. M. (PICT 2019–2943).

References

Abualreesh, M., J.N. Myers, J. Gurbatow, A. Johnson, D. Xing, J. Wang et al. 2021. Development of an effective cryopreservation protocol for blue catfish oogonia. N. Am. J. Aquac. 83: 336–345.

Ahammad, M.M., D. Bhattacharyya and B.B. Jana. 1998. Effect of different concentrations of cryoprotectant and extender on the hatching of Indian major carp embryos (*Labeo rohita, Catla catla* and *Cirrhinus mrigala*) stored at low temperature. Cryobiology 37: 318–324.

Ahammad, M.M., D. Bhattacharyya and B.B. Jana. 2003a. Hatching of common carp (*Cyprinus carpio* L.) embryos stored at 4 and –2°C in different concentrations of methanol and sucrose. Theriogenology 60: 1409–1422.

Ahammad, M.M., D. Bhattacharyya and B.B. Jana. 2003b. Stage-dependent hatching responses of rohu (*Labeo rohita*) embryos to different concentrations of cryoprotectants and temperatures. Cryobiology 46: 2–16.

Alam, M.A., S.M. Rahman, Y. Yamamoto, R.S. Hattori, T. Suzuki, M. Watanabe et al. 2018. Optimization of protocols for microinjection-based delivery of cryoprotective agents into Japanese whiting *Silliago japonica* embryos. Cryobiology 85: 25–32.

Anil, S., D. Rawson and T. Zhang. 2018. Development of molecular markers for zebrafish (*Danio rerio*) ovarian follicle growth assessment following *in-vitro* culture in cryopreservation studies. Cryobiology 83: 75–83.

Beirão, J., V. Robles, M.P. Herráez, C. Sarasquete, M.T. Dinis and E. Cabrita. 2006. Cryoprotectant microinjection toxicity and chilling sensitivity in gilthead seabream (*Sparus aurata*) embryos. Aquaculture 261: 897–903.

Cao, Z., X. Mao and L. Luo. 2019. Germline stem cells drive ovary regeneration in zebrafish. Cell Rep. 26: 1709–1717.

Chalde, T., A. Gárriz, E.A. Sánches and L.A. Miranda. 2016. Age-related gamete quality, reproductive performance and plasma sex steroids of pejerrey (*Odontesthes bonariensis*) broodstocks throughout the breeding season. Aquac. Res. 47: 969–982.

Chalde, T. and L.A. Miranda. 2020. Reproductive parameters of the native pejerrey fish (*Odontesthes bonariensis*) reared under natural temperature and light conditions. Aquac. Res. 51: 1706–1711.

Chao, N.H. and I.C. Liao. 2001. Cryopreservation of finfish and shellfish gametes and embryos. Aquaculture 197: 161–189.

Cloud, J.G., R. Armstrong, P. Wheeler, P.A. Kucera and G.H. Thorgard. 2000. The nortwest salmonid germplasm repository. pp. 338–342. *In*: Tiersch, T.R. and P.M. Mazik [eds.]. Cryopreservation Aquatic Species. World Aquaculture Society, Baton Rouge, USA.

del Fresno, P.S. and L.A. Miranda. 2013. Desarrollo de técnicas de congelamiento de embriones de pejerrey (*Odontesthes bonariensis*). XV Jornadas de la Sociedad Argentina de Biología. Chascomús, Buenos Aires, Argentina, 4 al 6 de diciembre de 2013.

Diwan, A.D., S.N. Harke and N.P. Gopalkrishna Archana. 2020. Cryobanking of fish and shellfish egg, embryos and larvae: an overview. Front. Mar. Sci. 7: 1–18.

Elisio, M., T. Chalde and L.A. Miranda. 2014. Endocrine regulation of pejerrey (*Odontesthes bonariensis*) oogenesis. Influence of environmental cues on its natural reproductive cycle. Comp. Biochem. Physiol. 175: 102–109.

Elisio, M., T. Chalde and L.A. Miranda. 2015. Seasonal changes and endocrine regulation of pejerrey (*Odontesthes bonariensis*) spermatogenesis in the wild. Gen. Comp. Endocrinol. 221: 236–243.

Faria Paes, M.C., R.C. da Silva, N.F. do Nascimento, F.N. Valentin, J.A. Senhorini and L.S.O. Nakaghi. 2014. Hatching, survival and deformities of piracanjuba (*Brycon orbignyanus*) embryos subjected to different cooling protocols. Cryobiology 69: 451–456.

Fornari, D.C., R.P. Ribeiro, D. Streit, L.C. Godoy, P.R. Neves, D. de Oliveira et al. 2011. Effect of cryoprotectants on the survival of cascudo preto (*Rhinelepis aspera*) embryos stored at –8°C. Zygote 22: 58–63.

Fornari, D.C., R.P Ribeiro, D. Streit, L. Vargas, L.C. Godoy, C.A.L. Oliveira et al. 2012. Increasing storage capability of pacu (*Piaractus mesopotamicus*) embryos by chilling: development of a useful methodology for hatcheries management. Cryo-Lett. 33: 125–133.

Fornari, D.C., D. Streit, J.M. Galo, D. Oliveira, N.L. Barrero, J.A. Povh et al. 2013. Viability of embryos of *Salminus brasilliensis* (Cuvier) cooled in different cryoprotectors solutions. Pes. Agrop. Bras. 48: 1175–1178.

Franěk, R., T. Tichopád, C. Steinbach, X. Xie, J. Lujić, Z. Marinović et al. 2019. Preservation of female genetic resources of common carp through oogonial stem cell manipulation. Cryobiology 87: 78–85.

Godoy, L.C., D.P. Streit, T. Zampolla, A. Bos-Mikich and T. Zhang. 2013. A study on the vitrification of stage III zebrafish (*Danio rerio*) ovarian follicles. Cryobiology 67: 347–354.

Goetz, FW and G. Theofan. 1979. *In vitro* stimulation of germinal vesicle breakdown and ovulation of yellow perch (*Perca flavescens*) oocytes. Effects of 17a-hydroxy-20b-dihydroprogesterone and prostaglandins. Gen. Comp. Endocrinol. 37: 273–285.

Guan, M., D.M. Rawson and T. Zhang. 2008. Development of a new method for isolating zebrafish oocytes (*Danio rerio*) from ovary tissue masses. Theriogenology 69: 269–275.

Guan, M., D.M. Rawson and T. Zhang. 2010. Cryopreservation of zebrafish (*Danio rerio*) oocytes by vitrification. Cryo-Lett. 31: 230–238.

Hagedorn, M., F.W. Kleinhans, D.E. Wildt and W.F. Rall. 1997. Chill sensitivity and cryoprotectant permeability of dechorionated zebrafish embryos, *Brachydanio rerio*. Cryobiology 34: 251–263.

Hettiarachchi, D.U., V.N. Alston, L. Bern, M. Shang, J. Wang, D. Xing et al. 2022. Producing xenogenic channel catfish, *Ictalurus punctatus* with cryopreserved testes and ovarian tissues of blue catfish, *I. furcatus*. Aquaculture 561: 738691.

Iwamatsu, T., S.Y. Takahashi, N. Sakai, Y. Nagahama and K. Onitake. 1987. Induction and inhibition of *in vitro* oocyte maturation and production of steroids in fish follicles by forskolin. J. Exp. Zool. 241: 101–111.

Jalabert, B. 1976. *In-vitro* oocyte maturation and ovulation in rainbow trout (*Salmo gairdneri*), northern pika (*Essox lucius*) and goldfish (*Carassius auratus*). J. Fish. Res. Board. Can. 33: 974–988.

Janik, M., F.W. Kleinhans and M. Hagedorn. 2000. Overcoming a permeability barrier by microinjecting cryoprotectants into zebrafish embryos (*Brachydanio rerio*). Cryobiology 41: 25–34.

Khosla, K., J. Kangas, Y. Liu, L. Zhan, J. Daly, M. Hagedorn et al. 2020. Cryopreservation and Laser nanowarming of zebrafish embryos followed by hatching and spawning. Adv. Biosyst. 138: 1–11.

Kohli, G., E. Clelland and C. Peng. 2005. Potential targets of transforming growth factor-beta I during inhibition of oocyte maturation in zebrafish. Reprod. Biol. Endocrinol. 3: 53–63.

Kondo, T., T. Yanagawa, N. Yoshida and M. Yamashita. 1997. Introduction of cyclin B induces activation of the maturation-promoting factor and breakdown of germinal vesicle in growing zebrafish oocytes unresponsive to the maturation-inducing hormone. Dev. Biol. 190: 142–152.

Lahnsteiner, F. 2008. The effect of internal and external cryoprotectants on zebrafish (*Danio rerio*) embryos. Theriogenology 69: 384–396.

Lee, S., N. Katayama and G. Yoshizaki. 2016. Generation of juvenile rainbow trout derived from cryopreserved whole ovaries by intraperitoneal transplantation of ovarian germ cells. Biochem. Biophys. Res. Commun. 478: 1478–1483.

Lee, S., W.Y. Bang, H.S. Yang, D.S. Lee and H.Y. Song. 2021. Production of juvenile masu salmon (*Oncorhynchus masou*) from spermatogonia-derived sperm and oogonia-derived eggs via intraperitoneal transplantation of immature germ cells. Biochem. Biophys. Res. Commun. 535: 6–11.

Leibo, S.P. and P. Mazur. 1971. The role of cooling rates in low-temperature preservation. Cryobiology 8: 447–452.

Li, S., Z. Mao, W. Han, Z. Sun, W. Yan, H. Chen et al. 1993. *In vitro* oocyte maturation in the zebra fish, *Brachydanio rerio*, and the fertilization and development of the mature egg. Chin. J. Biotechnol. 9: 247–255.

Lichtenstein, G., M. Elisio and L.A. Miranda. 2010. Development of sperm cryopreservation techniques in pejerrey *Odontesthes bonariensis*. Aquaculture 306: 357–361.

Lopes, T.S., E. Romagosa, D. Streit, R.P. Ribeiro and M. Digmayer. 2011. Cooling of pacu (*Piaractus mesopotamicus*) embryos at various stages of development for 6 or 10 hours. Theriogenology 75: 570–576.

Lopes, T.S., E.A. Sanches, R.Y. Okawara and E. Romagosa. 2015. Chilling of *Steindachneridion parahybae* (Siluriformes: Pimelodidae) embryos. Theriogenology 84: 538–544.

Lubzens, E., G. Young, J. Bobe and J. Cerdà. 2010. Oogenesis in teleosts: How fish eggs are formed. Gen. Comp. Endocrinol. 165: 367–389.

Lujić, J., Z. Marinović, S. Sušnik Bajec, I. Djurdjevič, E. Kása, B. Urbányi et al. 2017. First successful vitrification of salmonid ovarian tissue. Cryobiology 76: 154–157.

Macoretta, C.L. and L.A. Miranda. 2020. Cooling of pejerrey *Odontesthes bonariensis* (Teleostei, Atherinidae) embryos at sub-zero temperatures. Theriogenology 149: 123–130.

Marques, L.S., A. Bos-Mikich, L.C. Godoy, L.A. Silva, D. Maschio, T. Zhang et al. 2015. Viability of zebrafish (*Danio rerio*) ovarian follicles after vitrification in a metal container. Cryobiology 71: 367–373.

Marques, L.S., A.A.N. Fossati, M.S. Leal, R.B. Rodrigues, R.A. Bombardelli and D.P. Streit. 2018. Viability assessment of primary growth oocytes following ovarian tissue vitrification of neotropical teleost pacu (*Piaractus mesopotamicus*). Cryobiology 82: 118–123.

Marques, L.S., A.A.N. Fossati, R.B. Rodrigues, H.T. Da Rosa, A.P. Izaguirry, J.B. Ramalho et al. 2019. Slow freezing versus vitrification for the cryopreservation of zebrafish (*Danio rerio*) ovarian tissue. Sci. Rep. 9: 15353.

Marques, L.S., T.R. Freitas, R.B. Rodrigues, N.S. Teixeira, M.P. Atehortúa, H.T.R. Silva et al. 2021. Vitrification protocol for immature *Brycon orbignyanus* ovarian tissue as an extinction escape strategy. Cryobiology 103: 116–122.

Miranda, L.A., M.C. Cassará and G.M. Somoza. 2005. Increase in milt production by hormonal treatment in the pejerrey fish *Odontesthes bonariensis* (Valenciennes 1835). Aquac. Res. 36: 1473–1479.

Miranda, L.A. and G.M. Somoza. 2009. Induction of ovulation and spawning of cultured pejerrey *Odontesthes bonariensis* by implantation of a sustained-release GnRH-a pellet. Aquac. Res. 41: 129–134.

Nagahama, Y., H. Kagawa and F. Tashiro. 1980. The *in vitro* effects of various gonadotropins and steroid hormones on oocyte maturation in amago salmon and rainbow trout. Fish. Sci. 46: 1097–1102.

Pang, Y. and W. Ge. 2002. Gonadotropin and activin enhance maturational competence of oocytes in the zebrafish (*Danio rerio*). Biol. Reprod. 66: 259–265.

Pessoa, N.O., J.A.S. Galvão, F.G.M. De Souza Filho, M.L.N.M. De Sousa and C.M.S. Sampaio. 2014. Cooling of pirapitinga (*Piaractus brachypomus*) embryos stored at –10°C. Zygote 23: 453–459.

Pšenička, M., T. Saito, M. Rodina and B. Dzyuba. 2016. Cryopreservation of early stage Siberian sturgeon *Acipenser baerii* germ cells, comparison of whole tissue and dissociated cells. Cryobiology 72: 119–122.

Robles, V., E. Cabrita and M.P. Herráez. 2004. Microinjection of the antifreeze protein type III (AFPIII) in turbot embryos. Cryobiology 49: 317–318.

Robles, V., E. Cabrita, L. Anel and M.P. Herráez. 2006. Microinjection of the antifreeze protein type III (AFPIII) in turbot (*Scophthalmus maximus*) embryos: Toxicity and protein distribution. Aquaculture 261: 1299–306.

Robles, V., V. Barbosa, M.P. Herráez, S. Martínez-Páramo and M.L. Cancela. 2007. The antifreeze protein type I (AFP I) increases seabream (*Sparus aurata*) embryos tolerance lo low temperatures. Theriogenology 68: 284–289.

Robles, V. and D.G. Valcarce. 2019. The use of antifreeze proteins in the cryopreservation of gametes and embryos. Biomolecules 9: 1–12.

Routray, P., T. Suzuki, C.A. Strüssmann and R. Takai. 2003. Factors affecting the uptake of cryoprotective agents by fish eggs and embryos. Fish. Sci. 68: 965–966.

Šćekić, I., Z. Marinović, J. Lujić, T. Müller, N. Kitanović, B. Urbányi et al. 2020. A novel strategy for conservation of European eel (*Anguilla anguilla*) genetic resources: Cryopreservation of ovarian stem cells. Cryobiology 95: 151–156.

Seki, S., T. Kouya, R. Tsuchiya, D.M. Valdez, B. Jin, T. Hara et al. 2008. Development of a reliable *in vitro* maturation system for zebrafish oocytes. Reproduction 135: 285–292.

Selman, K., R.A. Wallace, A. Sarka and X. Qi. 1993. Stages of oocyte development in the zebrafish, *Brachydanio rerio*. J. Morphol. 218: 203–224.

Selman, K., T.R. Petrino and R.A. Wallace. 1994. Experimental conditions for oocyte maturation in the zebrafish, *Brachydanio rerio*. J. Exp. Zool. 269: 538–550.

Silva, L.A., L.S. Marques, T. Zhang, R.B. Rodrigues, L.R. Eloy and D.P. Streit. 2018. Production of live larvae following *in vitro* maturation of zebrafish oocytes. Theriogenology 105: 115–119.

Somoza, G.M., L.A. Miranda, G.E. Berasain, D.C. Colautti, M. Remes Lenicov and C.A. Strüssmann. 2008. Historical aspects, current status and prospects of pejerrey aquaculture in South America. Aquac. Res. 39: 784–793.

Streit Jr, D.P., M. Digmayer, R. Pereira Ribeiro, R. Nardez Sirol, G. Vanini de Moraes and J. Minardi Galo 2007. Embrioes de pacu submetidos a diferentes protocolos de resfriamento. Pesquisa Agropecuária Brasileira 42: 1199–1202.

Strüssmann, C.A., H. Nakatsugawa, F. Takashima, M. Hasobe, T. Suzuki and R. Takai. 1999. Cryopreservation of isolated fish blastomeres: Effects of cell stage, cryoprotectant concentration, and cooling rate on postthawing survival. Cryobiology 39: 252–261.

Suzuki, T., H. Komada, R. Takai, K. Arii and T. Kozima. 1995. Relation between toxicity of cryoprotectant DMSO and its concentration in several fish embryos. Fish. Sci. 61: 193–197.

Tsai, S., D.M. Rawson and T. Zhang. 2009. Development of cryopreservation protocols for early stage zebrafish (*Danio rerio*) ovarian follicles using controlled slow cooling. Theriogenology 71: 1226–1233.

Tsai, S., D.M. Rawson and T. Zhang. 2010. Development of *in vitro* culture method for early stage zebrafish (*Danio rerio*) ovarian follicles for use in cryopreservation studies. Theriogenology 74: 290–303.

Tsai, S. and C. Lin. 2012. Advantages and applications of cryopreservation in fisheries science. Braz. Arch. Biol. Technol. 55: 425–434.

Vajta, G. 2000. Vitrification of the oocytes and embryos of domestic animals. Anim. Reprod. Sci. 60: 357–364.

Valdez, D., A. Miyamoto, T. Hara, K. Edashige and M. Kasai. 2005. Sensitivity to chilling of medaka (*Oryzias latipes*) embryos at various developmental stages. Theriogenology 64: 112–122.

Ye, H., C. Zhou, H. Yue, M. Wu, R. Ruan, H. Du et al. 2021. Cryopreservation of germline stem cells in American paddlefish (*Polyodon spathula*). Anim. Reprod. Sci. 224: 106667.

Yön, K.N.D., Y. Aytekin and R. Yüce. 2008. Ovary maturatıon stages and histological investigation of ovary of the Zebrafish (*Danio rerio*). Braz. Arch. Biol. Technol. 51: 513–522.

Yoshizaki, G., M. Ichikawa, M. Hayashi, Y. Iwasaki, M. Miwa, S. Shikina et al. 2010. Sexual plasticity of ovarian germ cells in rainbow trout. Development 137: 1227–1230.

Zampolla, T., E. Spikings, S. Srirangarajah, D.M. Rawson and T. Zhang. 2011. Impact of cryoprotectants and cryopreservation on metabolic activity and cytoskeleton proteins of zebrafish (*Danio rerio*) ovarian fragments. Cryo-Lett. 32: 525–536.

Zampolla, T., D.M. Rawson and T. Zhang. 2012. The use of fetal bovine serum for cryopreservation of stage III zebrafish (*Danio rerio*) ovarian follicles. Cryo-Lett. 33: 435–442.

Zhang, Y.Z., S.C. Zhang, X.Z. Liu, Y.J. Xu, J.H. Hu, Y.Y. Xu et al. 2005. Toxicity and protective efficiency of cryoprotectants to flounder (*Paralichthys olivaceus*) embryos. Theriogenology 63: 763–773.

Zhang, T., D.M. Rawson, L. Tosti and O. Carnevali. 2008. Cathepsin activities and membrane integrity of zebrafish (*Danio rerio*) oocytes after freezing to –196°C using controlled slow cooling. Cryobiology 56: 138–143.

CHAPTER 9

Laboratory Techniques for the Analysis of Cells and Reproductive Tissues of Neotropical Species

Lauren Dorneles Dipp,[1] *Lis Santos Marques,*[1,2] *Rômulo Batista Rodrigues,*[1,3] *Thales de Souza França,*[1,4] *Nathalia dos Santos Teixeira,*[1] *Thaiza Rodrigues De Freitas,*[1] *Raquel Santos Dos Santos,*[1] *Larise Lima,*[1] *Jhony Lisboa Benato*[1] and *Danilo P. Streit Jr.*[1,*]

INTRODUCTION

This chapter brings methods usually performed in the Laboratory of Production and Conservation of aquatic biodiversity species (AQUAM), localized at the Federal University of Rio Grande do Sul (UFRGS), to analyze the cell and tissue conditions in practical experiments. All the methods described here were validated and applied in South American Neotropical fish species, and it is essential to highlight that even other methods are available. Here, we emphasize the daily basis and published techniques.

[1] Laboratory of Production and Conservation of Aquatic Biodiversity Species (AQUAM) Federal University of Rio Grande do Sul (UFRGS), Porto Alegre – RS, Brazil.

[2] Centro de Modelos Biológicos Experimentais – CeMBE da PUCRS, Porto Alegre, RS, Brasil.

[3] Departamento de Zootecnia e Ciências Biológicas, Universidade Federal de Santa Maria - Campus de Palmeira das Missões, RS, Brazil.

[4] Universitat Politècnica de València (UPV), Valencia, Spain.

Emails: lauren.dippd@hotmail.com, nati_st@hotmail.com, thaiza.rfreitas@gmail.com, racheelhere@gmail.com, larise.carolinevet@gmail.com, jhonybenato@hotmail.com, lismarx@gmail.com, romulo.batista@ufsm.br, thalesfranca@gmail.com

* Corresponding author: danilo.streit@ufrgs.br

The improvement of cellular analysis is in development, as new equipment has been released, and it also can be seen in the animal field. On the other hand, for fish cells, this process is slow. It is important to mention that this scenario has changed in recent years due to the popularization of zebrafish as a laboratory animal model, but it still needs more attention. As for the South American fishes, the application of protocols for reproductive cell validation has followed an even slower path compared to the number of modern techniques. It is understandable that few laboratories have high technology compared to the United States, Canada, Australia, Japan, and European countries such as Spain, the United Kingdom, France, Holland, Germany, the Czech Republic, and Hungary.

There is no doubt that the countries mentioned previously have a long history of development and practice in reproductive cell analysis applied in well-known aquaculture species, especially the variety of carp and salmon. Consequently, studies in cryopreserved reproductive cells, as in those two fishes, had more than fifty years. In Brazil, in a very beginner way, it occurred in the 1980s with the work of Fogli da Silveira et al. (1981), and only at the end of the first decade of the 2000s a significant volume of studies began to be produced, especially with male reproductive cells. Currently, in Brazil, approximately fifteen established research groups regularly produce papers on the reproductive cells of South American Neotropical fishes. Therefore, the amount of information also improved, attracting the attention of researchers, especially from countries considered reference.

The efficient application of cryopreservation techniques in reproductive cells of South American Neotropical fishes has already been published in a sequence of studies. The development of the protocol for vitrification in fish ovarian tissue by the AQUAM laboratory is an example that reflects this process since the initial study by Marques et al. (2015) and the improvement in another study also published by Marques et al. (2019) using zebrafish as a model. This study resulted in a protocol that has been repeatedly cited in human reproduction papers and applied in ovarian tissue of native species such as *Piaractus mesopotamicus* (Marques et al. 2018) and *Brycon orbignyanus* (Marques et al. 2021) by de AQUAM laboratory. The *Rhamdia quelen* sperm cryopreservation protocol is another example of the use of cellular techniques that not only validate the results but also help to popularize the use of free CASA software. Since the first work published (Adames et al. 2015) by the Research Group of Universidade Estadual do Oeste do Paraná (UNIOESTE), the studies that followed by the same group (Neumann et al. 2019; Xavier et al. 2021; Krause et al. 2021) and by AQUAM (Corso et al. 2019; Da Costa et al. 2019a, 2019b; Pérez-Atehortúa et al. 2022; França et al. 2022) used the same *R. quelen* sperm cryopreservation protocol for.

The chapter was organized in three parts: techniques applied to cells and tissues of both sexes, techniques applied to female cells and tissues, and techniques applied to male cells and tissues. Each technique described was adapted to the cell type under analysis, so we bring a brief content of the work carried out by the group to ensure other laboratories can make use of such techniques and improve them.

1. Methods for Female and Male Reproductive Cells and Tissue

1.1 MTT Assay – Cell Viability by Mitochondrial Activity

The MTT assay is a simple method that makes obtaining the number of living cells in samples through mitochondria activity possible. This assay can be used for different purposes, such as cytotoxicity tests, evaluation of drugs' mechanism of action, and even biologically active compounds. MTT assay involves converting [3-(4,5-dimethylthiazol-2yl)-2,5-diphenyl tetrazolium] bromide into formazan crystals by mitochondria. The formazan crystals, which have a blue/purple color, are solubilized, and it is possible to quantify the color intensity of each sample through the absorbance at 570 nm. This result presents the number of living cells in the sample. In most darker samples, there is a greater number of cells with mitochondrial activity. This test is usually applied in mammalian cells, suspended or fixed to culture surfaces, always with a predetermined concentration. MTT represents a convenient, reliable, and quantitative technique as it quantifies cell proliferation and viable cells (Sylvester 2011). Another critical point that we can highlight is the considerably low budget since the fast sample processing and the materials and equipment are accessible.

We performed some adaptations and validations to use the MTT assay in fish ovarian tissue. The first necessary adaptation was to change the temperature from 37°C to 28°C since the zebrafish cells, used as a model, showed maximum activity at this temperature. In order to apply the technique in tissue fragments, which was not possible to determine the cell concentration, tests were carried out demonstrating that the weight of the fragment could be used as a reference for the number of cells in the sample. Thus, we concluded that heavier tissue fragments would increase the number of cells. This methodology allowed the MTT assay to be used for the viability evaluation of fish tissues. Mitochondrial activity analysis in cells present in biological samples is important to evaluate the efficiency of the treatments to which the samples were exposed. The presence of succinate dehydrogenase enzyme activity, a component of the Krebs cycle, is an essential factor for the analysis of mitochondrial cell viability, which is the main component of the MTT assay.

On the other hand, techniques such as flow cytometry and confocal fluorescent microscopy, which are also applied for the analysis of mitochondrial activity and cell viability, have limitations regarding their costs. For the application of flow cytometry, obtaining a flow cytometer and specific commercial kits for sample preparation is essential, thus increasing the analysis budget. While in fluorescent microscopy, the technician must know some proprieties (e.g., fluorophore choice, excitation intensity, emission filters) that involve the method to achieve good quality images (Lichtman and Conchello 2005).

The MTT assay appears to be a reliable and low-cost method for assessing cell viability in cryopreservation protocols applied to fish tissues. In Marques et al. (2018), the MTT assay showed high rates of mitochondrial activity in two (VS1 and VS4) of the five treatments applied in cryopreservation protocol for ovarian tissue from pacu (*Piaractus mesopotamicus*). Also, as observed in the study by Marques et al. (2021), the membrane of oocytes in early stages in ovarian tissue from Piracanjuba (*Brycon orbignyanus*) did not show significant differences with the membrane integrity and

MTT assays due to the ultra-fast freezing provided by the protocol and decreasing the ice crystals formation inside the cell.

As the evaluation of female gametes, the evaluation of male gametes is essential for the effectiveness of fertilization. Studies in different animal species using the MTT assay (Aziz et al. 2005; Aziz et al. 2006; Berg 2015) concluded the method's reliability in evaluating mitochondrial activity and cell viability. More recently, the same analysis was applied to the South American silver catfish (*R. quelen*) by standardizing each step for the type of semen in question (França et al. 2022; Atehortúa et al. 2022). Mitochondria are organelles sensitive to low temperatures (Tsvetkov and Naydenova 1987) and responsible for providing energy to sperm in the ATP form (Kholodny et al. 2020). Thus, the MTT assay is a tool to validate sperm quality after the cryopreservation process.

Other proprieties could be mentioned about the MTT assay:

- It provides the evaluation of more than one type of sample. This positive point is due to the use of specific reading plates for spectrophotometers, which are commercially available in different wall numbers;
- The reaction remains stable for a few hours at room temperature at the end of the test. This stability allows more than one observation to confirm results;
- In addition to the traditional version of the method, a "dry" version is also available, which is even more cost-effective. This new version observes the color change of the MTT reagent (yellow) to purple on specific paper when exposed to a sample that contains viable sperm cells.

In conclusion, the mitochondrial activity assessment by MTT assay represents a reliable and practical analysis and is accessible to many laboratories. This analysis provides cell viability at the organelles level before and after exposing the samples to different treatments, including the cryopreservation process, and validates the applied techniques.

1.2 Evaluation of DNA Damage by Comet Assay

The comet assay, also known as "Single Cell Gel Electrophoresis," is a sensitive method for detecting breaks in DNA strands, as it allows visualization of the DNA damage resulting in agarose gel electrophoresis. The damage results in a "tail" close to the core sample application in the gel, and its evaluation is based on scores and categories available in public or paid domain software. However, only observing the results without software visualization is enough to get conclusions. The comet assay, alkaline or enzymatic, has considerable advantages: provides data at the level of individual cells for robust statistical analysis; requires a small amount of sample to carry out the procedures (< 10,000 cells); sensitivity for detecting damage present in the sample; application in any eukaryotic cell population, *in vivo* or *in vitro*, including aquatic organisms exposed to toxic agents, for purposes of genotoxicological studies and environmental monitoring (Kumaravel et al. 2009). *In situ* hybridization can add more specificity to the comet assay by identifying the damaged DNA or gene sequence. However, the addition of this method will result in an increase in the costs due to specific materials and equipment, as described earlier in the chapter.

In the study published by Marques et al. (2021), a modified comet assay was applied to observe possible cryopreservation damages in Piracanjuba (*Brycon orbignyanus*) primary growth oocytes. The samples were submitted to five vitrification treatments, and the comet assay showed that the VS1 treatment was the one that resulted in less damage to the DNA. This assay can also be used to evaluate fish sperm DNA damage by our research group. In the Da Costa et al. (2019b) study, the comet assay assessed the impact on the post-thaw sperm genetic material of South American silver catfish (*R. quelen*). These gametes were submitted to cysteine amino acid supplementation in the cryoprotectant solution, and post-thaw sperm were evaluated. The samples cryopreserved with 2.5, 5, and 20 mM cysteine amino acid concentrations presented higher post-thawed sperm DNA damage than the control group (without cysteine). Thus, the comet assay is a tool of considerable value in cryobiology experiments. It provides the genetic material quality status before and after the cryopreservation processes.

1.3 Cell Membrane Integrity Evaluation by Fluorescent Dye

The rapid fluorometric method combining fluorescein diacetate (FDA) and propidium iodide (PI) can evaluate cell membrane integrity and viability in fish tissue. The analysis uses an inclusion dye (FDA) and an exclusion dye (PI) simultaneously. FDA is a non-polar ester capable of crossing the plasma membrane and is hydrolyzed by intracellular esterases to produce free fluorescein (bright green). In contrast, PI does not enter the cell with an integrity membrane. Thus, viable cells are stained bright green with FDA, and non-viable cells are stained bright red with PI.

The application of two fluorescent dyes provides the identification of intact cells in addition to the observation of esterase activity. This method can be performed by more specific analysis, such as flow cytometry and *in situ* hybridization. However, the addition of complex techniques or even the staining replacement cannot be accessible due to the costs and handling. Flow cytometry and *in situ* hybridization, which comprises probes with RNA or DNA for identifying cells and tissues by complementarity of sequences, require expensive inputs and equipment, which is inaccessible for many laboratories.

According to the study by Marques et al. (2018), the evaluated primary growth oocytes of *Piaractus mesopotamicus* did not show significant differences compared to the fresh sample, and this result was observed in all treatments using equilibrium solutions and vitrification solutions (VS1, VS2, VS3, VS4, and VS5). Here, the cell membrane integrity analysis by fluorescent dyes (FDA and PI) is a complete and accessible test with results that show that the cells did not suffer damage after being submitted to the treatments, compared to the control. Finally, the method can be used to assess the occurrence of damage in fish reproductive cells after being subjected to different treatments.

The evaluation of cell viability by staining with fluorescent dyes can also be applied to male reproductive cells. The application of the method in both cell populations is carried out in the same way by incubating the sample with the dyes involved in the process. The PI and SYBR-14 dyes were applied in this protocol, as validated and described in Benato et al. (2023). Thus, the budget involved in the process does not vary significantly, maintaining this method's advantage compared to complex analyses such as flow cytometry and *in situ* hybridization.

2. Methods for Female Reproductive Cells and Tissue

2.1 Trypan Blue Test for Cellular Membrane Integrity

The trypan blue test is based on the principle that living cells have an intact membrane and would not allow the vital dye trypan blue to enter the cytoplasm. Therefore, cells stained in blue after exposure to the trypan blue test are considered dead, and those that remain unstained are considered alive. The trypan blue test for cellular membrane integrity quantifies live and dead cells in a sample.

The trypan blue test represents a practical tool for directly assessing cellular membrane damage. Enhanced techniques such as electron microscopy and flow cytometry are used as complements or substitutes, or both for trypan blue. However, these enhanced techniques had fastidious steps in sample preparation. Electron microscopy aims to produce detailed images of intracellular content from the sample prepared in paraffin slices and exposed to electron beams. However, this technique presents the same points that make the sample preparation and analysis difficult, such as the fixation step, which must be carried out carefully not to modify the dimensions of the cells in the sample, as the sectioning step, which may result in a small, thin slice of the sample (Winey et al. 2014).

In previous studies by the AQUAM group (Marques et al. 2018; Marques et al. 2021), it is observed that the trypan blue analysis provided the membrane integrity visualization of the early stages oocyte samples of *Piaractus mesopotamicus* and *Brycon orbignyanus* without significant differences in relation to the control (fresh tissue). This result supports the analysis protocol, firstly applied to zebrafish (*Danio rerio*) (Marques et al. 2015) and then to ovarian tissue vitrification of Neotropical fishes.

3. Methods for Male Reproductive Cells and Tissue

3.1 Sperm Concentration Analysis

Sperm concentration is a quantitative analysis and can be related to fertilization rate. Thus, it is important to know the spermatozoa number in a semen sample. The counting is performed using a hemocytometer, a Neubauer chamber or another cell counting chamber, an optical microscope, and a cell counter. The semen should be diluted in a fixative solution (i.e., 10% buffered formalin saline) to provide an appropriate cell concentration in the observation field.

Other methodologies for sperm concentration measurement are available, including flow cytometry, fluorescence microscopy, and spectrophotometry (Ciereszko et al. 2020). However, the costs matter, which include equipment and materials involved in these methodologies, are still an issue when choosing which techniques will be applied.

The concentration sperm measurement can be used to report a semen characteristic (Da Costa et al. 2022; Gallo et al. 2022) and to calculate the ideal inseminating dose for *in vitro* fertilization. Consequently, using the same number of spermatozoa per oocyte in all treatments is possible. Evidencing the effects of post-thawing dilution (França et al. 2022) and density gradient separation (Atehortúa et al. 2022) on *R. quelen* post-thawed sperm. Thus, the sperm concentration analysis using the Neubauer chamber count is valuable and reliable for experimental procedures.

3.2 Sperm Kinetics Analysis by CASA Software

Sperm kinetics is a qualitative analysis of semen; it also provides information to understand how oocyte fertilization occurs and about the reproductive strategy of the species. Relative sperm motility, one of the kinetic parameters, represents the number of sperm cells that move in relation to the static sperm cells. Several studies show a high correlation with fertilization, which is why it is considered one of the main biomarkers of fish semen quality (Gallego and Asturiano 2018a).

The observation of sperm kinetics can be performed in two ways; the first is subjective, which consists of visualizing the sperm movement directly under an optical microscope, thus obtaining the percentage of moving/static sperm ratio. Another parameter that could be estimated is vigor, where the cell speed is observed and classified on a scale of 1 to 5 (1 = slowest movements, 5 = fastest movements). The second way to evaluate sperm kinetics is using the objective method. Here, a camera coupled to an optical microscope and a computer can record videos of sperm movements. The Computer-Assisted Sperm Analyzer (CASA) software can be used to evaluate the videos, and it is possible to observe the following parameters: motility time, motility percentage, curvilinear velocity, average path velocity, straight-line velocity, straightness index, wobble, progression, and beats per seconds.

The limiting factors of the sperm kinetics analysis, mainly in the subjective form, lead to low reproducibility and difficulty in comparing results in different publications (Gallego et al. 2018b), which led to the rejection of this method by scientific journals with a high impact factor. On the other hand, it is a low-cost and practical analysis that can be used when it is not necessary to obtain precise data about sperm kinetics. In the form of objective analysis, there is more than one CASA software; most of them are available under payment, but there is public domain CASA software for free. The difference between the two software is the speed of obtaining the results and the manual editing and analysis. Despite this difference between the paid and non-paid versions, the accuracy of the results is the same. The analysis of sperm kinetics through CASA has high accuracy, and because it evaluates parameters that are impossible to assess through subjective analysis, it is the most accepted analysis method by scientific journals with a high impact factor.

Our group has already used subjective motility analysis (Corso et al. 2019; Da Costa et al. 2020). However, nowadays, sperm kinetic parameters evaluation is performed with the free CASA plug-in software from ImageJ, seeking greater accuracy analysis. In our studies, semen samples from South American silver catfish (França et al. 2022; França et al. 2023; Athuértua et al. 2022) and *Brycon hillari* (Da Costa et al. 2022) were observed using objective analysis.

3.3 Evaluation of Sperm Membrane Integrity by Eosin Y and Nigrosin Dye

Evaluating sperm membrane integrity by eosin Y and nigrosin dye allows the identification and quantification of viable sperm cells in a sample. Eosin Y is a supravital dye that does not penetrate cells with an intact plasma membrane, but when damaged, the cell becomes stained in pink color. Nigrosin is responsible for the darker background contrast of the slide, thus allowing the visualization of unstained cells. So, dead spermatozoa are visualized in pink while live cells remain unstained.

As well as the membrane integrity tests applied to the evaluation of oocytes, the staining tests for the evaluation of sperm membranes are equally essential. According to Maria et al. (2010), sperm viability and morphological abnormalities are directly related to controlled reproduction and sperm conservation of Neotropical fishes.

As previously mentioned, cost and handling issues arise despite the availability of more recent techniques. Equipment and materials related to these techniques, including flow cytometry and *in situ* hybridization, have costs beyond the reach of many laboratories and research teams. These techniques do not provide applicability in the field, where it is necessary to use reagents that do not require refrigeration or other more specific attention, as well as easily transported equipment.

3.4 Evaluation of Sperm Cells Morphology by Rose Bengal Stain

Sperm morphology is the evaluation of the physical structure of spermatozoa, from which morphological changes and abnormalities in cells can be identified. Observing the head and middle piece and tail morphology allows an evaluation of the fertilizing capacity of the gamete. These alterations may reflect the fish's physiological, sanitary, or nutritional stage and environmental effects. Modifications may occur in the spermatogenesis process due to malnutrition, diseases, and stress, which are classified as primary alterations. On the other hand, secondary alterations originate during semen manipulation at the time of collection, for example, excessive pressure applied to the fish's coelomic cavity. The morphological analysis and other methods may predict the male reproductive performance in the management of fish farming. This helps increase the effectiveness of reproduction biotechnologies applications, such as sperm cryopreservation.

Acid Red 94, also known as "Rose Bengal," was discovered in the 19th century. It has since become known for its application in microscopy and for being a financially accessible, biocompatible, and available dye (Demartis et al. 2021). Rose Bengal is applied in morphology and viability analysis of male gametes in different species of animals (Herman and Swanson 1941; Silva et al. 2019; Kitiyanant et al. 2000), which contributes to the observation of the fertility properties of semen. In fish, sperm evaluation is essential in determining the best techniques involving the gamete cryopreservation process to assist in conserving endangered species and fish farming. Although laboratories evaluate these properties of male gametes through electronic microscopy, the use of Rose Bengal dye in optical microscopy is straightforward and can be used in field conditions due to its simplicity (Streit Jr et al. 2004).

According to Streit Jr et al. (2004), the Rose Bengal is the most effective dye for observing fish sperm morphology. In this study, using pacu (*Piaractus mesopotamicus*) sperm samples, a comparison was made between three different histological dyes combined with three fixation solutions. They concluded that using Rose Bengal with formalin buffered saline is the best method for evaluating sperm morphology because it does not modify the cell structure and allows a precise observation. This same staining method, already standardized, was applied to the piraputanga (*Brycon hilarii*) sperm, also a Neotropical species that suffers from population decline due to human intervention and excessive fishing (Da Costa et al. 2022).

References

Adames, M.S., C.P.R. de Toledo, G. Neumman, A.H. Buzzi, C.N. Buratto, P.A. Piana et al. 2015. Optimization of the sperm: oocyte ratio and sperm economy in the artificial reproduction of *Rhamdia quelen* using fructose as a sperm motility modulator. Anim. Reprod. Sci. 161: 119–128.

Atehortúa, M.P., A.G. Galuppo, R.B. Rodrigues, T.S. França, N.S. Teixeira, T.R. Freitas et al. 2022. The use of differential separation and density gradient with AllGrad 90% after thawing improves the sperm quality of South American catfish (*Rhamdia quelen*). Aquaculture 553: 1–10.

Aziz, D.M., L. Ahlswede and H. Enbergs. 2005. Application of MTT reduction assay to evaluate equine sperm viability. Theriogenology 64: 1350–1356.

Aziz, D.M. 2006. Assessment of bovine sperm viability by MTT reduction assay. Anim. Reprod. Sci. 92: 1–8.

Benato, J.L., D. Streit Jr, N.S. Teixeira, R.B. Rodrigues, T.R. de Freitas, M. Okomato et al. 2023. *Pogonias courbina* spermatozoa characteristics in its first reproductive season. PeerJ (in press).

Berg, B.M.V. 2015. Microscopic analysis of MTT stained boar sperm cells. Open Vet. J. 5: 58–63.

Ciereszko, A., S. Judycka, J. Nynca, M. Słowińska and M.A. Dietrich. 2020. Factors influencing milt quality in fishes and its usefulness to cryopreservation. pp. 25–67. *In*: Betsy, J. and S. Kumar [eds.]. Cryopreservation of Fish Gametes. Springer Nature Singapore Pte Ltd., Singapore, Singapore.

Corso, M.N., L.S. Marques, L.F.G. Gracia, R.B. Rodrigues, L.J.G. Barcellos and D.P. Streit Jr. 2019. Effects of different doses of eugenol on plasma cortisol levels and the quality of fresh and frozen-thawed sperm in south American catfish (*Rhamdia quelen*). Theriogenology 125: 135–139.

Da Costa, B.B., L.S. Marques, P.G. Lassen, R.B. Rodrigues and D.P Streit Jr. 2019a. Cryopreservation-induced morphological changes in the sperm of South American silver catfish (Rhamdia quelen). J. Appl. Ichthyol. 34: 1–7.

Da Costa, B.B., L.S. Marques, P.G. Lassen, R.B. Rodrigues, H.T.R. Silva, J.C.F Moreira et al. 2019b. Effects of cysteine supplementation on the quality of cryopreserved sperm of South American silver catfish. Aquac. Res. 51: 455–464.

Da Costa, B.B., J.A. Povh, E.A. Sanches, L.N. Spica, R.B. Rodrigues, N.S. Teixeira et al. 2022. Characterization of sperm quality in *Brycon hilarii*: How does morphology affect sperm movement? Theriogenology Wild 1: 1–9.

Demartis, S., A. Obinu, E. Gavini, P. Giunchedi and G. Rassu. 2021. Nanotechnology-based rose Bengal: A broad-spectrum biomedical tool. Dyes Pigm. 188: 109236.

Fogli da Silveira, W., E.T Kavamoto and M.Y. Narabara. 1981. Avaliação quali-quantitativa e criopreservação em forma de pellets do sêmen do bagre *Rhamdia hilarii*. Ciên. Cult. 33: 620.

França, T.S., I.C. Gomes, E.A. Sanches, M.P. Atehortúa, N.S. Teixeira and R.B. Rodrigues. 2022. Post-thaw dilution of *Rhamdia quelen* sperm improves the reproductive success. Anim. Reprod. Sci. 243: 1–10.

De Souza França, T., I.C. Gomes, E.A. Sanches, M. Pérez Atehortúa, N.S. Teixeira, R.B. Rodrigues et al. 2023. Fish sperm cryopreservation using biodegradable containers: New low-cost and environment-friendly methodology. Reproduction 166: 89–97.

Gallego, V. and J.F. Asturiano. 2018a. Sperm motility in fish: technical applications and perspectives trough CASA-Mot systems. Reprod. Fert. Dev. 30: 820–832.

Gallego, V., J.G.H. Jusdado, C. Rozenfeld, L. Pérez and J.F. Asturiano. 2018b. Subjective and objective assessment of fish sperm motility: When the technique and technicians matter. Fish Fisiol. Biochem. 40: 1457–1467.

Gallo, J.M., R.B. Rodrigues, D.C. Fornari, J.A. Povh, R.P. Ribero, T. Zhang et al. 2022. Does semen quality of *Colossoma macropomum* change the productivity of larvae during the reproductive period? Aquaculture 558: 738376.

Herman, H.A. and E.W. Swanson. 1941. Variations in dairy bull semen with respect to its use in artificial insemination. Res. Bull. 326.

Kholodnyy, V., S. Boryshpolets, B. Dzyuba and J. Cosson. 2020. Energetics of fish spermatozoa. pp. 69–116. *In*: Betsy, J. and S. Kumar [eds.]. Cryopreservation of Fish Gametes, ed 1, Singapore: Springer Nature Singapore Pte Ltd.

Kitiyanant, Y., M.J. Scmidt and K. Pavasuthipaisit. 2000. Evaluation of sperm acrosome reaction in the Asiatic elephant. Theriogenology 53: 887–896.

Krause, R.A., G. Neuman, A.C.O. Pedreira, M.A. Syperreck, A.M. Malacarne, M.N.F. Braga et al. 2021. *Rhamdia quelen* semen pre-cryopreservation processing and sperm viability time after thawing. Braz. J. Dev. 7: 43131–43147.

Kumaravel, T.S., B. Vilhar, S.P. Faux and A.N. Jha. 2009. Comet assay measurements: a perspective. Cell. Biol. Toxicol. 25: 53–64.

Lichtman, J.W. and J.A. Conchello. 2005. Fluorescence microscopy, Nat. Methods 2: 910–919.

Maria, A.N., H.C. Azevedo, J.P. Santos, C.A. Silva and P.C.F Carneiro. 2010. Semen characterization and sperm structure of the Amazon tambaqui *Colossoma macropomum*. J. Appl. Ichtyol. 26: 779–783.

Marques, L.S., A. Bos-Mikich, L.C. Godoy, L.A. Silva, D. Maschio, T. Zhang et al. 2015. Viability of zebrafish (*Danio rerio*) ovarian follicles after vitrification in a metal container. Cryobiology 71: 367–373.

Marques, L.S., A.A.N. Fossati, M.S.L. Leal, R.B. Rodrigues, R.A. Bombardelli and D.P. Streit. 2018. Viability assessment of primary growth oocytes following ovarian tissue vitrification of neotropical teleost pacu (*Piaractus mesopotamicus*). Cryobiology. 82: 118–123.

Marques, L.S., A.A.N. Fossati, R.B. Rodrigues, H.T. Da Rosa, A.P. Izaguirry, J.B. Ramalho et al. 2019. Slow freezing versus vitrification for the cryopreservation of zebrafish (*Danio rerio*) ovarian tissue. Sci. Rep. 9: 15353.

Marques, L.S., T.R. de Freitas, R.B. Rodrigues, N.S. Teixeira, M.P. Atehortúa, H.T.R. Silva et al. 2021. Vitrification protocol for immature *Brycon orbignyanus* ovarian tissue as an extinction escape strategy. Cryobiology 103: 116–122.

Morris, R. 2015. Spectrophotometry. Curr. Protoc. Essential Lab. Tech. 11: 2.1.1–2.1.30.

Neumann, G., P.V. Sanches and R.A. Bombardelli. 2019. Effects on fertility of motile sperm to egg ratio with use of cryopreserved *Rhamdia quelen* semen at different post-activation times. Anim. Reprod. Sci. 201: 84–92.

Pérez-Atehortúa, M., A.G. Galuppo, R.B. Rodrigues, T.S. França, N.T. Santos, T.F. Rodrigues et al. 2022 The use of differential separation and density gradient with AllGrad 90% after thawing improves the sperm quality of South American catfish (*Rhamdia quelen*). Aquaculture 553: 738072.

Silva, H.V.R., T.G.P. Nunes, L.R. Ribeiro, L.A. de Freitas, M.F. de Oliveira, A.C. Assis Neto et al. 2019. Morphology, morphometry, ultrastructure, and mitochondrial activity of jaguar (*Panthera onca*) sperm. Anim. Reprod. Sci. 203: 84–93.

Streit Jr, D.P., G.V. de Moraes, R.P. Ribeiro, J.A. Povh, E.D. de Souza and C.A.L. de Oliveira. 2004. Avaliação de diferentes técnicas para coloração de sêmen de peixes. Arq. Ciên. Vet. Zool. UNIPAR 7: 157–162.

Sylvester, P.W. 2011. Optimization of the Tetrazolium Dye (MTT) colorimetric assay for cellular growth and viability. Methods. Mol. Biol. 716: 157–168.

Tsvetkov, T. and Z. Naydenova. 1987. Activity of ATP synthetase complex after low temperature treatment or freeze-drying of mitochondria isolated from skeletal muscles. Cryobiology 24: 280–284.

Winey, M., J.B. Meehl, E.T. O'Toole and T.H. Giddings Jr. 2014. Conventional transmission electron microscopy. Mol. Biol. Cell. 25: 319–323.

Xavier, A.M.M., G. Neumann, E.A. Sanches, S.U. Cardoso and R.A. Bombardelli. 2021. Extenders with vitamin C and E applied to *Rhamdia quelen* sperm cryopreservation. Braz. J. Dev. 7: 118046–118060.

CHAPTER 10

Sperm Cryopreservation Protocols for Neotropical South American Species

Danilo P. Streit Jr.,[1,2,*] *Rômulo Batista Rodrigues,*[3] *Eduardo Antônio Sanches,*[4] *Jayme Aparecido Povh,*[5] *Diógenes Henrique de Siqueira-Silva,*[6] *Ana Carina Nogueira Vasconcelos*[7] and *Tiantian Zhang*[8]

INTRODUCTION

Countless technological tools are used, with different objectives, to ensure the preservation of animal genes. In this sense, semen cryopreservation is a biotechnique considered classic and initially developed for mammals by Polge et al. (1949). Since the first study published on cryopreservation with fish semen by Blaxter (1953), more than 200 species have been studied (Tsai and Lin 2012). The first trial with the cryopreservation of semen from South American Neotropical fish with *Rhamdia*

[1] Aquam Research Group, Animal Science Research Program, Federal University of Rio Grande do Sul, Porto Alegre, RS, Brazil.
[2] Veterinary Science Program, Federal University of Rio Grande do Sul, Porto Alegre, RS, Brazil.
[3] Departamento de Zootecnia e Ciências Biológicas, Universidade Federal de Santa Maria - Campus de Palmeira das Missões, RS, Brazil.
[4] Universidade Estadual Paulista "Julio de Mesquita Filho" – UNESP, FCAVR and CAUNESP. Av. Nelson Brihi Badur 430, Vila Tupy, 11900-00, Registro, São Paulo, Brazil.
[5] Faculty of Veterinary Medicine and Animal Science, Federal University of Mato Grosso do Sul, Campo Grande-MS, Brazil.
[6] Faculty of Biology, Federal University of the South and Southeast of Pará, Marabá-PA, Brazil.
[7] JBS Biotech Innovation Center, Florianopolis, SC, Brazil.
[8] Faculty of Science and Technology, Bournemouth University, Poole, United Kingdom.
Emails: romulo.batista@ufsm.br, eduardo.sanches@unesp.br, jayme.povh@ufms.br, diogenessilva@unifesspa.edu.br, anacarina.bio@gmail.com, tzhang@bournemouth.ac.uk
* Corresponding author: danilo.streit@ufrgs.br

Figure 1. History of publications with South American Neotropical species recorded in sperm cryopreservation studies. The number of publications in the respective year is indicated in parentheses when it is greater than one publication. [1]Flogi da Silveira et al. (1981), [2]Coser et al. (1984), [3]Coser et al. (1987), [4]Kavamoto et al. (1989), [5]Fogli da Silveira et al. (1990), [6]Farias et al. (1999), [7]Murgas et al. (2001), [8]Murgas et al. (2003), [9]Carolsfeld et al. (2003), [10]Ribeiro and Godinho (2003), [11]Cruz-Casallas et al. (2004), [12]Navarro et al. (2004), [13]Fresneda et al. (2004), [14]Arciniegas et al. (2005), [15]Velasco-Santamaría et al. (2006), [16]Ninhaus-Silveira et al. (2006), [17]Melo and Godinho (2006), [18]Cruz-Casallas et al. (2006), [19]Maria et al. (2006a), [20]Maria et al. (2006b), [21]Streit Jr. et al. (2006), [22]Medina-Robles et al. (2007); [23]Oliveira et al. (2007), [24]Murgas et al. (2007), [25]Viveiros et al. (2008), [26]Taitson et al. (2008), [27]Menezes et al. (2008), [28]Viveiros et al. (2009b), [29]Viveiros et al. (2009a), [30]Streit Jr. et al. (2009), [31]Felizardo et al. (2010), [32]Viveiros et al. (2010), [33]Nascimento et al. (2010), [34]Ramirez-Merlano et al. (2011a), [35]Galo et al.

Figure 1 contd. ...

hilarii (synonym of *R. quelen*) was published as a congress summary by Fogli da Silveira et al. (1981), and later, it was completed in the mid-1980s (Fogli da Silveira et al. 1985). The expressive wealth of the South American ichthyofauna currently represented by 739 genera, 69 families, and 20 orders, about 1/3 of all freshwater fish on the planet (Reis et al. 2016), has aroused interest in the preservation of genetic resources in the South American continent, using seminal cryobiology, both for application in fish production and in environmental compensation programs.

Existing scientific studies involving semen cryopreservation from South American Neotropical fish were searched using the main scientific databases (Scopus, Web of Science, Pub Med, and Science Direct). The eligibility criteria of the articles included the use of semen cryopreservation of native South American species, publication in Portuguese, English, and Spanish, and no limitations on the date of publication.

The development of semen cryopreservation techniques with South American migratory species began in the 1980s (Figure 1), when the first complete scientific article was published with *Prochilodus scrofa* (synonym of *Prochilodus lineatus*) and *Salminus maxillosus* (synonym of *Salminus brasiliensis*) (Coser et al. 1984). However, the first assay using the technique in Brazil was performed with the semen of *Rhamdia hilarii* (synonym of *R. quelen*) and published in 1981 as a congress summary (Fogli da Silveira et al. 1981). The full article was only published in the mid-1980s (Fogli da Silveira et al. 1985).

...Figure 1 contd.

(2011), [36]Miliorini et al. (2011), [37]Felizardo et al. (2011), [38]Viveiros et al. (2011), [39]Ramirez-Merlano et al. (2011b), [40]Viveiros et al. (2012b), [41]Viveiros et al. (2012a), [42]Viveiros et al. (2012c), [43]Nascimento et al. (2012), [44]Nascimento et al. (2012), [45]Paula et al. (2012), [46]Martínez et al. (2012a), [47]Martínez et al. (2012b), [48]Carneiro et al. (2012), [49]Varela Jr. et al. (2012a), [50]Paulino et al. (2012), [51]Varela Jr. et al. (2012b), [52]Atencio et al. (2013), [53]Martínez et al. (2013), [54]Zanandrea et al. (2014), [55]Navarro et al. (2014), [56]Viveiros et al. (2014), [57]Carvalho et al. (2014), [58]Atencio-Garcia et al. (2014), [59]Viveiros et al. (2015a), [60]López et al. (2015), [61]Viveiros et al. (2015b), [62]Weingartner et al. (2015), [63]Lopes et al. (2015), [64]Atencio-Garcia et al. (2015), [65]Viveiros et al. (2015b), [66]Vasconcelos et al. (2015), [67]Varela Jr. et al. (2015a), [68]Viveiros and Leal (2016), [69]Garcia et al. (2015), [70]Maria et al. (2015), [71]Melo-Maciel et al. (2015), [72]Varela Jr. et al. (2015b), [73]Adames et al. (2015), [74]Pardo-Carrasco et al. (2015), [75]Zanandrea et al. (2016), [76]Pinheiro et al. (2016b), [77]Viveiros et al. (2016), [78]Nunes et al. (2016), [79]Mello et al. (2016), [80]Oliveira et al. (2016), [81]Pinheiro et al. (2016a), [82]Restrepo-Betancur et al. (2017), [83]Di Chiacchio et al. (2017), [84]Atencio-Garcia et al. (2017), [85]Monteiro et al. (2017), [86]Do Nascimento et al. (2017), [87]Di Chiacchio et al. (2017), [88]Viveiros et al. (2017), [89]Gallego et al. (2017), [90]Mello et al. (2017), [91]Goes et al. (2017), [92]Lenz et al. (2018), [93]Lopes et al. (2018), [94]Pereira et al. (2018) [95]Pires et al. (2018), [96]Perry et al. (2019), [97]Viveiros et al. (2019), [98]Paula et al. (2019), [99]Assis et al. (2019), [100]Galo et al. (2019a), [101]Galo et al. (2019b), [102]Medina-Robles et al. (2019), [103]Gheller et al. (2019), [104]Pereira et al. (2019), [105]Neumann et al. (2019), [106]Da costa et al. (2019b), [107]Corso et al. (2019), [108]Da Costa et al. (2019a), [109]Herrera-Cruz et al. (2019), [110]Da Costa et al. (2020), [111]Acosta et al. (2020), [112]Pereira et al. (2020), [113]Teodozia et al. (2020), [114]Borges et al. (2020), [115]Duarte-Trujilo et al. (2021), [116]Marques et al. (2021), [117]Krause et al. (2021), [118]Pérez-Atehortúa et al. (2021), [119]Xavier et al. (2021), [120]Motta et al. (2022), [121]Torres et al. (2022), [122]Paula et al. (2022), [123]Pérez-Atehortúa et al. (2022), [124]França et al. (2022), [125]Velarde et al. (2023), [126]Atencio-Garcia et al. (2023), [127]Medina-Robles et al. (2021), [128] Medina-Robles et al. (2023a), [129]Medina-Robles et al. (2023b), [130]Calizaya and Calizaya (2022), [131]Egger et al. (2021), [132]Andrade et al. (2014), [133]Felizardo et al. (2016), [134]França et al. (2020), [135]Palhares et al. (2020), [136]Perry et al. (2020), and [137]Palhares et al. (2021).

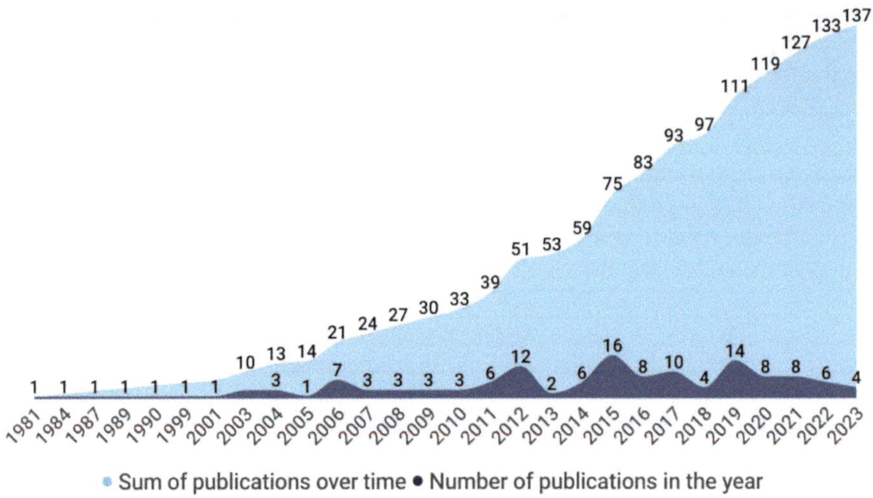

Figure 2. Annual growth trend of articles from 1981 to 2023 on cryopreservation of Neotropical fish sperm.

In the '90s, the only remarkable study was carried out again by Fogli da Silveira et al. (1990), focusing on semen from *Piaractus mesopotamicus*. It was only from the 2000s that a significant increase in the number of studies occurred, highlighted by the extensive study by Carolsfeld et al. (2003), which developed protocols for four species of fish, and the review by Viveiros and Godinho (2009) on the cryopreservation of semen from South American Neotropical species (Figure 2).

1. Cryoprotectants and Extenders

Our review shows that 34 South American Neotropical species have been studied for the development of cryopreservation protocols over four decades. Although some species showed a single study, *Prochilodus scrofa* (synonym of *P. lineatus*) has been the most studied over the years, with 25 published articles and a well-established cryopreservation protocol. In total, 126 publications were considered for this review, which was distributed according to the different fish families: Serrasalmidae (37), Prochilodontidae (37), Bryconidae (27), Pimelodidae (23), Salminae (6), and Anastomidae (4). Here, we highlight that the difference between the number of published studies (129) and the sum of studies published with each family (134) is related to some studies that carried out tests with different families of fish within the same study, as we saw in Carolsfeld et al. (2003) and Di Chiacchio et al. (2017), for example.

The bibliographic survey identified a total of 219 cryoprotectant analyses with ten different types of internal cryoprotectants. DMSO was used in 42.47% of the studies and is the main intracellular cryoprotectant tested, followed by methanol with 18.72% (Figure 3), which has been tested in almost all fish families except for Salminae. Methylglycol (17.35%) was consolidated as an intracellular cryoprotectant only in

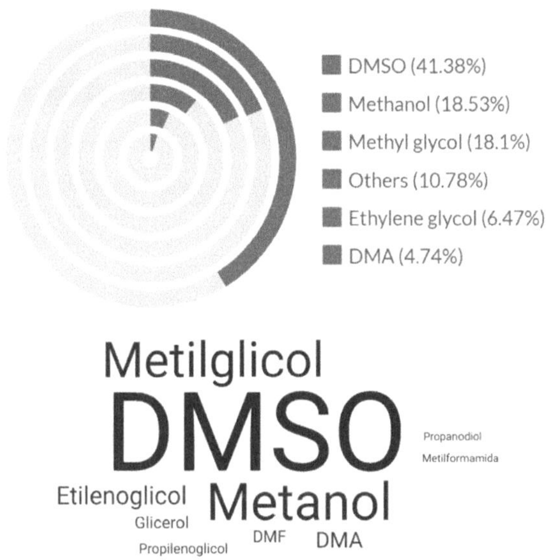

Figure 3. Percentages of use of intracellular cryoprotectants tested in different species of South American Neotropical fish and word cloud. We analyzed 129 scientific papers published between 1981 and 2023. Dimethylsulfoxide – DMSO (C_2H_6OS); methanol (CH_3OH); methylglycol ($C_3H_8O_2$); ethylene glycol ($C_2H_6O_2$); dimethylacetamide – DMA (C_4H_9NO); glycerol ($C_3H_8O_3$); propylene glycol ($C_3H_8O_2$); dimethylformamide – DMF (C_3H_7NO); methylformamide – MF (C_2H_5NO).

the late 2000s, and recent studies showed two new cryoprotectants have been tested in South American species – dimethylformamide (DMF) and methylformamide (MF).

Figure 4 shows the percentage of use of internal cryoprotectants in the protocols developed for the different South American Neotropical fish families. We observed that, except for the Bryconidae family, which presented a higher percentage of tests using two cryoprotectants (DMSO and methylglycol), the other families showed the predominance of only one cryoprotectant. The DMSO was the preferred compound for the families of Salminae, Anastomidae, Serrasalmidae, and Prochilodontidae, while methanol was used for catfishes in the Pimelodidae family. The cryoprotectant analysis for seminal cryopreservation of the Serrasalmidae family was the most explored, totaling 66 tests divided into nine internal cryoprotectants, while the Salminae family showed the least number of studies (ten), testing only three different compounds.

Although glycerol was the first successful intracellular cryoprotectant for fish semen (Blaxter 1953), DMSO became widely used after successful studies with salmonids (Martínez-Páramo et al. 2017) and carp semen (Suquet et al. 2000); since 1984 (Coser et al. 1984), it has been used to cryopreserve semen from native Brazilian species.

As with intracellular cryoprotectants, a range of seminal extenders has already been tested in native South American species, and so far, new diluents are still being tested. Here, we consider extenders in the composition of the cryoprotective solution components used as external cryoprotectants, such as glucose, powdered milk, and

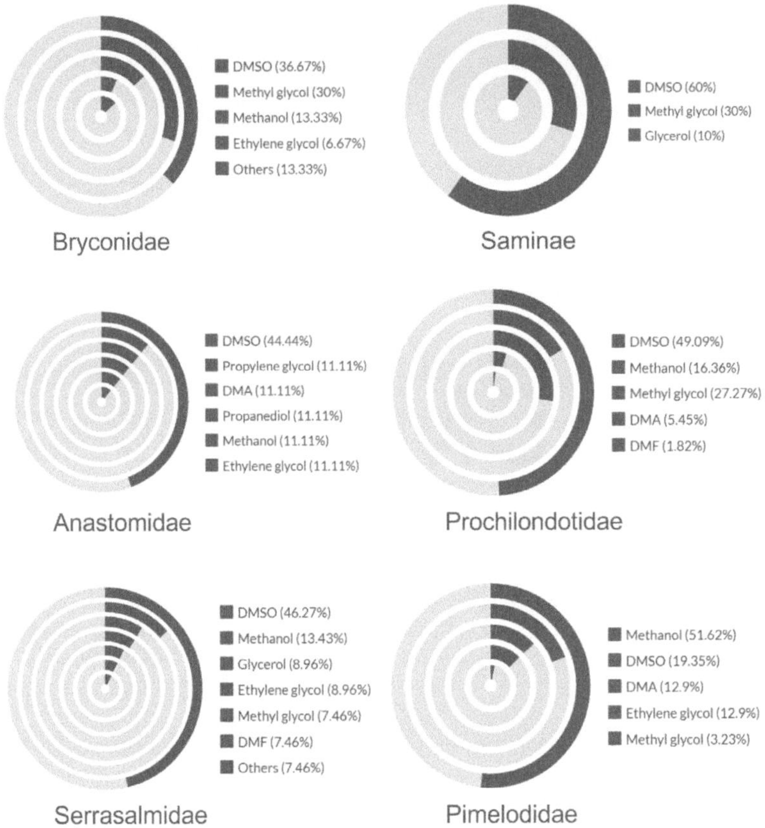

Figure 4. Cryoprotectants were tested on different families of South American Neotropical fish, listed in 129 articles published between 1981 and 2023. Number of cryoprotectant tests for each family: Bryconidae (n = 49); Prochilodontidae (n = 54); Salminae (n = 10); Anastomidae (n = 9); Serrasalmidae (n = 66); Pimelodidae (n = 31). Dimethylsulfoxide – DMSO (C_2H_6OS); methanol (CH_3OH); methylglycol ($C_3H_8O_2$); ethylene glycol ($C_2H_6O_2$); dimethylacetamide – DMA (C_4H_9NO); glycerol ($C_3H_8O_3$); propylene glycol ($C_3H_8O_2$); dimethylformamide – DMF (C_3H_7NO); methylformamide – MF (C_2H_5NO).

egg yolk (in nature). In the literature, there is no standardization as to the terms used to refer to these components, especially sugars (glucose, trehalose, sucrose, etc.) considered both extenders' solutions and external cryoprotectants. In this case, glucose was the most present component (28.36%), followed by egg yolk (18.18%) and Beltsville Thawing Solution-BTS (14.91%) (Figure 5). Although glucose is a regular component in different extenders, here, we considered it as an extender only when it was used in its isolated form.

The diluent BTS was originally formulated to conserve swine semen, and it has been efficiently adapted to cryopreserve semen of South American Neotropical fish. Initial studies with these fish species had solutions like Courtland and Hank's Balanced Saline Solution (HBSS) as the preferred extenders; Figure 6 shows the percentage of use of extenders and/or external cryoprotectants in protocols developed for seminal cryopreservation of South American Neotropical fish families. We

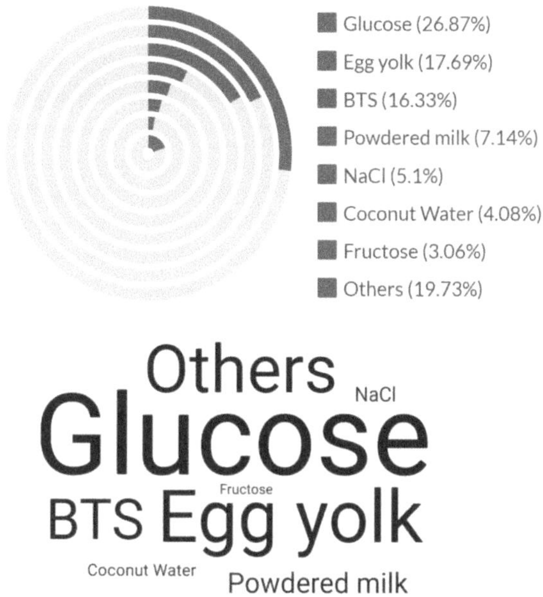

Figure 5. Percentage of extenders used in sperm cryopreservation from different families of South American Neotropical fish and word cloud with the main extenders tested. We analyzed 129 scientific papers published between 1981 and 2023.

can observe that glucose was the most used component in all the families studied, except for Pimelodidae, which showed more efficiency when powdered milk was used. Different components were tested for each fish family, and we observed a higher diversity of extenders for the species of Serrasalmidae (n = 0), followed by Prochilodontidae (n = 15), Pimelodidae (n = 12), Bryconidae (n = 10), Salminae (n = 6), and Anastomidae (n = 4).

2. Species

As the families belonging to the orders Characiformes and Siluriformes gather the largest number of studies, we analyzed the efficiency of the cryoprotectants and diluting solutions used and suggested protocols within each family.

2.1 Characiformes

The order Characiforme is the largest and most complex among Neotropical fish (Nelson 2016). It includes the Bryconidae, Salminae, Serrasalmidae, Prochilodontidae, and Anastomidae families, of which seminal cryopreservation studies were found in 25 species.

2.1.1 Bryconidae

The high number of studies with the Brycon genus is related to restocking river research financed by companies in the hydroelectric sector. The research focus is related to the ecological importance of these species since they are considered

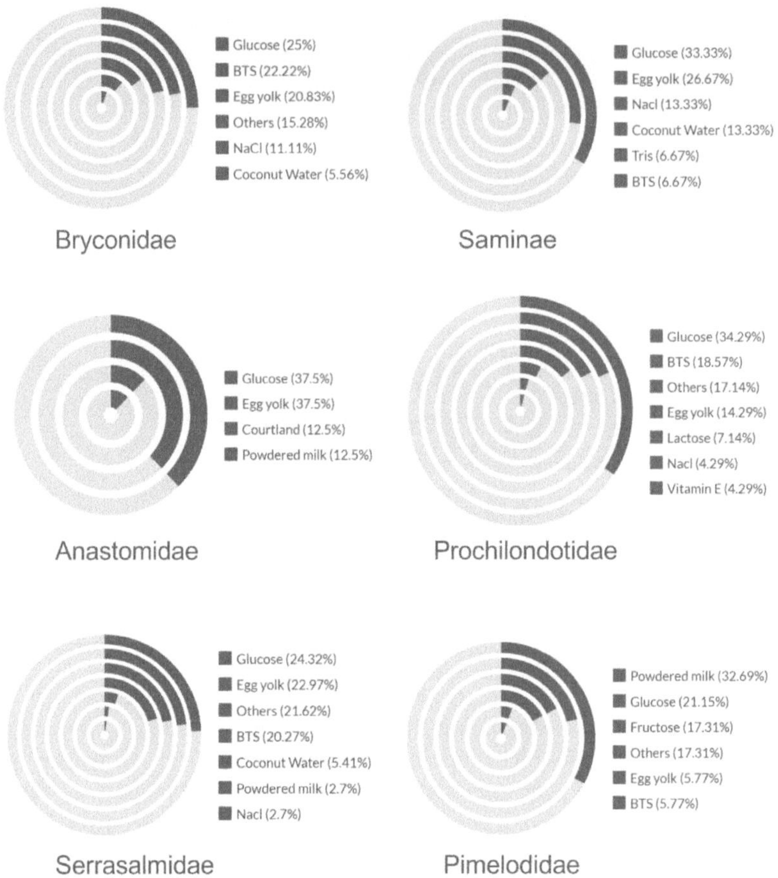

Figure 6. Extenders tested on different South American Neotropical fish families, listed in 129 articles published between 1981 and 2023. A number of tests of different extenders and/or external cryoprotectants for each family include Bryconidae (n = 61), Prochilodontidae (n = 69), Salminae (n = 15), Anastomidae (n = 8), Serrasalmidae (n = 71), and Pimelodidae (n = 52).

environmental indicators, and the construction of dams puts strong pressure on fish stocks. According to the Brazilian Red Book of Threatened Species of Fauna (ICMBIO 2018), *B. orbignyanus*, and *B. insignis* are endangered, while *B. nettereri* and *B. opalinus* are classified as vulnerable.

Eight intracellular cryoprotectants were tested for the genus: DMSO, methanol, methyl glycol, ethylene glycol, dimethylformamide (DMF), dimethylacetamide (DMA), N-methylformamide (NMF), and propylene glycol. *Brycon orbignyanus* was the most studied species in this family, with nine publications identified, and served as a basis for developing studies for other species of Bryconidae. We can divide the seminal cryopreservation protocols into two groups—the species that had the cells most efficiently cryopreserved with DMSO and the ones that showed better results with methyl glycol.

The initial studies with Bryconidae date back to the beginning of 2000, initially by Murgas et al. (2001) and Murgas et al. (2003), and the study by Carolsfeld et al. (2003) is considered a milestone for the cryopreservation of several Neotropical South American species, compiling several experiments with different species, including *B. orbignyanus*. The basic protocol proposed by the authors—DMSO and egg yolk diluted in glucose solution—was extensively tested in numerous species of the genus Brycon, varying the yolk concentration between 5% and 12%, as well as glucose between 5% and 6%, and seminal activation by sodium bicarbonate (1%). This protocol was also used by Medina-Robles et al. (2007) to validate the protocol for *B. amazonicus*, and added for the species the use of 4 mL macro-straws and thawing of the semen at 35°C/90s. Galo et al. (2011) correlated the increase in spermatic morphological changes with low sperm motility after cryopreservation in Bryconidae. The last published data using the protocol with DMSO, egg yolk, and glucose analyzed the sperm cell of Colombian *B. moorei* (Atencio-Garcia et al. 2017).

In the mid-2000s, a protocol using 10% methyl glycol (intracellular cryoprotectant) began to be investigated for the *Brycon* genus by Maria et al. (2006) using the species *B. orbignyanus*. It is important to note that the protocol with methyl glycol has been adjusted and validated in four species of Bryconidae (*B. orbignyanus, B. nattereri, B. opalinus*, and *B. insignis*). Unlike in the protocol development using DMSO diluted only in glucose solution, the methyl glycol was tested with at least five extensors, varying in molar concentrations. Thus, except for the study by Maria et al. (2006) using Merck III® and Viveiros et al. (2008) using coconut water as an extender to dilute methyl glycol, the other studies evaluated the effectiveness of NaCl, glucose, or BTS in the cryopreservation of sperm cells.

While developing the protocol using methyl glycol, some components and procedures were modified, besides adjusting the extensors. The antibiotic gentamicin was suggested in the study by Maria et al. (2006) and regularly incorporated into the methyl glycol protocol. Seminal cryopreservation protocol for *Brycon* species cannot be hypoosmotic (< 270 mOsm/kg) (Nascimento et al. 2011); besides suggesting an equilibration time between 15–30 minutes, semen storage in straws of 0.5 mL or 4 mL and thawing process at 60°C for 8s (Viveiros et al. 2012b), the sugar or ionic compound (between 0 and 200 mOsm/kg) should be present in the Bryconidae sperm activating solution (Viveiros et al. 2019). When compared to other intracellular cryoprotectants (DMSO and methanol), methyl glycol has provided better structural protection of Brycon sperm cells (Viveiros et al. 2015a, b), and it did not compromise the efficiency of the cells during reproduction (Di Chiacchio et al. 2017).

A recent study (Perry et al. 2019) testing alternative cryoprotectants—methylformamide (MF), dimethylformamide (DMF), and dimethylacetamide (DMA)—suggested a new protocol for *Brycon* genus using 7.5% DMF as an intracellular cryoprotectant. The large number of data published allows a safe and effective sperm cryopreservation of the species of *Brycon* genus.

2.1.2 Salminae

The Salminae family has always aroused the interest of Brazilian researchers due to their economic and social importance in sport and artisanal fishing, respectively. According to Freitas (2010), *S. brasiliensis* is the senior synonym of *S. maxillosus*. The researchers' interest in this species is related to the fact that they are large migratory fish, and the wild populations suffer from anthropogenic actions in the basin watersheds where they are found.

In Brazil, the first study with semen cryopreservation was with Salminus (Coser et al. 1984). Although very simple compared to the current one, this study is valuable, and the authors suggested a protocol with 10% diluted DMSO in the NaCl extender (0.8%). However, almost two decades later, Carolsfeld et al. (2003) defined a new protocol for the species with DMSO (10%) associated with egg yolk (10%) and glucose (5%). The DMSO appears to be the intracellular cryoprotectant with the best results for Salminus semen, and it was confirmed by Viveiros et al. (2009b) after testing DMSO, methylglycol and glycerol diluted in different concentrations of NaCl (154, 200 mM and 200 mM-tris), glucose (277 mM) and BTS (4%). Considering the publications, the protocol by Carolsfeld et al. (2003) seems to be more suitable for preserving the semen of *Salminus*, and its efficiency regarding fertilization was shown by Weingartner et al. (2015). Cryopreservation of *Salminus* semen was alternatively analyzed using methylglycol (10%) and ACP (3.4%) (Zanandrea et al. 2014, 2016), and the authors did not find any difference when compared to 10% DMSO with 10% egg yolk and 5% glucose. Although the Salminae family was one of the initial targets for semen cryobiology research in the 1980s, more efficient results have not been observed since the protocol proposed by Carolsfeld et al. (2003).

2.1.3 Serrasalmidae

Three species of two genera in the Serrasalmidae family are known as "round fish" such as *Colossoma macropomum*, *Piaractus mesopotamicus, Piaractus brachypomus*, and *Piaractus orinoquensis*. These species are particularly important to aquaculture, and 163,721.406 tons were produced in Brazil in 2022 (IBGE 2023). For being so economically important, several studies involving semen cryopreservation of these species have been developed over the decades.

The first study involving the reproduction of *P. mesopotamicus* was published in 1977 (Godinho et al. 1977) and in 1990 (Fogli da Silveira et al. 1990), a cryopreservation protocol based on a published study for carp semen (Stein and Bayrle 1978) was tested, using DMSO (10%) diluted in the V2e extender. More than a decade later, Carolsfeld et al. (2003) proposed the use of DMSO (10%), egg yolk (10%), and glucose (5%) for the species, and based on the same protocol, Streit et al. (2009) found a significant increase in the percentage of primary abnormalities in sperm morphology after the cryopreservation. The same was confirmed later by Galo et al. (2019b), who attributed primary pathologies as the main factor for reducing the fertilization rate and hatching of *P. mesopotamicus* using this protocol. A new combination of cryoprotectants was suggested by Teodozia et al. (2020) using only glucose (5%) associated with DMSO (10%), together with the use of a dry-shipper or programmable freezer during the cryopreservation process.

Based on the basic protocol of Carolsfeld et al. (2003), other researchers tested alternative extenders trying to replace the original ones—glucose and/or egg yolk. Streit Jr et al. (2006) tested BTS and ZOR (both at 10%), but they were not recommended by the authors. On the other hand, Paulino et al. (2012) adjusted the concentration of the BTS extender to 5%, proposed a protocol using DMSO and methanol, and thawing the semen at 40 or 60°C for 12 or 8 seconds, respectively, to minimize the incidence of morphological changes in the sperm. With BTS (5%), Carvalho et al. (2014) tested caffeine to enhance the efficiency of DMSO (10%); however, there was no success with this additive. Then, a new component—100 mM raffinose—was added by Pires et al. (2018) to the DMSO (10%) diluted in BTS (5%) (osmolality adjusted to 330 mOsm/kg) to make it more efficient.

Since the second half of the 2000s, there has been a significant increase in the number of studies with seminal cryopreservation of *C. macropomum*—tambaqui, making it the second most studied South American species on this subject. Since it represents more than 80% of the native fish produced in Brazil, developing a cryopreservation protocol for this species focuses on animals' zootechnical potential. The first study involving the preservation of tambaqui semen was published in 1999 (Farias et al. 1999), suggesting a protocol with 5% glycerol diluted in coconut water. Later, Menezes et al. (2008) recommended a protocol with propylene glycol or ethylene glycol; however, there is no information regarding the concentration of these cryoprotectants or the extender used for dilution. Pereira et al. (2020) tested different sulfated polysaccharides from different natural sources to cryopreserve the semen of the species using the protocol of 10% DMSO diluted in coconut water, but this addition did not convince them. Varela Jr et al. (2012a) proposed the use of DMSO (10%) as the intracellular cryoprotectant and BTS as the extender for dilution. Later, Varela Jr et al. (2015a) suggested using a thermal box for the cooling process prior to freezing. However, Oliveira et al. (2016) recommended using a programmable cooling freezer to cryopreserve the species' semen, using glucose (5%) as an extender.

Semen cryopreservation protocols using intracellular cryoprotectants other than DMSO were also investigated. Varela Jr et al. (2012b) recommended a protocol with DMA (8%) or DMF (8% and 11%) when compared to traditional intracellular cryoprotectants (DMSO, methanol and methylglycol) and thawing 38°C/8s. This protocol with 8% DMF diluted in 92% BTS was not recommended by Marques et al. (2021) to cryopreserve the sperm of the species contaminated with 10% or more of urine, feces, or blood. On the other hand, Carneiro et al. (2012) recommended the use of methyl glycol (10%), egg yolk (5%), and glucose-290 mOsm (5%). Finally, Gallego et al. (2017), using this protocol, identified sperm subpopulations in cryopreserved semen from *C. macropomum*. The volume of straws was the objective of two studies using different cryopreservation protocols. Maria et al. (2015) suggested the use of macropalettes of 1.6 mL and 4.5 mL cryotubes, using the protocol with methyl glycol (10%), egg yolk (5%), and glucose-290 mOsm (5%) and thawing the samples for 90s and 60°C. In a recent study, a macropalette was also recommended by Medina-Robles (2019); however, with 2.0 mL and a protocol with 10% methanol and glucose (5.5%).

Based on the protocols of Varela Jr et al. (2012a,b), Carneiro et al. (2012), and Maria et al. (2015), other researchers produced information not with the focus of proposing a complete and exhaustively tested protocol but adding some additive or even testing other intracellular cryoprotectants and extenders, and the consequences of its use for the sperm of the species. In this sense, Melo-Maciel et al. (2015) suggested adding aloe vera to enhance the efficiency of a protocol when using DMSO; Garcia et al. (2015) used enriched glucose (different salts and powdered coconut water) associated with DMSO; Pinheiro et al. (2016a) vetoed the use of egg yolk when associated with DMSO; Mello et al. (2016) proposed a protocol to evaluate the degree of toxicity of cryoprotectants DMF, DMSO, methanol, ethylene glycol, and glycerol on sperm DNA; Mello et al. (2017) determined that DMSO and ethylene glycol produced less DNA methylation than the other cryoprotectants tested; Lopes et al. (2018) suggested adding vitamin E and taurine (1 mM both) to the protocol with DMSO (10%) diluted in glucose (5%). In this same line, adding antioxidants, Pereira et al. (2018) recommended using 4 Mm reduced glutathione to DMSO base protocol (10%), and Lenz et al. (2018) used as a base the solution composed of egg yolk (20%), glucose (5%), and NaCl (65%), recommended methanol (10%) after comparing it with DMSO, DMF, and glycerol. Galo et al. (2019a) stated that proteins weighing less than 50 kDa can be used as post-cryopreservation seminal quality markers for the species.

Considering all round fish species, *P. brachypomus* has the least number of scientific studies, and the regular use of its semen in the production of hybrids for Brazilian fish farming reinforces the importance of studying the effect of cryopreservation on the *P. brachypomus* semen. In an initial study, Fresneda et al. (2004) suggested using two protocols for the species—the first with DMSO (5%) and powdered milk (15%) diluted in a glucose solution (5.4%) with NaCl (0.8%), and the second protocol using methanol (8%) diluted in skim powdered milk solution (15%). In the same year, Navarro et al. (2004), after testing different combinations with cryoprotectants and extenders, concluded that the best protocols should be composed of DMSO (10%) combined with egg yolk (7%) or methanol (10%) associated with powdered milk (15%). Later, another protocol was recommended by Nascimento et al. (2010), which contained methyl glycol (10%) diluted in glucose (5%). A complete protocol was published later (Ramirez-Merlano et al. 2011a), with the authors suggesting the use of DMSO or methanol (10%) and egg yolk (15%) diluted in glucose solution (5.5%). To date, there are only four studies published, and a definitive protocol was not achieved. On the other hand, the recommendation to use macrotubes (2.5 mL and 5 mL) (Ramirez-Merlano et al. 2011a) allows for a more practical operation, as there are efforts to produce the species on a large scale in Brazil and Colombia. In 2023, two studies were published by the same research group (Medina-Robles et al. 2023a, b) with a new Serrasalmidae species, *Piaractus orinoquensis*. The species was reclassified by Escobar et al. (2019) from the species *Piaractus brachypomus*. In the study by Medina-Robles et al. (2023a), the authors concluded that semen samples of the species cryopreserved with DMSO (10%), glucose (5.5%), and egg yolk (12%) in 4 mL macro-straws, maintained the same fertilizing capacity after seven years cryopreserved storage than fresh semen.

In the publication by Medina-Robles et al. (2023b), it was found that the storage time of cryopreserved samples negatively affected sperm quality, especially in sperm morphology and DNA integrity.

So far, the protocols developed for round fish species are in different stages of development, certainly due to their importance in the Brazilian aquaculture sector. On the one hand, studies involving *C. macropomum* are at a very advanced stage, while the other two species (*P. brachypomus* and *P. mesopotamicus*) still need studies that reinforce the protocols investigated so far.

2.1.4 Prochilodontidae

The Prochilodontidae family is found along the Brazilian territory with different species inhabiting the hydrographic basins, and since the first studies of their seminal cryopreservation, these species have been used in developing protocols for Neotropical species.

In the 1980s, *P. lineatus*, formerly known as *P. scrofa*, was the target of initial studies of seminal cryobiology of South American Neotropical species. In 1984, the first study (Coser et al. 1984) tested a protocol using DMSO (10%) diluted in NaCl (0.8%), and a few years later, Kavamoto et al. (1989) diluted DMSO (10%) in another extender called V2e, with inconclusive results. From the mid-2000s onwards, studies were resumed to develop an efficient protocol for the species of Prochilodontidae. Thus, two protocols were developed based on two intracellular cryoprotectants—DMSO and methyl glycol—and, to date, they are still being tested.

The protocol that was developed based on the DMSO was initially proposed by Carolsfeld et al. (2003), using DMSO (10%) and glucose (5%). The study by Murgas et al. (2007) defined a protocol containing 10% of DMSO or methanol diluted in BTS (4.5%). Other authors tried to improve the results from this protocol, suggesting some adjustments in the concentration of cryoprotectants. Felizardo et al. (2010) proposed the use of methanol (10%) diluted in BTS (5%); Miliorini et al. (2011) suggested diluting methanol (5%) and DMSO (7.5%) in BTS (5%); Felizardo et al. (2011) indicated the efficiency of using activators with high osmolarities (≥ 293 mOsm/kg) when using methanol or DMSO; Vasconcelos et al. (2015) proposed the use of lactose (8%) associated with 10% DMSO or methanol diluted in BTS (5%); Varela Jr et al. (2015a) tested different concentrations of DMSO and concluded that 5% diluted in BTS (8%) was the best combination, while Paula et al. (2019) determined that the cooling curve for cryopreservation should be 30°C/min for the same protocol. Recently, Acosta et al. (2020) recommended the replacement of the traditional intracellular cryoprotectant (DMSO) for amides (DMF 2–8%) diluted in BTS (318 mOsm/kg).

Another internal cryoprotectant was also studied to develop a protocol for *P. lineatus*. Studies involving the test of methyl glycol (10%) and glucose (5%) (Viveiros et al. 2009a) or coconut water (300 mOsm/kg) (Viveiros et al. 2010) were followed by several publications trying to improve the protocol. Nascimento et al. (2012) used methyl glycol (10%) diluted in BTS or glucose, while Viveiros et al. (2014) tested DMSO, methanol, and methyl glycol diluted in glucose or BTS and concluded that methanol and methyl glycol combined with either of the two extenders are indicated for the species. Then Viveiros and Leal (2016) recommended using the 1:10 ratio (semen: cryoprotective solution), adding more information to the protocol with 10% methyl glycol diluted in 5% glucose. Using the same protocol, Viveiros et al. (2016) suggested the use of NaCl or glucose with an osmolarity of 100–150 mOsm/kg to activate the cryopreserved semen, and Di Chiacchio et al. (2017) observed no difference in the quality of cryopreserved semen during the entire reproductive period of the species when using 10% methyl glycol diluted in 5% glucose. Viveiros et al. (2017) recommended the cryopreservation of the semen within 3 hours after collection. In a recent study, Motta et al. (2022) were not definitive in recommending the addition of melatonin in the protocol with 10% methyl glycol diluted in 5% glucose.

In order to improve and enhance the suggested protocols, some studies were conducted by different researchers, testing additives in the cryopreservation protocols of *P. lineatus*. Thus, Paula et al. (2012) tested vitamin E or reduced glutathione in the protocol with methanol (10%) and lactose (15%) but did not recommend adding them. On the other hand, Navarro et al. (2014) recommended adding 0.1 μg of vitamin C to the protocol with DMSO (5%) diluted in BTS (5%), and Carvalho et al. (2014) analyzed the addition of different concentrations of caffeine to the same components, but there was no improvement in post-thaw results. The addition of melatonin was tested by Assis et al. (2019) to improve the protocol with DMSO (8%) diluted in BTS (5%) but also did not find significant gains in seminal quality after thawing.

Prochilodus magdalenae is a species originally from the Magdalena River basin in Colombia, and researchers from that country have conducted some studies. It should be noted that the investigated protocols were based on studies with another species of the same genus—*P. lineatus*. Initially, Martínez et al. (2012a) proposed a protocol containing DMSO (5–10%) diluted in glucose (5–6%) after having tested different combinations of intracellular cryoprotectants and extenders. Martínez et al. (2012b) complemented the protocol by adding egg yolk (12%), recommending filling 0.5 mL straws. Then, the protocol was further refined by Martínez and Pardo-Carrasco (2013) with DMSO (10%) and egg yolk (12%) and diluted in glucose (6%) adjusted to 333 mM. Finally, Atencio-Garcia et al. (2015) adjusted glucose to 5.5% and validated the protocol, suggesting 160,000 sperm cells per oocyte when using cryopreserved semen. The only alternative protocol investigated for the species was suggested by Atencio et al. (2013), who used DMA (8%), egg yolk (12%), and glucose (6%). Recently, another species of Prochilodontidae (*P. mariae*), in this case from the Orinoco River basin in Colombia, was studied by Duarte-Trujillo et al. (2021). The focus of the study was whether or not to add 12% egg yolk, curiously diluted in 6% glucose, and 10% methanol instead of DMSO or DMA, reported as efficient for the protocol of another species from this region, *P. magdalenae*. Another study was published in 2021 by Medina-Robles et al. (2021) with *P. mariae*. The authors concluded that the semen of the species was successfully cryopreserved using a programmable freezer or dry-shipper with a cryoprotectant solution containing 10% methanol.

From the São Francisco River basin, in the northeast of Brazil, *P. brevis* is the other species of the Prochilodontidae family that is also a target for seminal cryopreservation studies. Therefore, the protocol suggested by Lopes et al. (2015) comprised 5% glucose and 10% methylglycol, diluted between 1:3 and 1:6 semen and cryoprotective solution. Pinheiro et al. (2016b) completed the protocol recommending using a dry-shipper or programmable freezer for the cooling curves. However, the protocol suggested by Nunes et al. (2016) recommended diluting the DMSO (10%) in glucose (5%). The latest studies with the species have tried optimizing the protocol results using DMSO (10%) as a permeable cryoprotectant. There was an attempt to replace glucose (5%) with powdered coconut water to dilute DMSO (Do Nascimento et al. 2017), but no significant improvement was achieved. The same happened when Monteiro et al. (2017) added vitamin C (0.0001 mg), trying to optimize the protocol by Nunes et al. (2016), and no improvements were obtained. Recently, Torres et al. (2022), when testing different non-permeable cryoprotectants, added 5% egg yolk to the protocol with 10% DMSO and 5% glucose.

All articles considered for this review showed that the semen cryopreservation protocols for the Prochilodontidae family are in different stages of development, certainly due to the importance/distribution of these species within the South American continent. While for the species *P. magdalenae* and *P. brevis,* the cryopreservation protocols have been little improved, studies of *P. lineatus* cell cryopreservation defined two optimal protocols, whose fact is associated with the wide distribution of the species and the great interest in technological development.

Another important fact that should be emphasized as an explanation for the studies of *P. lineatus* is using its fish larvae as food for other species. In the articles studied here, the protocols tested for *P. lineatus* are associated with developing some species of the *Brycon* genus, for example, which also certainly contributed to configuring the South American Neotropical species with the largest number of studies developed with seminal cryobiology.

2.1.5 Anastomidae

Seminal cryopreservation protocols with the species of this family date back to the 1980s, with more recent ones emerging in the 2000s. In total, for three species of this family, seminal cryopreservation protocols have already been proposed, and until recently, all of them belonged to the genus Megaleporinus, formerly Leporinus. After revision studies, the species *L. silvestrii, L. macrocephalus, L. obtusidens*, and *L. elongatus* were reclassified and renamed by ichthyological studies. In this case, *L. silvestrii* and *L. obtusidens* are synonym of *Megaleporinus obtusidens*; *L. elongatus* is a synonym of *Megaleporinus elongatus*, and *L. macrocephalus* is a synonym of *Megaleporinus macrocephalus*. Here, we adopted the scientific name used in the publication for originality related to the mentioned document.

In the article by Coser et al. (1987), the authors reported that although *L. silvestrii* was abundant in rivers in southwestern Brazil, at that time, including the species in restocking programs due to the construction of large hydroelectric dams was necessary. This was the main reason for the authors to study a protocol for this species. The extender media tested, as well as the intracellular cryoprotectants, served as a basis for other studies developed in the 2000s for species of the Anostomidae family.

Coser et al. (1987) tested the extenders NaCl and the modified Courtland, developed by Truscott et al. (1968). When the study with *L. silvestrii* was developed in the 1980s, knowledge about the biochemical composition of semen of native South American species was unknown; therefore, the use of diluting media developed for salmonids was common, which is the case of the modified Courtland. Interestingly, the final protocol suggested for this species contains DMSO (5–10%), but there was no reference to a preference for either tested extender. Although the motility result was less than 10%, the limited use of techniques and knowledge about the processes of fish cryobiology in Brazil in the 1980s make the information in that article valuable. Common in sport fishing, the species *L. elongatus* and *L. obtusidens* were the target of restocking programs for population conservation due to their reproductive migratory characteristics. In the classic study by Carolsfeld et al. (2003), the authors recommended a protocol composed of DMSO (10%) and egg yolk (10%) diluted in glucose (5%) for *L. elongatus*. Subsequently, Taitson et al. (2008) completed the protocol, in this case using *L. obtusidens* and suggesting that the semen:cryoprotectant ratio should be 1:5 in 5 mL straws and use $NaHCO_3$ (119 mM) as a sperm cells activator and fertilizing in a proportion of 112,000 sperm cells per oocyte.

The interest in the species *L. macrocephalus* occurred strongly in the first decade of the 2000s due to its use in fish farming and recreational activities (sport fishing). Thus, the justification for the use of *L. macrocephalus* in fish farming led Ribeiro and Godinho (2003) to propose a cryopreservation protocol from semen collected directly from the gonads of animals kept in fish farming, since the species produces a reduced volume of semen-making collection complicated. The protocol recommended by the authors contains 8.9% of one of the three intracellular cryoprotectants (DMSO, DMA, or propylene glycol), egg yolk (8.9%), and diluted in glucose (5%).

After the study by Taitson et al. (2008), no more publications were found for the Anostomidae family. We attribute this fact to the low production (2,818.844 ton) of the species in the national aquaculture IBGE (2023), the reduction of the recreational fishing trades in the country, and the change of focus of the restocking programs in Brazil for the species of the Bryconidae family.

2.2 Siluriforme

Fish of the Siluriform Order are popularly known as catfish, and their main characteristics are the absence of scale on the body, the skin being smooth or covered with dermal plates, the presence of spindles on the dorsal and pectoral fins and tactile barbels around the mouth. In addition, they have predominantly nocturnal habits. The studies with cryobiology of South American catfish are of genera in the family Pimelodidae, belonging to the order Siluriforme, where a total of nine species were investigated.

Recently, Fricke et al. (2019) reclassified *R. hilarii* as *R. quelen*, but we will adopt the original name mentioned in the article when the study was developed. Research on cryopreservation of South American catfish semen began in the 1980s in a study with *Rhamdia hilarii* (Fogli da Silveira et al. 1981). In the study, the authors proposed a cryopreservation protocol by diluting the semen in V2e

(1:3 ratio), dripping the semen directly onto an ice block, and then the pellets were cryopreserved. *R. quelen* semen was also used as a model study for cryopreserve submitted to different anesthetic protocols before collection by Corso et al. (2019), using, in this case, the extender for swine semen, BTS (5%) associated with 10% methanol.

The importance of *R. quelen* for fish farming in southern Brazil might have contributed to further developing a semen cryopreservation protocol among South American catfish. More than 30 years after the first study, Adames et al. (2015) developed new research with *R. quelen* and established an efficient protocol for cryopreservation of semen composed of methanol (10%), powdered milk (5%) diluted in a fructose solution (5%). Recently, Krause et al. (2021) adjusted for an exact concentration of the methanol cryoprotectant from 10% to 11.66%, in addition to recommending keeping it at 25°C for up to 30 minutes after thawing without qualitative loss of semen. The success of this protocol motivated further research on other indicators to qualify the efficiency of the protocol on fertilization rate (Goes et al. 2017; Neumann et al. 2019) and the sperm morpho pathologies caused by cryopreservation (Da Costa et al. 2019a). To improve the protocol, Da Costa et al. (2019b) added different concentrations of cysteine in the cryoprotective solution; however, it did not improve the results after cryopreservation. Xavier et al. (2021) recommended 4 mg/mL of vitamin C as an antioxidant to cryopreserve *R. quelen* semen. However, it is important to point out that the authors replaced fructose with glucose in the original protocol developed by Adames et al. (2015), which may have been the reason for the average post-cryopreservation motility of only 38.2% for the vitamin C concentration that was recommended in the study.

Three recent articles suggest a significant improvement in the protocol Adames et al. (2015) developed for *R. quelen*. Pérez-Atehortúa et al. (2021a) established a density gradient protocol (AllGrad 45%) to be used after seminal cryopreservation. Then Pérez-Atehortúa et al. (2021b) recommended a simple centrifugation to optimize the sperm quality of the cryopreserved sample. However, the simple methodological adjustment proposed by França et al. (2022), diluting the semen right after thawing the sample in NaCl extender (325 mOsm kg^{-1}; pH 7.6; 24°C), is essential to obtain fertilization and hatching rates equal to fresh semen. Based on this methodological adjustment, it can be stated that the protocol for *R. quelen* is complete.

The first study involving the genus *Pseudoplatystoma* was developed in 2003 (Carolsfeld et al. 2003), and for *P. corruscans*, a protocol using methanol (10%) diluted in powdered milk (15%) was tested and suggested by the authors. Interestingly, DMSO showed no promising results for *P. corruscans* or *P. fasciatum* (Arciniegas et al. 2005) even when combined with different concentrations of other extenders (glucose, egg yolk, and Stein and Bayrle's solution). On the other hand, Ramirez-Merlano et al. (2011b) suggested two protocols for *P. metaense*: the first containing DMSO (10%), powdered milk (5%) diluted in glucose (5.5%), and the second containing methanol (12%) and powdered milk (5.5%) diluted in glucose (5.5%). Recently, Herrera-Cruz et al. (2019) proposed for the species *P. magdaleniatum*, DMSO, DMA, or ethylene glycol as intracellular cryoprotectants in a concentration of 5% to 10% diluted in an extender solution containing powdered milk (3%), glucose (6%) and vitamin E (0.4%). DMA (8% and 10%) combined with 5% glucose seems to be a promising cryoprotectant for the genus *Pseudoplatystoma*, as recently attested by Velarde et al. (2023) with *P. corruscans*. It was decisive in this context to protect the DNA of gametes better.

Popularly known as white catfish, *S. cuspicaudus* is a species of great fishing importance in Colombia in the basins of the rivers Sinu and Magdalena. The species was studied by Atencio-Garcia et al. (2014), who proposed a protocol with good viability after thawing containing ethylene glycol (5%) powdered milk (5%) diluted in glucose (6%). The replacement of the intracellular cryoprotectant ethylene glycol by 8% DMA in the protocol with powdered milk (5%) diluted in glucose (6%) was the recommendation of Pardo-Carrasco et al. (2015), who also recommended filling the samples in 5 mL macro-straws. In the recent study by the same Colombian research group with the cryopreservation of *S. cuspicaudus* semen, Atencio-Garcia et al. (2023) increased the concentration of ethylene glycol (6%, 8%, and 10%) and did not find a concentration considered more efficient.

Recently, studying the semen of *Steindachneridion scriptum*, Pereira et al. (2019) did not recommend the use of methanol, and the suggested protocol was composed of methyl glycol (7.5–10%) diluted in BTS without determining the concentration of the extender. As for the species *Leiarius marmoratus*, Gheller et al. (2019) obtained average motility between 27.7% and 35.9% from a simple protocol with BTS enriched with trehalose (100 mM and 200 mM). However, Borges et al. (2020) opted for a protocol for *L. marmoratus* using glucose as an extender to dilute 10% of three intracellular cryoprotectants—methanol, DMSO, and ethylene glycol. The average motility obtained by the authors was low—8.30% (methanol), 6.86% (DMSO), and 0.65% (ethylene glycol)—especially if compared with the study by Gheller et al. (2019). From this information, we can interpret that for *L. marmoratus*, a protocol using a more complex sugar and the absence of an intracellular cryoprotectant should be prioritized in the formulation of a protocol. In these studies with Brazilian catfish, the productive efficiency of cryopreserved semen was not tested, such as the fertilization and hatching rate.

The semen cryopreservation protocols of South American Siluriformes are in different stages of development. If, on the one hand, *R. quelen* has a consolidated protocol, the other species' protocols need improvement in their stage, which is considered very incipient. This favorable condition for *R. quelen* is certainly related to its importance for fish farming in southern Brazil, in addition to the fact that most research centers that develop studies with cryobiology are in that region of the country.

3. Final Considerations

For almost 40 years, semen cryopreservation studies in South American fish focused on species with production potential, such as *R. quelen* and *P. lineatus*, and since the 2000s, they have been extended to species at risk of extinction, especially of the Brycon genus. However, these studies still have a long way to reach a consolidation stage for the wide practical application of seminal cryopreservation protocols.

Some factors are inherent in the methodology, and the technical limitations cover the practical application in species. However, we realize that most studies focus on sperm motility as a decisive factor in the efficiency of the process, and little progress is made on the frontier of knowledge, such as the physical-chemical aspects involved in cryopreservation processes from the use of cryoprotectants and extender solutions. It is strictly necessary that the cryopreserved sperm cells are capable of fertilizing the oocytes and resulting in viable embryos and individuals from the morphological and genetic point of view. Most of the reviewed bibliography lacks information to explain in more detail what happens in cryopreserved sperm at the morphological, physiological, and molecular levels. Consequently, efficient protocols for most studied species have not yet been developed. The species *P. lineatus* and *B. orbignyanus* have developed protocols with good results and efficiency in generating larvae. As for the species *R. quelen*, it is possible to state that the protocol is complete and sufficiently validated to be used efficiently, being the first South American species with this status.

Using commercial germplasm banks to produce juveniles destined for fish production farms is not yet a reality in Brazil. Perhaps this scenario changes from a breeding program for species such as *C. macropomum, L. marmoratus*, and the genus *Pseudoplatystoma*. In this case, a germplasm bank would be widely used in the composition of programs to explore the zootechnical potential of these species, whether pure or hybrid. As for the germplasm banks that are used for environmental purposes, they are generally maintained if there is a project linked to some company, usually in the hydroelectric sector, as they are expensive to maintain, and the universities and public research centers do not have resources to keep them viable for a long time.

Finally, we believe that there should be a greater alignment of researchers as to the focus of study for the development and improvement of semen cryopreservation protocol, to evolve in the propositions concretely and not just speculate with information that contributes little to an advancement of the productive and/or environmental sector.

Acknowledgments

We thank Dr. Vagner Leonardo M. dos Santos for his help in updating the scientific names of native South American species.

References

Acosta, I.B., C.D. Corcini, S.M.M. Gheller, C.R.C. Brito, T.L.S. Goulart and A.S. Varela-Jr. 2020. Effect of amide on semen cryopreservation of Curimba (*Prochilodus Lineatus*). Cryo-Lett. 41: 1–5.

Adames, M.S., C.P.R. de Toledo, G. Neumman, A.H. Buzzi, C.N. Buratto, P.A. Piana et al. 2015. Optimization of the sperm:oocyte ratio and sperm economy in the artificial reproduction of *Rhamdia quelen* using fructose as a sperm motility modulator. Anim. Reprod. Sci. 161: 119–128.

Arciniegas, P.S.M., J.E. Mojica-Rodríguez and P.E. Cruz-Casallas. 2005. Ensayos preliminares sobre crioconservación de semen de bagre rayado (*Pseudoplatystoma fasciatum* Linnaeus, 1766). Orinoquia 9: 28–37.

Assis, I.L.A., P.C. Palhares, G.J. Machado, J.G.S. Souza, T.S. França and V.O. Felizardo. 2019. Effect of melatonin on cryopreserved sperm of *Prochilodus lineatus* (Characiformes). Cryo-Lett. 40: 152–158.

Atencio, V.F., E.J. Pérez, J.A. Espinosa and S.C. Pardo. 2013. Evaluación de dimetilacetamida como crioprotetor para la crioconservación de semen de bocachico *Prochilodus magdalenae*. Arch. Med. Vet. 45: 151–158.

Atencio-García, V.J., M. Dorado, E. Navarro, F. Pérez, B. Herrera, J. Movilla et al. 2014. Evaluation of ethylene glycol as a cryoprotectant in the sperm cryopreservation of trans-andean shovelnose catfish (*Sorubim cuspicaudus*, Pimelodidae). Acta Biol. Colomb. 19: 271–279.

Atencio-García, V.J., J.A. Espinosa, J.G. Martínez and S.C.P. Carrasco. 2015. Insemination of bocachico fish (*Prochilodus magdalenae*) with fresh or cryopreserved semen. Rev. Colomb. Cienc. Pec. 28: 347–355.

Atencio-García, V.J., M.D. Longas, C.M. Petro, M. Prieto-Guevara and J. Espinosa-Araujo. 2017. Cryopreservation of dorada *Brycon moorei* sperm with dimethyl sulfoxide. Rev. Colomb. Biotecnol. 19: 87–94.

Atencio-García, V., D. Padilla-Izquierdo, J. Robles-González, M. Prieto-Guevara, S. Pardo-Carrasco and J. Espinosa-Araujo. 2023. Damage to *Sorubim cuspicadus* sperm cryopreserved with ethylene glycol. Animal 13: 235.

Blaxter, J.H.S. 1953. Sperm storage and cross-fertilization of spring and autumn spawning herring. Nature 172: 1189–1190.

Borges, A.M., K.O. Araújo, I. Pivato and R.D. Navarro. 2020. Ultraestrutura e criopreservação de sêmen de jundiá amazônico (*Leiarius marmoratus*) em cativeiro. Arq. Bras. Med. Vet. Zootec. 72: 253–262.

Carneiro, P.C.F., H.C. Azevedo, J.P. Santos and A.N. Maria. 2012. Cryopreservation of tambaqui (*Colossoma macropomum*) semen: extender, cryoprotectants, dilution ratios and freezing methods. Cryo-Lett. 33: 385–393.

Carolsfeld, J., H.G. Godinho, E. Zaniboni Filho and B.J. Harvey. 2003. Cryopreservation of sperm in Brazilian migratory fish conservation. J. Fish Biol. 63: 472–489.

Carvalho, A.F.S., M.R.F. Machado, E.S. Andrade, L.D.S. Murgas, M.G. Zangeronimo, T.C. Barros et al. 2014. Effect of caffeine added to the activating solution on sperm motility of fresh and thawed semen of pacu, *Piaractus mesopotamicus*, and curimba, *Prochilodus lineatus*. J. World Aquac. Soc. 45: 75–81.

Corso, M.N., L.S. Marques, L.F.G. Gracia, R.B. Rodrigues, L.J.G. Barcellos and D.P. Streit Jr. 2019 Effects of different doses of eugenol on plasma cortisol levels and the quality of fresh and frozen-thawed sperm in South American catfish (*Rhamdia quelen*). Theriogenology 125: 135–139.

Coser, A.M., H.P. Godinho and D. Ribeiro. 1984. Cryogenic preservation of spermatozoa from *Prochilodus scrofa* and *Salminus maxillosus*. Aquaculture 37: 387–390.

Coser, A.M.L., H.P. Godinho and V.C. Torquato. 1987. Criopreservação do sêmen do peixe Piau *Leporinus silvestrii* (Boulenger, 1902). Arq. Bras. Med. Vet. Zootec. 98: 37–42.

Cruz-Casallas, P.E., S.C. Pardo-Carrasco, J.A. Arias-Castellanos, P.E. Lombo-Castellanos, D.A. Lombo-Rodriguez and J.E. Pardo-Mariño. 2004. Cryopreservation of yamu *Brycon siebenthalae* milt. J. World Aquac. Soc. 35: 529–535.

Cruz-Casallas, P.E., V.M. Medina-Robles and Y.M. Velasco-Santamaria. 2006. Evaluación de diferentes crioprotectores para la crioconservación de espermatozoides de yamú (*Brycon amazonicus*). Ver. Colomb. Cienc. Pec. 19: 152–159.

Da Costa, B.B., L.S. Marques, P.G. Lassen, R.B. Rodrigues and D.P. Streit Jr. 2019a. Cryopreservation-induced morphological changes in the sperm of South American silver catfish (*Rhamdia quelen*). J. Appl. Ichthyol. 35: 987–993.

Da Costa, B.B., L.S. Marques, P.G. Lassen, R.B. Rodrigues, H.T.R. Silva, J.C.F. Moreira et al. 2019b. Effects of cysteine supplementation on the quality of cryopreserved sperm of South American silver catfish. Aquac. Res. 51: 455–464.

Da Costa, B.B., L.S. Marques, P.G. Lassen, R.B. Rodrigues, H.T. Da Rosa-Silva, J.C.F. Moreira et al. 2021. Effect of glutamine and cysteine supplementation on quality of cryopreserved sperm of South American silver catfish. Aquac. Res. 52: 2173–2181.

Di Chiacchio, I.M., I.L.G. Almeida, M.C. Leal and A.T.M. Viveiros. 2017. Sperm quality and its freezing ability throughout the spawning season in *Prochilodus lineatus* and *Brycon orbignyanus*. Theriogenology 90: 284–288.

Do Nascimento, R.V.L., L.V. Leite-Castro, A.R. Montenegro, M.S. Oliveira-Araújo, J.T. Lopes, P.S. Almeida-Monteiro et al. 2017. Influência do tempo de resfriamento e soluções diluentes sobre a congelabilidade do sêmen de *Prochilodus brevis*. Acta Sci. Vet. 45: 1–9.

Duarte-Trujillo, A.S., M.B. Ardila-Artunduaga, D.N. Guaje-Ramírez and V.M. Medina-Robles. 2021. Cryoprotective effect of egg yolk in semen of *Prochilodus mariae* (Characiformes: Prochilodontidae). Biotec. Sector Agropec. Agroindustrial 19: 191–205.

Escobar, L., M.D., R.P. Ota, A. Machado-Allison, I.P. Farias and T. Hrbek. 2019. A new species of *Piaractus* (Characiformes: Serrasalmidae) from the Orinoco Basin with a redescription of *Piaractus brachypomus*. J. Fish Biol. 95: 411–427.

Farias, J.O., J.F. Nunes, M.A.M. Carvalho and C.C.M. Salgueiro. 1999. Avaliação "*in vitro*" e "*in vivo*" do sêmen de Tambaqui (*Colossoma macropomum*) conservado a temperatura ambiente e criopreservado em água de coco. Rev. Cient. Prod. Animal 1: 44–58.

Felizardo, V.O., R.A. Mello, L.D.S. Murgas, E.S. Andrade, M.M. Drumond and P.V. Rosa. 2010. Effect of cryopreservant combinations on the motility and morphology of curimba (*Prochilodus lineatus*) sperm. Anim. Reprod. Sci. 122: 259–263.

Felizardo, V.O., L.D.S. Murgas, R.D. Navarro, A.C.S. Gonçalves and M.S. Paulino. 2011. Osmolaridade e taxa de diluição na ativação do sêmen criopreservado de *Prochilodus lineatus*. Arch. Zootec. 60: 1255–1262.

Fogli da Silveira, W., E.T. Kavamoto and M.Y. Narabara. 1981. Avaliação quali-quantitativa e criopreservação em forma de pellets do sêmen do bagre *Rhamdia hilarii*. Ciên. Cult. 33: 620.

Fogli da Silveira, W., E.T. Kavamoto, M.G. Rigolino, L.A. Penteado and A.C. Carvalho Filho. 1985. Primeiros resultados de fertilização com sêmen congelado de truta arco-íris, *Salmo irideus* Gibbons, no Brasil. Bol. Inst. Pesca 11: 131–136.

Fogli da Silveira, W., E.T. Kavamoto, M.A. Cestarolli, H.P. Godinho, S.M. Ramos and A.N. Silveira. 1990. Avaliação espermática, preservação criogênica e fertilidade do sêmen do pacu, *Piaractus mesopotamicus* (HOLMBERG, 1887), proveniente de reprodução induzida. Bol. Inst. Pesca 17: 1–13.

França, T.S., I.C. Gomes, E.A. Sanches, M. Pérez-Atehortúa, N.S. Teixeira, R.B. Rodrigues et al. 2022. Post-thaw dilution of *Rhamdia quelen* sperm improves the reproductive success. Anim. Reprod. Sci. 243: 107018.

Freitas, L.A.C.M. 2010. Diversidade genética de Dourado (*Salminus brasiliensis* Curvier, 1816), uma espécie de grande interesse comercial no Pantanal Mato-Grossense. Ph.D. Thesis, Federal University of São Carlos, São Carlos, Brazil.

Fresneda, A., G. Lenis, E. Agudelo and M. Olivera-Ángel. 2004. Espermiación inducida y crioconservación de semen de cachama blanca (*Piaractus brachypomus*). Rev. Colomb. Cienc. Pec. 17: 46–52.

Fricke, R., W.N. Eschmeyer and R. Van der Laan. 2019. Eschmeyer's catalog of fishes: genera, species, references [Internet]. San Francisco: California Academy of Sciences.

Gallego, V., S.S. Cavalcante, R.Y. Fujimoto, P.C.F. Carneiro, H.C. Azevedo and A.N. Maria. 2017. Fish sperm subpopulations: changes after cryopreservation process and relationship with fertilization success in tambaqui (*Colossoma macropomum*). Theriogenology 87: 16–24.

Galo, J.M., D.P. Streit Jr., R.N. Sirol, R.P. Ribeiro, M. Digmayer, V.X.L. Andrade et al. 2011. Spermatic abnormalities of piracanjuba *Brycon orbignyanus* (Valenciennes, 1849) after cryopreservation. Braz. J. Biol. Sci. 71: 693–699.

Galo, J.M., D.P. Streit Jr., C.A. Oliveira, J.A. Povh, D.C. Fornari, M. Digmayer et al. 2019a. Quality of fresh and cryopreserved semen and their influence on the rates of fertilization, hatching and quality of the larvae of *Piaractus mesopotamicus*. Braz. J. Biol. Sci. 79: 438–445.

Galo, J.M., D.P. Streit Jr., C.D. Corcini, A.S. Varela Jr, R.D. Jardim and R.P. Ribeiro. 2019b. Protein profile as a quality indicator of cryopreserved semen from Tambaqui *Colossoma macropomum* (Cuvier, 1818). Braz. J. Biol. 80: 752–762.

Garcia, R.R.F., A.C.N. Vasconcelos, J.A. Povh, E.R. Oberst, A.S. Varela Jr., C.D. Corcini et al. 2015. Functional integrity of *Colossoma macropomum* (Cuvier, 1816) sperm cryopreserved with enriched extender solutions. Neotrop. Ichthyol. 13: 599–606.

Gheller, S.M.M., C.D. Corcini, C.R.C. de Brito, I.B. Acosta, G.C. Tavares, S.L. Soares et al. 2019. Use of trehalose in the semen cryopreservation of Amazonian catfish *Leiarius marmoratus*. Cryobiology 87: 74–77.

Godinho, H.M., N.A. Fenerich, M.Y. Narahara and J.M.B. Barker. 1977. Sobre a reprodução induzida do pacu *Cobeeoma mitrei* (Berg, 1895). Ciênc. Cult. 29: 796–797.

Goes, M.D., E.S.R. Goes, R.P. Ribeiro, N.M. Lopera-Barrero, P.L. Castro, T.S. Bignotto et al. 2017. Natural and artificial spawning strategies with fresh and cryopreserved semen in *Rhamdia quelen*: Reproductive parameters and genetic variability of offspring. Theriogenology 88: 254–263.

Herrera-Cruz, E.E., J.L. Aristizabal-Regino, J.A. Yepes-Blandón, A.L. Estrada-Posada, J.A. Espinosa-Araujo and V.J. Atencio-García. 2019. Criopreservación de semen de bagre rayado *Pseudoplatystoma magdaleniatum* con tres diferentes crioprotectores. Rev. Colomb. Biotecnol. 21: 55–62.

IBGE - Instituto Brasileiro de Geografia e Estatística. 2023. Available at https://sidra.ibge.gov.br/home/ipca/brasil Accessed in: January 2024.

ICMBIO - Instituto Chico Mendes de Conservação da Biodiversidade. 2018. Livro Vermelho da Fauna Brasileira Ameaçada de Extinção: Volume VII - Invertebrados. ICMBio/MMA, Brasília, Brasil.

Kavamoto, E.T., W. Fogli da Silveira, H.M. Godinho and E. Romagosa. 1989. Fertilização em *Prochilodus scrofa* Steindachner 1881, com sêmen criopreservado em nitrogênio líquido. Bol. Inst. Pesca 16: 29–36.

Krause, R.A., G. Neumann, A.C.O. Pedreira, M.A. Syperreck, A.M. Malacarne, M.N.F. Braga et al. 2021. *Rhamdia quelen*'s semen pre-cryopreservation processing and sperm viability time after thawing. Braz. J. Dev. 7: 43131–43147.

Lenz, D.R., M.C.C. Lima, T.F. Prado, A.M. Victorio, A.G. de Paula, M.L.G. Meirinhos et al. 2018. Avaliação de diferentes crioprotetores e tempos de descongelamento de sêmen de tambaqui *Colossoma macropomum*. Rev. Bras. Reprod. Anim. 42: 36–41.

Lopes, J.T., J.P.S. Pinheiro, L.T. Nunes, R.R.R. Pinheiro, M.E.M. Souza, O.S. Almeida et al. 2015. Evaluation of different crioprotectants and dilution ratios on *Prochilodus brevis* sperm cryopreservation. Rev. Bras. Reprod. Anim. 38: 170.

Lopes, J.T., M.S. Oliveira-Araújo, R.V. do Nascimento, Y.M. Ferreira, A.R. Montenegro and C.S.B. Salmito-Vanderley. 2018. Efeito de vitaminas e aminoácidos como suplementação da solução crioprotetora para a criopreservação do sêmen de tambaqui (*Colossoma macropomum*). Acta Sci. Vet. 46: 1–8.

López, D.I., M.C. Leal and A.T.M. Viveiros. 2015. Extender composition and osmolality affects post-thaw motility and velocities of piracanjuba *Brycon orbignyanus* (Valenciennes, 1850) (Characiformes) sperm. J. Appl. Ichthyol. 31: 114–118.

Maria, A.N., A.T.M. Viveiros, R.T.F. Freitas and A.V. Oliveira. 2006a. Extenders and cryoprotectants for cooling and freezing of piracanjuba (*Brycon orbignyanus*) semen, an endangered Brazilian teleost fish. Aquaculture 260: 298–306.

Maria, A.N., A.T.M. Viveiros, L.H. Orfão, A.V. Oliveira and G.F. Moraes. 2006b. Effects of cooling and freezing on sperm motility of the endangered fish piracanjuba *Brycon orbignyanus* (Characiformes, Characidae). Anim. Reprod. 3: 55–60.

Maria, A.N., A.C.M. Carvalho, R.V. Araújo, J.P. Santos, P.C.F. Carneiro and H.C. Azevedo. 2015. Use of cryotubes for the cryopreservation of tambaqui fish semen (*Colossoma macropomum*). Cryobiology 70: 109–114.

Marques, L.S., L.F.G. Gracia, M. Pérez-Atehortúa, R.B. Rodrigues, N.T. dos Santos, E.R. Oberst et al. 2021. Urine, feces, and blood contamination of frozen and fresh tambaqui (*Colossoma macropomum* Cuvier, 1818) sperm. Cryobiology 102: 121–126.

Martínez, J.G., V.A. García and S.P. Pardo-Carrasco. 2012a. DNA fragmentation and membrane damage of bocachico *Prochilodus magdalenae* (Ostariophysi: Prochilodontidae) sperm following cryopreservation with dimethylsulfoxide and glucose. Neotrop. Ichthyol. 10: 577–586.

Martínez, J.G., A.M. Tarazona-Morales and S.C. Pardo-Carrasco. 2012b. Sperm cryopreservation of freshwater fish bocachico (*Prochilodus magdalenae*) in DMSO and glucose and its effects on fertilization and hatching efficiency. Anim. Reprod. 9: 19–26.

Martínez, J.G. and S.C. Pardo-Carrasco. 2013. Effect of freezing and thawing rates on sperm motility in Bocachico *Prochilodus magdalenae* (Pisces, Characiformes). Rev. MVZ Cordoba 18: 3295–3303.

Martínez-Páramo, S., Á. Horváth, C. Labbé, T. Zhang, V. Robles, P. Herráez et al. 2017. Cryobanking of aquatic species. Aquaculture 472: 156–177.

Medina-Robles, V.M., Y.M. Velasco-Santamaria and P.E. Cruz-Casallas. 2007. Effects of packaged volume on freezing and thawing rates and the fertility of cryopreserved sperm of yamú (*Brycon amazonicus*). Arch. Med. Vet. 39: 229–237.

Medina-Robles, V.M., D.N. Guaje-Ramírez, L.C. Marin-Cossio, L.Y. Sandoval-Vargas and P.E. Cruz-Casallas. 2019. Crioconservação seminal de *Colossoma macropomum* como estratégia de produção e conservação na Orinoquia Colombiana. Orinoquia 23: 15–24.

Medina-Robles, V.M., L.Y. Sandoval-Vargas, D. Guaje-Ramírez, L.C. Marín-Cossio, I. Valdebenito-Isler and P.E. Cruz-Casallas, P.E. 2021. Cryopreservation of coporo (*Prochilodus mariae*) milt using three permeating cryoprotectant agents and two freezing systems. Aquaculture Res. 52: 6760–6769.

Medina-Robles, V.M., R.O. Suárez-Martínez, B. Baldisserotto and P.E. Cruz-Casallas. 2023a. Cryopreserved semen of *Piaractus orinoquensis* (Serrasalmidae): Post-thaw storage times and activating solutions. Acta Biol. Colomb. 28: 86–94.

Medina-Robles, V.M., L.Y. Sandoval-Vargas, R.O. Suárez-Martínez, E. Gómez-Ramírez, D.N. Guaje-Ramírez and P.E. Cruz-Casallas. 2023b. Cryostorage of white cachama (*Piaractus orinoquensis*) sperm: Effects on cellular, biochemical and ultrastructural parameters. Aquac. Rep. 29: 101477.

Melo, F.C.S.A. and H.P. Godinho. 2006. A protocol for cryopreservation of spermatozoa of the fish *Brycon orthotaenia*. Anim. Reprod. 3: 380–385.

Melo-Maciel, M., L. Leite-Castro, J. Leite, M. Oliveira, P. Almeida-Monteiro, J. Nunes et al. 2015. Aloe vera na criopreservação do sêmen de tambaqui (*Colossoma macropomum*). Arq. Bras. Med. Vet. Zootec. 67: 945–949.

Mello, F., J.S. Garcia, A. Depince, C. Labbé and D.P. Streit Jr. 2016. DNA Isolation from fresh and cryopreserved semen of Tambaqui (*Colossoma macropomum*). J. FisheriesSciences.com 10: 21–24.

Mello, F., J.S. Garcia, L.C. Godoy, A. Depince, C. Labbé and D.P. Streit Jr. 2017. The effect of cryoprotectant agents on DNA methylation patterns and progeny development in the spermatozoa of *Colossoma macropomum*. Gen. Comp. Endocrinol. 245: 94–101.

Menezes, J.T.B., L.J. Queiroz, C.D.R. Costa and J.R.J.B. Menezes. 2008. Avaliação espermática pós-descongelamento em tambaqui, *Colossoma macropomum* (Cuvier, 1818). Acta Amaz. 38: 365–368.

Miliorini, A.B., L.D.S. Murgas, P.V. Rosa, G. Oberlender, G.J.M. Pereira and D.V. Costa. 2011. A morphological classification proposal for curimba (*Prochilodus lineatus*) sperm damages after cryopreservation. Aquac. Res. 42: 177–187.

Monteiro, P.S.A., M.S. Oliveira-Araújo, R.R.R. Pinheiro, J.T. Lopes, Y.M. Ferreira, A.R. Montenegro et al. 2017. Influence of vitamins C and E on the quality of cryopreserved semen *Prochilodus brevis* (Prochilodontidae, Teleostei). Semina: Ciênc. Agrár. 38: 2669–2680.

Motta, N.C., R.C. Egger, K.S. Monteiro, A.V. de Oliveira and L.D.S. Murgas. 2022. Effects of melatonin supplementation on the quality of cryopreserved sperm in the neotropical fish *Prochilodus lineatus*. Theriogenology 179: 14–21.

Murgas, L.D.S., A. Gualhanone, M.O.B. Silva, C.B. Mello, R.T.F Freitas and M.G. Zangeronimo. 2001. Calidad seminal del pez piracanjuba (*Brycon orbignyanus*) post-descongelación. An. Vet. Murcia 17: 3–10.

Murgas, L.D.S., R.T. Franciscatto and A.G.O. Santos. 2003. Avaliação Espermática Pós-Descongelamento em Piracanjuba (*Brycon orbignyanus*, Vallenciennes, 1849). R. Bras. Zootec 32: 1810–1814.

Murgas, L.D.S., A.B. Miliorini, R.T.F. Freitas and G.J.M. Pereira. 2007. Criopreservação do sêmen de curimba (*Prochilodus lineatus*) mediante adição de diferentes diluidores, ativadores e crioprotetores. R. Bras. Zootec. 36: 526–531.

Nascimento, A.F., A.N. Maria, N.O. Pessoa, M.A.M. Carvalho and A.T.M. Viveiros. 2010. Out-of-season sperm cryopreserved in different media of the Amazonian freshwater fish pirapitinga (*Piaractus brachypomus*). Anim. Reprod. Sci. 118: 324–329.

Nascimento, A.F., A.C.S. Gonçalves, R.V. Reis Neto, M.C. Leal and A.T.M. Viveiros. 2012. Extender composition, osmolality, cryoprotectant and equilibration time effects on fresh sperm motility of two Characiformes fish: piracanjuba (*Brycon orbignyanus*) and streaked prochilod (*Prochilodus lineatus*). Anim. Reprod. 9: 103–110.

Nascimento, N.F., M. Pereira-Santos, L.H. Piva, B. Manzini, T. Fujimoto, J.A. Senhorini et al. 2017. Growth, fatty acid composition, and reproductive parameters of diploid and triploid yellowtail tetra *Astyanax altiparanae*. Aquaculture 471: 163–171.

Navarro, O.J., Y.M. Velasco-Santamaría and P. Cruz-Casallas. 2004. Evaluación de cinco crioprotectores para la crioconservación de semen de cachama blanca (*Piaractus brachypomus*). Rev. Colomb. Cienc. Pec. 17: 53–59.

Navarro, R.D., F.K.S.P. Navarro, V.O. Felizardo, L.D.S. Murgas and E.S. Andrade. 2014. Semen quality of Curimba (*Prochilodus lineatus*) cryopreserved with vitamins. Acta Sci. Technol. 36: 55–60.

Nelson, J.S. 2016. Fishes of the World. John Wiley, New York.

Neumann, G., P.V. Sanches and R.A. Bombardelli. 2019. Effects on fertility of motile sperm to egg ratio with use of cryopreserved *Rhamdia quelen* semen at different post-activation times. Anim. Reprod. Sci. 201: 84–92.

Ninhaus-Silveira, A., F. Foresti, R. Veríssimo-Silveira and J.A. Senhorini. 2006. Seminal analysis, cryogenic preservation, and fertility in matrinxã fish, *Brycon cephalus* (Günther, 1869). Braz. Arch. Biol. Technol. 49: 651–659.

Nunes, L.T., M.S. Oliveira, J.T. Lopes, M.E.M. Souza, R.R.R. Pinheiro, C.C. Campello et al. 2016. Cryopreservation of *Prochilodus brevis* semen: freezing media and thawing rates. Semina: Ciênc. Agrár. 37: 1643–1654.

Oliveira, A.V., A.T.M. Viveiros, A.N. Maria, R.T.F. Freitas and Z.A. Izaú. 2007. Sucesso do resfriamento e congelamento de sêmen de pirapitinga *Brycon nattereri*. Arq. Bras. Med. Vet. Zootec. 59: 1509–1515.

Oliveira, M.S., P.S. Almeida-Monteiro, L.T. Nunes, F.R.A. Linhares, J.P.S. Pinheiro, R.R.R. Pinheiro et al. 2016. Cryopreservation of tambaqui semen using a dry shipper and a programmed freezing machine. Semina: Ciênc. Agrár. 37: 2167–2180.

Pardo-Carrasco, M.V.Z., J. Salas-Villalva, L. Reza-Gaviria, J. Espinosa-Araujo and V. Atencio-Garcia. 2015. Cryopreservation of Trans-Andean shovelnose catfish (*Sorubim cuspicaudus*) semen using dimethylacetamide. Rev. CES Med. Vet. Zootec. 10: 122–131.

Paula, D.A.J., E.S. Andrade, L.D.S. Murgas, V.O. Felizardo, E.U. Winkaler, W. Zeviani et al. 2012. Vitamin E and reduced glutathione in *Prochilodus lineatus* (curimba) semen cryopreservation (Characiformes: Prochilodontidae). Neotrop. Ichthyol. 10: 661–665.

Paula, D.A.J., L.D.S. Murgas, T.F.D. Castro, I.L. Assis, R.V. Reis Neto and S. Marcussi. 2019. Effects of cooling rates on the quality of *Prochilodus lineatus* (Valenciennes, 1836) sperm. Reprod. Domest. Anim. 54: 1034–1043.

Paulino, M.S., L.D.S. Murgas, V.O. Felizardo and R.T.F. Freitas. 2012. Anormalidades espermáticas de *Piaractus mesopotamicus* após descongelamento utilizando diferentes metodologias. Arq. Bras. Med. Vet. Zootec. 64: 1591–1596.

Pereira, F.A., C.D. Corcini, A.C. da Silva, D.M. Pires, J.R. Pereira, S.M.M. Gheller et al. 2018. Reduced Glutathione and ATP in the seminal cryopreservation of tambaqui. Cryo-Lett. 39: 371–379.

Pereira, J.R., F.A. Pereira, C.T. Perry, D.M. Pires, J.R.E. Muelbert, J.R.E. Garcia et al. 2019. Dimethylsulfoxide, methanol and methylglycol in the seminal cryopreservation of Suruvi, *Steindachneridion scriptum*. Anim. Reprod. Sci. 200: 7–13.

Pereira, V.A., D.B. de Alencar, I.W.F. de Araújo, J.A.G. Rodrigues, J.T. Lopes, L.T. Nunes et al. 2020. Supplementation of cryodiluent media with seaweed or Nile tilapia skin sulfated polysaccharides for freezing of *Colossoma macropomum* (Characiformes: Serrasalmidae) semen. Aquaculture 528: 735553.

Pérez-Atehortúa, M., A.G. Galuppo, R.B. Rodrigues, N.S. Teixeira, T.R. de Freitas, T.S. França et al. 2021. Sperm selection of cryopreserved milt of catfish (*Rhamdia quelen*) by density gradient centrifugation with AllGrad® 90%. MethodsX 8: 101422.

Pérez-Atehortúa, M., A.G. Galuppo, R.B. Rodrigues, T.S. França, N.S. Teixeira, T.R. de Freitas et al. 2022. The use of differential separation and density gradient with AllGrad 90% after thawing improves the sperm quality of South American catfish (*Rhamdia quelen*). Aquaculture 553: 738072.

Perry, C.T., C.D. Corcini, A.N. Anciuti, M.V. Otte, S.L. Soares, J.R.E. Garcia et al. 2019. Amides as cryoprotectants for the freezing of *Brycon orbignyanus* sperm. Aquaculture 508: 90–97.

Pinheiro, J.P.S., L.V. Leite-Castro, F.C.E. de Oliveira, F.R.A. Linhares, J.T. Lopes and C.S.B. Salmito-Vanderley. 2016a. Quality of tambaqui (*Colossoma macropomum*) semen cryopreserved in different concentrations of egg yolk. Ciênc. Anim. Bras. 17: 267–273.

Pinheiro, J.P.S., M.A.P. Melo-Maciel, F.R.A. Linhares, J.T. Lopes, P.S. Almeida-Monteiro, R.R.R. Pinheiro et al. 2016b. Use of glucose or BTS®TM combined with DMSO or methylglycol under two different freezing protocols for the cryopreservation of sperm from the common curimatã (*Prochilodus brevis*). Anim. Reprod. 13: 779–786.

Pires, D.M., C.D. Corcini, A.C.D. Silva, S.M.M. Gheller, F.A. Pereira, J.R. Pereira et al. 2018. Association between DMSO and Sugars in the sperm cryopreservation of pacu. Cryo-Lett. 39: 121–130.

Polge, C., A.U. Smith and A.S. Parkes. 1949. Revival of spermatozoa after vitrification and dehydration at low temperatures. Nature 164: 4172.

Ramirez-Merlano, J.A., Y.M. Velasco-Santamaría, V.M. Medina-Robles and P.E. Cruz-Casallas. 2011. Cryopreservation effects on the sperm quality of cachama blanca *Piaractus brachypomus* (Cuvier 1818). Aquac. Res. 42: 738–745.

Reis, R.E., J.S. Albert, F. Di Dario, M.M. Mincarone, P. Petry and L.A. Rocha. 2016. Fish biodiversity and conservation in South America. J. Fish Biol. 89: 12–47.

Restrepo-Betancur, G., J.D.M. Páez and L.A. Chacón. 2017. Evaluación de dos crioprotectores y tres curvas de congelación programable en la criopreservación de semen de *Brycon henni* (Pisces: Characidae). Rev. Investig. Vet. Peru 28: 597–605.

Ribeiro, R.I.M.A. and H.P. Godinho. 2003. Criopreservação do sêmen testicular do teleósteo piauçu *Leporinus macrocephalus*. Arq. Bras. Med. Vet. Zootec. 55: 75–79.

Stein, H. and H. Bayrle. 1978. Cryopreservation of sperm of some freshwater teleosts. Ann. Biol. Anim. Biochim. Biophys. 18: 1073–1076.

Streit Jr., D.P., C. Benites, G.V. de Moraes, R.P. Ribeiro, E.S. Sakaguti and R.F. Caldieri. 2006. Sêmen de pacu (*Piaractus mesopotamicus*) criopreservado com diluentes utilizados para sêmen de suínos. Ciênc. Anim. Bras. 7: 289–297.

Streit Jr., D.P., A.C. de Oliveira, R.P. Ribeiro, R.N. Sirol, G.V. de Moraes, J.M. Galo et al. 2009. Motilidade, vigor e patologias seminal in natura e pós criopreservação de *Piaractus mesopotamicus*. Bol. Inst. Pesca. 35: 159–167.

Suquet, M., C. Dreanno, C. Fauvel, J. Cosson and R. Billard. 2000. Cryopreservation of sperm in marine fish. Aquac. Res. 31: 231–243.

Taitson, P.F., E. Chami and H.P. Godinho. 2008. Gene banking of the neotropical fish *Leporinus obtusidens* (Valenciennes, 1836): a protocol to freeze its sperm in the field. Anim. Reprod. Sci. 105: 283–291.

Teodozia, E.R.P., M.S.O. Araújo, J.T. Lopes, A.R. Montenegro and C.S.B. Salmito-Vanderley. 2020. Semen cryopreservation in *Piaractus mesopotamicus*: The effect of different diluents and freezing equipment. Rev. Ciênc. Agronômica 51: e20196754.

Torres, T.M., P.S. Almeida-Monteiro, R.V. Nascimento, V.A. Pereira, Y.M. Ferreira, J.S. Lobato et al. 2022. Sperm cryopreservation of *Prochilodus brevis* using different concentrations of non-permeable cryoprotectants. Anim. Reprod. 19: e20210083.

Truscott, B., D.R. Idler, R.J. Hoyle and H.C. Freeman. 1968. Sub-zero preservation of Atlantic salmon sperm. J. Fish. Res. Board Can. 25: 363–372.

Tsai, S. and C. Lin. 2012. Advantages and applications of cryopreservation in fisheries science. Braz. Arch. Biol. Technol. 55: 425–434.

Varela Jr., A.S., C.D. Corcini, D.P. Streit Jr., G. Rizzoto, R.D. Jardim, T. Lucia Jr. et al. 2012a. Efeito crioprotetor de diferentes concentrações do dimetilsulfóxido no congelamento de sêmen de Tambaqui *Colossoma macropomum*. Atlântica 34: 129–137.

Varela Jr., A.S., C.D. Corcini, S.M.M. Gheller, R.D. Jardim, T. Lucia Jr. and D.P. Streit Jr. 2012b. Use of amides as cryoprotectants in extenders for frozen sperm of tambaqui, *Colossoma macropomum*. Theriogenology 78: 244–251.

Varela Jr., A.S., E.F. Silva, T.F. Cardoso, E.Y. Namba, R.D. Jardim, D.P. Streit Jr. et al. 2015a. The role of dimethyl sulfoxide in the cryopreservation of Curimba (*Prochilodus lineatus*) semen. Semina: Ciênc. Agrár. 36: 3471–3480.

Varela Jr, A.S., K.L. Goularte, J.P. Alves, F.A. Pereira, E.F. Silva, T.F. Cardoso et al. 2015b. Methods of cryopreservation of Tambaqui semen, *Colossoma macropomum*. Anim. Reprod. Sci. 157: 71–77.

Vasconcelos, A.C.N., V.O. Felizardo, A.F.S. de Carvalho, R.R.F. Garcia, S.E. Ramos and L.D.S. Murgas. 2015. Cryopreservation of *Prochilodus lineatus* semen: effect of cryoprotectants combination. Bol. Inst. Pesca 41: 817–824.

Velasco-Santamaría, Y.M., V.M. Medina-Robles and P.E. Cruz-Casallas. 2006. Cryopreservation of yamú (*Brycon amazonicus*) sperm for large scale fertilization. Aquaculture 256: 264–271.

Velarde, J.M.C., N.M. Bastos, L. Carneiro-Leite, L.P. Borges, E.G. Vieira, R. Veríssimo-Silveira et al. 2023. Dimethyl acetamide and dimethyl sulfoxide associated at glucose and egg yolk for cryopreservation of *Pseudoplatystoma corruscans* semen. Neotrop. Ichthyol. 21: e220071.

Viveiros, A.T.M., A.N. Maria, L.H. Orfão, M.A.M. Carvalho and J.F. Nunes. 2008. Powder coconut water (ACP) as extender for semen cryopreservation of Brazilian migratory fish species. Cybium 32: 215.

Viveiros, A.T.M. and H.P. Godinho. 2009. Sperm quality and cryopreservation of Brazilian freshwater fish species: a review. Fish Physiol. Biochem. 35: 137–150.

Viveiros, A.T.M., L.H. Orfão, A.N. Maria and I.B. Allaman. 2009a. A simple, inexpensive and successful freezing method for curimba *Prochilodus lineatus* (Characiformes) semen. Anim. Reprod. Sci. 112: 293–300.

Viveiros, A.T.M., A.V. Oliveira, A.N. Maria, L.H. Orfão and J.C. Souza. 2009b. Sensibilidade dos espermatozoides de dourado (*Salminus brasiliensis*) a diferentes soluções crioprotetora. Arq. Bras. Med. Vet. Zootec. 61: 883–889.

Viveiros, A.T.M., A.F. Nascimento, L.H. Orfão and Z.A. Isaú. 2010. Motility and fertility of the subtropical freshwater fish streaked prochilod (*Prochilodus lineatus*) sperm cryopreserved in powdered coconut water. Theriogenology 74: 551–556.

Viveiros, A.T.M., T.B. Amaral, L.H. Orfão, Z.A. Isaú, D. Caneppele and M.C. Leal. 2011. Sperm cryopreservation of tiete tetra *Brycon insignis* (Characiformes): effects of cryoprotectants, extenders, thawing temperatures and activating agents on motility features. Aquac. Res. 42: 858–865.

Viveiros, A.T.M., A.N. Maria, T.B. Amaral, L.H. Orfão, Z.A. Isaú and R. Veríssimo-Silveira. 2012a. Spermatozoon ultrastructure and sperm cryopreservation of the Brazilian dry season spawner fish pirapitinga, *Brycon nattereri*. Aquac. Res. 43: 546–555.

Viveiros, A.T.M., L.H. Orfão, A.F. Nascimento, F.M. Corrêa and D. Caneppele. 2012b. Effects of extenders, cryoprotectants and freezing methods on sperm quality of the threatened Brazilian freshwater fish pirapitinga-do-sul *Brycon opalinus* (Characiformes). Theriogenology 78: 361–368.

Viveiros, A.T.M., Z.A. Isaú, D. Caneppele and M.C. Leal. 2012c. Sperm cryopreservation affects post-thaw motility, but not embryogenesis or larval growth in the Brazilian fish *Brycon insignis* (Characiformes). Theriogenology 78: 803–810.

Viveiros, A.T.M., T.R. Taffarel and M.C. Leal. 2014. Osmolality and composition of the extender during the cold storage of *Prochilodus lineatus* (Characiformes: Prochilodontidae) sperm. Neotrop. Ichthyol. 12: 643–648.

Viveiros, A.T.M., A.F. Nascimento, M.C. Leal, A.C.S. Gonçalves, L.H. Orfão and J. Cosson. 2015a. Methyl glycol, methanol and DMSO effects on post-thaw motility, velocities, membrane integrity and mitochondrial function of *Brycon orbignyanus* and *Prochilodus lineatus* (Characiformes) sperm. Fish Physiol. Biochem. 41: 193–201.

Viveiros, A.T.M., A.C.S. Gonçalves, A.F. Nascimento and M.C. Leal. 2015b. Fresh, equilibrated and post-thaw sperm quality of *Brycon orbignyanus* (Valenciennes, 1850) and *Prochilodus lineatus* (Valenciennes, 1837) treated with either salmon GnRHa and domperidone or pituitary extract. Neotrop. Ichthyol. 13: 157–164.

Viveiros, A.T.M. and M.C. Leal. 2016. Sperm dilution ratio affects post-thaw motility rate and velocity of *Prochilodus lineatus* (Characiformes) sperm. Zygote 24: 662–667.

Viveiros, A.T.M., M.C. Leal, T.S. França, I.L.G. Almeida and Z.A. Isaú. 2016. Osmolality and composition of the activating solution affects motility of fresh and frozen *Prochilodus lineatus* sperm differently. Anim. Reprod. Sci. 173: 73–79.

Viveiros, A.T.M., I.M. Di Chiacchio, I.L.G. Almeida, T.R. Taffarel and M.C. Leal. 2017. Storage and transportation of *Prochilodus lineatus* (Characiformes) sperm prior to cryopreservation. Gen. Comp. Endocrinol. 245: 84–88.

Viveiros, A.T.M., N.C. Motta, Z.A. Isaú, I.L.G. Almeida and M.C. Leal. 2019. Ions and osmolality on post-thaw sperm motility activation of the endangered *Brycon insignis* (Characiformes). J. Appl. Ichthyol. 35: 739–746.

Weingartner, M., A.C.V. Zanandrea and E. Zaniboni-Filho. 2015. Cryopreserved sperm for oocyte fertilization of dourado *Salminus brasiliensis*. Ciênc. Rural 45: 892–897.

Xavier, A.M.M., G. Neumann, E.A. Sanches, S.U. Cardoso and R.A. Bombardelli. 2021. Extenders with vitamins C and E applied to *Rhamdia quelen* sperm cryopreservation. Braz. J. Dev. 7: 118046–118060.

Zanandrea, A.C.V., M. Weingartner and E. Zaniboni-Filho. 2014. Use of ACP-104® at different dilutions for sperm cryopreservation of dourado, *Salminus brasiliensis*. J. World Aquac. Soc. 45: 82–87.

Zanandrea, A.C.V., M. Weingartner and E. Zaniboni-Filho. 2016. Induced reproduction of dourado (*Salminus brasiliensis*): fertilization with sperm cryopreserved in ACP-104®. Acta Sci. Anim. Sci. 38: 17–22.

CHAPTER 11

Stem Germ Cells in Fish, the Methods for Cryopreservation in Neotropical South American Species and Its Potential for Fish Conservation and Restoration

Saynara Souza dos Santos Miranda,[1,2] *Hadda Tercya,*[1,3] *Taiju Saito*[4] and *Diógenes Henrique Siqueira-Silva*[2,5,*]

INTRODUCTION

Of the vertebrate groups, fish are the predominant category in the aquatic environment, with over 35,000 described species (Fricke et al. 2020) and many others that have not yet been cataloged. Among them, the largest known freshwater ichthyofauna is located in the neotropics, which has approximately 5,160 species (Froese and Pauly 2015; Reis et al. 2016), of which more than 3,600 are concentrated

[1] Research Group of Studies on the Reproduction of Amazon fish (GERPA/LANEC), University of South and Southern of Pará, Marabá, Brazil.

[2] PPG in Animal Reproduction in the Amazon (ReproAmazon), University of Pará, Belém, 66075-110, Brazil.

[3] Instituto Politécnico de Bragança. Campus de Santa Apolónia, Bragança, 5300-253, Portugal.

[4] Uchidomari, Ainan, Minamiuwa-district, Ehime, 798-4206, Japan. South Ehime Fisheries Research Center (SEFREC), Ehime University.

[5] Research Group of Studies on the Reproduction of Amazon fish (GERPA/LANEC), University of South and Southern of Pará, Marabá, 68507-590.

Emails: narabiol@gmail.com, htercya@gmail.com, taiju76@gmail.com

* Corresponding author: diogenessilva@unifesspa.edu.br

in Brazil (ICMBIO/MMA 2018). The orders Siluriformes and Characiformes are those with the greatest diversity and distribution in the Neotropical region, with the Siluriformes, represented by catfishes, counting 1,915 species and Characifomes with 1,700 fish species (Castro and Vari 2003; Nelson 2006; Reis et al. 2016). Many of these are species of great socioeconomic importance for aquaculture worldwide, and some others are listed as threatened with extinction (Nakatani et al. 2001; Alves et al. 2014; Froese and Pauly 2015; ICMBIO/MMA 2018).

In spite of the great diversity of fish species, as well as the investments made in aquaculture production, several species have been found to be threatened with extinction, mainly due to anthropic actions such as habitat degradation, the introduction of exotic species, deforestation, and production of domestic, industrial and agricultural pollution (Agostinho et al. 2005; Miranda 2012; Tolussi et al. 2018). Such actions have intensified the increasing number of teleost species in the Red Book of Endangered Species, which includes 52 species of the order Characiformes and 91 species of the Siluriformes order (ICMBIO/MMA 2018).

Based on these factors responsible for compromising and decreasing fish biodiversity, biotechnologies have been developed to contribute to both the conservation and reconstitution of threatened species and to leverage the economic chain related to fish production. Thus, fish conservation strategies such as the creation of *ex-situ* gene banks through the cryopreservation of several cell types, such as primordial germ cells (Kobayashi et al. 2007; Higaki et al. 2010; Kawakami et al. 2010; Yasui et al. 2011), stem oogonia (Higaki et al. 2018), and stem spermatogonia (Lee and Yoshizaki 2016a; Franěk et al. 2019a) have been employed aiming the use in surrogate broodstock technique.

Surrogate broodstock has been proven to be an important tool not only in the preservation of endangered species (Orwig and Schlatt 2005; Lacerda et al. 2006b; Higuchi et al. 2011; Yoshizaki et al. 2011; Morita et al. 2012) but also the study of reproductive biology and regulatory mechanisms of gametogenesis (Nóbrega 2009; Schulz et al. 2010; Lacerda et al. 2012), such in the interaction Sertoli-germ cell (Lonnie et al. 2000), sexual plasticity of germ cells (Yoshizaki et al. 2010; Yoshizaki et al. 2012), as well as in the maintenance of economically important species (Okutsu et al. 2006b; Yazawa et al. 2013) and genetic material of improved animals, thus enabling the increase of genetic diversity in the short term (Lacerda et al. 2006b).

This technique is based on the removal of stem germ cells (PGCs, spermatogonia, or oogonia) from the gonad of a donor individual and their transfer to the blastoderm of an embryo (Saito et al. 2008) to the coelomic cavity of newly hatched larvae (Takeuchi et al. 2003; Okutsu et al. 2006a; Pšenička et al. 2015; Zhou et al. 2021) or to the gonad of a sterile adult recipient (Lacerda et al. 2010a; Nóbrega et al. 2010; Yoshizaki et al. 2010; Majhi et al. 2014; Silva et al. 2016; Siqueira-Silva et al. 2019). The surrogate broodstock technique can also be performed by PGC extraction from an embryo blastoderm and their subsequent transfer to a host embryo (Yamaha et al. 2003; Saito et al. 2010). Another method for surrogate broodstock, first performed by Nagler et al. (2001) and recently replicated by Kayo et al. (2022), concerns the allogeneic testis transplantation between adult individuals. In Nagler et al. (2001) research, the left testis from the host species was removed and, in its place, another testicle was transplanted. They observed the transplanted testis maturate and also produced viable and fertile sperm. Kayo et al. (2022), in turn, grafted part of the donor testis into a host from medaka (*Oryzias latipes*) partially castrated. Those animals were then submitted to natural mating, and the authors could follow the production of viable offspring derived from the sperm produced in the allografted testis for at least three months. Regardless of the performed method, in common, those techniques culminate to originate mature gametes with genetic characteristics from the donor species (Izadyar et al. 2000; MClean 2005; Lacerda 2006a; Nóbrega et al. 2009; Yoshizaki et al. 2019; Jin et al. 2021).

1. Surrogate Broodstock Approach in Fish

In fish, the surrogate broodstock approach was first reported in zebrafish (Lin et al. 1992). In this study, blastomeres of wild-type zebrafish were aspirated into a glass needle and then transplanted into the host embryos of albino zebrafish at the same developmental stage (BT). Donor-derived germ cell transmission to the next generation was evaluated by pigmentation in offspring. A similar approach has been applied in studies using medaka (Wakamatsu et al. 1993), trout (*Oncorhynchus mykiss*) (Takeuchi et al. 2001), loach (*Misgurnus anguillicaudatus*) (Nakagawa et al. 2002), and pikeperch (*Sander lucioperca*) (Güralp et al. 2016). However, the efficiency of donor germ cell transfer was low and varied among experiments: 11.9–17.9% in zebrafish, 29.9% in medaka, 31.6% in trout, and 25% in loach and pikeperch. However, Li et al. (2016) optimized the blastomeres transplantation technique in medaka by using PGC-boosted donors and sterile hosts, improving efficiency to nearly 100%. Their method has not been tested in other fish species yet.

Yoon et al. (1997) identified the location of PGCs in each embryonic stage in zebrafish by using whole-mount *in situ* hybridization (WISH) with Vasa transcripts (Yoon et al. 1997). This analysis demonstrated that PGCs are located at the marginal region of the blastodisc at the blastula stage. With this information, Yamaha et al. (2001) transferred the PGC-containing lower part of the blastodisc of a donor embryo into a host blastula-stage embryo (blastoderm transplantation [BdT]) to improve the efficiency of PGC transfer (Yamaha et al. 2001). The BdT approach demonstrated the production of donor-derived gametes from all analyzed chimeras. This report also showed the first evidence of xenogeneic gamete production in fish (between crusian carp, *Carassius*, and goldfish, *Carassius auratus*). Using the BdT technique with sterile hybrid recipients, Yamaha et al. (2003) confirmed mono-sex and donor-sole gamete production (Yamaha et al. 2003). In this study, goldfish PGC-containing blastula fragments (genetically female, XX) were transplanted into sterile male (XY) hybrid embryos from a cross between female (XX) goldfish and super-male (YY) common carp; donor-derived sperm was produced following maturation of the transplanted embryos. The donor goldfish PGCs (XX) underwent sex reversal and differentiated into sperm in the male hybrids.

However, the BT and BdT methodologies cannot be used to produce surrogate broodstock when the donor and host species are distantly related. Saito et al. (2010) demonstrated that zebrafish embryos transplanted with goldfish or loach blastomeres showed severe malformations because of donor cell aggregation; this effect may be due to differences in cell affinity between donor and host cells (Saito et al. 2010). To overcome this problem, Saito et al. (2008) transplanted a single PGC, which is visualized by injecting GFP-*nos3* 3'UTR mRNA into donor eggs into a sterilized recipient embryo under a fluorescence stereomicroscope (Saito et al. 2008). With this single PGC transplantation (SPT) approach, germline chimeras were produced between unrelated species—for example, pearl danio (*Danio albolineatus*) to zebrafish—and between different genera (goldfish to zebrafish) and families (loach to zebrafish). In all combinations, a transplanted PGC could migrate to the genital ridge of zebrafish embryos without inducing malformations. These xenogeneic germline chimeras also developed normally with high PGC transfer efficiency (more than 40%).

Moreover, donor-derived sperm was obtained in all combinations tested. The result showed that a single PGC was sufficient for gonad formation and gametogenesis when transplanted into a sterile host. Similar transplantation experiments have since been performed for many fish species combinations with various phylogenetic relationships (Saito et al. 2011, 2014). However, the methodologies have some limitations. BT, BdT, and SPT methods must be undertaken during a short window in embryonic development, and they can only be used for species for which embryonic manipulation techniques have been developed, such as model fish species (e.g., zebrafish, medaka, goldfish, and loach). In many fish species, dechorionation and embryonic manipulation are extremely difficult.

Takeuchi and colleagues (2004) developed an innovative approach to transplant PGCs that avoided the complexity of the above methods (Takeuchi et al. 2004). They isolated PGCs from the genital ridge of newly hatched trout embryos and transplanted the cells into the peritoneal cavities of host salmon embryos at the same stage as the donor. The transplanted donor germ cells colonized and proliferated in the developing gonads of recipients, and they obtained 13.5% female and 17% male normal germline chimeras that produced donor-derived gametes. This approach is called peritoneal cavity transplantation (PCT). Although the success rate of surrogate broodstock production using this method was not as high as that previously reported for BT, BdT, and SPT, PGC, transplantation into hatched larvae has considerable advantages, such as the high developmental rate after manipulation and the use of comparatively simple equipment and procedures, particularly in salmonid species that have larvae that are large enough for this type of manipulation. Subsequently, it was shown that spermatogonia and oogonia could be used as donor cell sources for PCT (Okutsu et al. 2006a; Wong et al. 2011). The significance of using gonial cells isolated from adult fish is the simplicity of obtaining these cells from adult gonads and the ease of use for various cell operations, such as cryopreservation and cell culture. Surprisingly, a transplantable quantity and quality of gonial cells can be isolated from frozen testes (Lee et al. 2013; Pšenička et al. 2016), frozen ovary (Lee et al. 2016b), and even whole frozen fish in a freezer (Lee et al. 2015).

Furthermore, recently, it has been shown that spermatogonia retrieved from a fish that has been dead for 12 hours in culture water can efficiently colonize and proliferate in the gonads of recipients after transplantation (Yang et al. 2022). From these advantages, spermatogonial or oogonial transplantation by PCT is considered the optimal strategy to produce germline chimeras for the surrogate broodstock technique. However, applying PCT can be technically very difficult for species with small and fragile larvae, such as many marine fish species. The survival rate from hatchling to larvae in marine species is generally low; the additional mortality induced by experimental manipulation means that for some species, the number of chimeras produced by PCT will be insufficient for the practical usage of surrogate production in aquaculture.

To overcome the problems posed by small larvae, Franca and colleagues developed a new technique in which donor germline cells are transplanted into the gonads of an adult fish via the genital pore. Genital pore transplantation (GPT) is a variation of the method for spermatogonial transplantation into seminiferous tubules that was established in mammals by Brinster and collaborators in the 1990s (Brinster and Zimmermann 1994). Lacerda et al. (2006a) isolated germ cells from the testes of tilapia and labeled them with the fluorescent dye PKH26. Recipient adult fish were treated with busulfan and high temperatures to induce the degeneration of endogenous germ cells prior to transplantation. The isolated tilapia germ cells were then transplanted into the spermatic ducts of the testis of the recipient via the genital papillae. Analysis of the recipient testes confirmed the proliferation of the transplanted germ cells. In a subsequent study, they performed GPT with spermatogonia from other tilapia species and obtained donor-derived offspring (6.3%; 2 out of 32) using semen from germline chimeras (Lacerda et al. 2010b). GPT has also been used in varied fish species (Nóbrega et al. 2010; Xu et al. 2020). The most significant advantage of this technique is the low level of physical damage to the recipient; additionally, the technique requires only simple equipment, such as a syringe. GPT can be conducted easily without intensive practice using micromanipulation devices, as in the case of other transplantation methods (Siqueira-Silva et al. 2018). Donor-derived gametes can be obtained in a comparatively short period; compared to other methods in which transplantation is performed in the embryonic or early developmental stages, the latter method needs to raise the chimera to adulthood to obtain gametes. However, GPT requires many isolated donor cells before transplantation, as adult recipient fish have large spaces in their gonads. In the tilapia model, 10^7 isolated gonial cells in 1 ml were injected into a single fish (Lacerda et al. 2010a), whereas 10^4 gonial cells were enough for inducing germline chimera in PCT in trout (Okutsu et al. 2006b), and only a single PGC is sufficient in SPT in zebrafish (Saito et al. 2008). It has also been shown that endogenous germ cells survive even after sterilization treatment in some adult recipients. The incomplete recipient preparation might be a disadvantage for GPT.

2. Host Sterilization for the Surrogate Broodstock

One of the most important bottlenecks for the surrogate broodstock technique, however, is the achievement of sterile hosts in order to both obtain gonads containing the germinal epithelium free of endogenous germ cells, opening available niches for donor germ cell colonization and proliferation, as well as avoid donor and endogenous germ cells competition, which, inevitably, ends up with donor germ cells elimination

(Shinohara et al. 2002; Brinster et al. 2003; Lacerda 2006a; Golpour et al. 2016). A batch of methods has been employed, differing according to the research objective.

The most used techniques for host sterilization are employed on eggs through chromosome manipulation techniques, such as triploidization (Okutsu et al. 2007; Adamov et al. 2017; Hattori et al. 2019; Franěk et al. 2019b; Zhou et al. 2021) and hybridization (Yamaha et al. 2003; Wong et al. 2011). In common, both techniques have the advantage to be applied on a large scale, since the triploidization involves thermal shock by cold (Piferrer et al. 2003) or heat water (Adamov et al. 2017; Sato et al. 2020), hydrostatic pressure shock (Xu et al. 2008), or chemical treatment (Johnstone et al. 1989) in the whole offspring, which may contain thousands of eggs at once. The hybridization may also be performed in the whole offspring as well (Kawamura et al. 2020). Both of the techniques show some disadvantages as well. The triploidization may not be effective to generate 100% of triploid animals, as shown by Sato et al. (2020) and Adamov et al. (2017) in the Neotropical fishes *Colossoma macropomum* and *A. altiparanae,* respectively.

In spite of hybrid sterility being a common phenomenon for several eukaryotic interspecies hybrids (Kawamura et al. 2020), it is not 100% sure for fish species, as in the case of some South American catfishes (Hashimoto et al. 2012). Martino (1994) proved hybrid females from the Neotropical Characiformes *Colossoma macropomum* and *Piaractus brachypomus* were able to produce mature oocytes, which could be fertilized and developed in normal offspring. Another huge problem with those techniques involves the absence of available niches for the donor germ cell colonization since, although infertility can be achieved, many hybrids still produce endogenous germ cells that occupy the germinal epithelium (Okutsu et al. 2007) and some others are even able to produce gametes (Nascimento et al. 2017) (Figure 1), therefore preventing the donor cells from colonizing the epithelium. In order to overcome this problem and achieve the so-called "ideal host" (Golpour et al. 2016), Piva and collaborators (2018) crossed some Neotropical tetra species. Following, the hybrid eggs were submitted to thermal shock, aiming for embryo triploidization. They obtained a hybrid triploid interspecies between females of *A. altiparanae* and males of *A. fasciatus* owning the germinal epithelium totally depleted of germ cells and containing intact Sertoli cells. This is the ideal gonad for the reception of donor-transplanted cells (Figure 1). However, just males were achieved in that study, and thus, since the somatic environment determines the production of gametes, only sperm could be produced using those animals as hosts.

Still using eggs, Saito and colleagues (2018) eliminated primordial germ cells (PGCs) in sturgeon embryos using ultraviolet (UV) irradiation. Despite the practice of this method, it is only possible in embryos whose PGCs are specified at the egg vegetal pole, such as in the case of sturgeon species. All the Neotropical species whose PGC specifications were studied showed their synthesis in the animal pole, which impossibilities the use of the technique because UV rays will damage the DNA of the cells from the developing embryos. In addition, this technique did not ensure 100% efficiency. Until this moment, the unique technique that ensures 100% efficiency in the production of sterile animals when eggs are treated is the knockdown of the dead-end gene (dnd), a vertebrate-specific gene that is crucial for the migration and

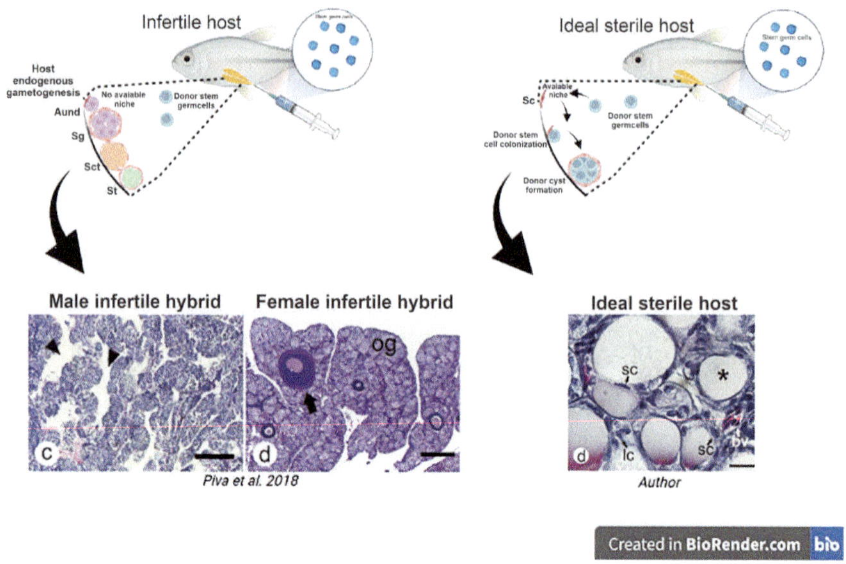

Figure 1. Schematic illustration showing an infertile host that does not produce sperm but is unable to receive donor stem germ cells by transplantation due to the presence of spermatogenic cysts into the seminiferous tubule (a–b), and the "ideal host" in which only Sertoli cells (SC) are present in the germinal epithelium, being, therefore, an ideal environment to receive donor stem germ cells through transplantation (c). Aund: a undifferentiated spermatogonia; bv: blood vessel; lc: Leydig cells; Og: Oogonia; Sc: Sertoli cells; Sct: Spermatocyte; Sg: Spermatogonia; St: Spermatid; Arrow: primary growth oocyte; arrowhead: opened lumen; Asterisk: empty tubular lumen.

survival of PGCS (Rushton et al. 2003). It can be performed using the morpholino oligonucleotide antisense (MO) (Fujimoto et al. 2010; Linhartová et al. 2015; Yoshizaki et al. 2016) or CRISPR/Cas9 (Balock et al. 2019). In spite of its efficiency, this sterilization method cannot be applied on a large scale since the embryos must be injected one by one. Furthermore, the availability of this method depends on the MO injection in embryos before they achieve the 2-cell cleavage stage, which is very short in Neotropical fish, such as in the Siluriformes *Pseudopimelodus mangurus* (40 minutes on average at room temperature) and *Pimelodus maculatus* (30 minutes on average at room temperature) (Arashiro et al. 2018).

Besides the techniques applied to eggs, sterilization can also be performed in adult fish. In this case, the most commonly employed method is performed by the association between the cytostatic chemotherapeutic drug busulfan with high water temperatures (Lacerda 2006; Lacerda et al. 2008, 2010a; Majhi et al. 2009b; Nóbrega et al. 2010; Siqueira-Silva et al. 2015; Majhi et al. 2017). Busulfan works by alkylating DNA bases during cell division, culminating with cell death. As elevated temperature accelerates fish metabolism, germ cell divisions occur more quickly, improving busulfan action in killing germ cells and, consequently, the depletion of gonads. Furthermore, Siqueira-Silva and collaborators (2015) described the high water temperature by itself could cause depletion of the testes in the Neotropical species *A. altiparanae*. This technique can also be applied on a large scale. However, it presents the disadvantage of the transient state of gonadal sterility since the stem spermatogonia, which stay quiescent during the treatment, are not affected and then proliferates to repopulate the depleted gonads (Siqueira-Silva et al. 2015).

Independently, the chosen germ cell transplantation technique and its approach allows a promising scenario in the area of biotechnology, genetic conservation, and production in aquaculture (Lacerda et al. 2006b; Siqueira-Silva et al. 2018; Goto and Saito 2019; Yoshizaki and Yzawa 2019). However, it is increasingly necessary to study and understand the process of gametogenesis from the donor and host species, including the role and physiology of the stem gem cells, to process the surrogate broodstock technology better.

3. Types of Germ Cells Used in the Surrogate Broodstock Technique

3.1 Primordial Germ Cells

The Primordial Germ Cells (PGCs) originate early in embryonic development, generally at a different location from where the gonad develops (Linhartová et al. 2014; Saito et al. 2014). Such distance indicates that after establishing a germ lineage, PGCs migrate from where they are initially specified toward the developing gonads (Raz 2003; Coelho et al. 2019; Grimaldi and Raz 2019). This migration process is in charge of the chemoattractive system SDF – 1a (stromal-derived factor) and their cell membrane receptor, the transmembrane proteins CXCR4 and CXCR7. The PGCs migration pathway is strictly related to the dynamic patterns of mRNA expression by the chemokine SDF-1a (Raz 2003; Raz et al. 2004; Raz and Reichman-Fried 2006; Boldajipour et al. 2008). Doitsidou et al. (2002) showed that the knockdown of the CXCR4 receptor activity and its ligand SDF-1a resulted in severe defects to the migration of zebrafish PGCs, clarifying that despite the PGCs movement being independent, their direction depends on the activation of the receptor system.

The PGC can be differentiated from somatic cell types by their structural properties during early embryonic development. It is histologically characterized as a rounded or elliptical cell with a relatively large size (10–20 μm), evident, large, and voluminous nucleus (6–10 μm), and clear nuclear membrane, which presents electron-dense granular structures with an irregular outline and a poorly basophilic cytoplasm (Braat et al. 1999; Nagai et al. 2001; Yamaha et al. 2010; Yön and Akbulut 2015; Mazzoni et al. 2015). In spite of the common morphology for the species, the PGCs may present morphological alterations, which will depend on the migration phase in which they are observed. During the migration process, these cells can have an amoeboid shape, an irregularly contoured nucleus, and cytoplasmic projections called pseudopodia (Braat et al. 1999; Doitsidou et al. 2002; Yön and Akbulut 2015; Saito et al. 2015; Grimaldi and Raz 2019).

The use of PGCs in the surrogate broodstock approach for the conservation and reconstitution of endangered species and in the maintenance of species with high economic value has become a very promising approach since these cells are the precursor for all germ cell lineage and formed gametes. Additionally, the PGCs are able to preserve maternal constituents, such as germplasm and mitochondrial DNA (Yoshizaki et al. 2003; Yamaha et al. 2007; Siqueira-Silva et al. 2018; Goto and Saito 2019), which means to say that they are responsible for transmitting genetic inheritance to future generations, thus originating new organisms through fertilization (Ciruna et al. 2002; Takeuchi et al. 2003; Yoshizaki et al. 2003; Saito et al. 2008; Yasui et al. 2011; Robles et al. 2017). Saito and collaborators (2008) showed that one single PGC is able to restore gametogenesis in zebrafish hosts, whose endogenous gametogenesis was eliminated using morpholino oligonucleotide (MO). However, only spermatogenesis was restored. This result shows PGC numbers are as important as the somatic part of the gonadal ridge (Fujimoto et al. 2010) for teleost sex differentiation, evidencing that single PGC transplantation cannot be used for conservation programs.

3.2 Oogonia and Spermatogonia

In most teleost fish, reproduction occurs cyclically and seasonally, where there is a resting period interspersed by periods of sexual activity that are finalized with the emergence of new offspring (Quagio-Grassiotto et al. 2013; Andrade et al. 2015). In both female and male teleost, gametogenesis begins respectively in the germinal epithelium of the ovarian lamellae and the epithelium of the seminiferous tubules from the oogonia and spermatogonia, respectively (Grier 2002; Grier et al. 2016).

The undifferentiated oogonia, which is surrounded by the cytoplasmic processes of the prefollicular epithelial cells (Nakamura et al. 2010, 2011; Quagio-Grassiotto et al. 2011), as well as the A undifferentiated spermatogonia (Aund* and Aund), which is surrounded by the cytoplasmic processes of the Sertoli cells (Schulz et al. 2010), have a discontinuous distribution in the germinal epithelium and are not connected via cytoplasmic bridge to clone members. Their self-renewal via mitoses ensures the continuous production of eggs and sperm, respectively (Schulz et al. 2010; Quagio-Grassiotto et al. 2011).

Undifferentiated spermatogonial cells are usually divided into two morphological types: type A undifferentiated* (Aund*) and type A undifferentiated (Aund) spermatogonia (Leal et al. 2009; Schulz et al. 2010). Those are the largest cells from the spermatogenesis of any fish species with a large and well-evident nucleus (e.g., around 8 cm in diameter in the Neotropical tetra *Astyanax altiparanae*) (Rodrigues et al. 2015) and in zebrafish *Danio rerio* (Leal et al. 2009), containing poorly condensed chromatin and one or two compact nucleoli. The most distinct difference is the nuclear envelope, which is irregular for the type Aund* (Leal et al. 2009). Studying those cells' topographical distribution in zebrafish, Nóbrega et al. (2010) observed their preference to reside in a microenvironment close to the interstitial compartment, where blood vessels and Leydig Cells, among other connective tissue elements, are found. Siqueira-Silva et al. (2013) and Siqueira-Silva et al. (2019) have already shown this aspect in the Neotropical cichlid species *Cichla kelberi* (Siqueira-Silva et al. 2013) and *C. intermedia* (Vilela et al. 2003), in which although the A undifferentiated spermatogonia be distributed along the entire germinal epithelium, there is a clear preference by the blind end of the seminiferous lobules near the tunica albuginea (the connective tissue that involves the testes).

Both types of A undifferentiated spermatogonia are considered stem cells since they can self-renewal. Another important role of the A undifferentiated spermatogonia is its capacity to colonize the gonadal germinal epithelium (Higuchi et al. 2011). Moreover, their differentiation in the type A differentiated spermatogonia always involves mitotic activity and is not influenced by other germ cells (Lacerda et al. 2014). Therefore, once in a germ-cell-empty epithelium, those cells can form a new population of germ cells by proliferation, giving rise to mature spermatozoa. A undifferentiated spermatogonia was also proved to be able to migrate toward and colonize the gonadal ridge (Yoshizaki et al. 2011). Another very interesting characteristic of the A undifferentiated spermatogonia is their ability to transdifferentiate in oogonia and produce viable, fertile oocytes when in a female gonadal environment (Nóbrega et al. 2010; Hattori et al. 2019; Octavera and Yoshizaki 2020). This aspect is important since one species' population can be fully restored from one single donor male fish through A undifferentiated transplantation into hosts from other species.

Although less studied, the oogonia were already proven to present similar physiological characteristics of A undifferentiated spermatogonia (Wong et al. 2011; Lee et al. 2016a; Franěk et al. 2019c), even the high sexual plasticity (Yoshizaki et al. 2010), proving that the sexual differentiation of those germ cells is controlled solely by the somatic microenvironment, rather than being cell autonomous. Altogether, those characteristics enable both, the A undifferentiated spermatogonia and oogonia as very interesting transplantable cells for the surrogate broodstock method. Understanding the biological and physiological role of stem germ cells is essential for their wide application in biotechnologies such as the surrogate broodstock approach.

4. Surrogate Broodstock in Neotropical Fish

The surrogate broodstock approach in Neotropical fish was performed for the first time by Majhi's team (2009a). That experiment was performed between two pejerrey species the *Odontesthes bonariensis* (as donor) and *O. hatcheri* (as host). The authors isolated spermatogonial stem cells from the testes of young individuals of *O. bonariensis* and then labeled them using the CFDA-SE Cell Linker in order to track their behavior and development into the host seminiferous tubules. Following, germ cell transplantation was performed into the gonads, via urogenital papilla, of sexually mature specimens of the Patagonian pejerrey *O. hatcheri.* Host endogenous gametogenesis was depleted by incubation of the sexually mature animals in heated water (25°C), being injected with two doses of busulfan at a four-week interval. Four weeks after transplantation, in two out of three transplanted animals, stem spermatogonia had colonized the cortical region of the testes and in 6 weeks (1 host animal), they proliferate and form a monolayer network in the host germinal epithelium. PCR analysis showed the presence of donor-derived spermatozoa in 20% of host specimens. Their milt was then used to fertilize *O. bonariensis* eggs, resulting in 1.2 to 13.3% of pure *O. bonariensis*, and other hybrids from the two used species. Those results suggest that donor-derived germ cells could differentiate into fully functional spermatozoa in the xenogeneic recipients.

Following Lacerda (2010b), transplanted stem spermatogonia, previously stained with the PKH26 dying, from the yellow peacock bass (*Cichla monoculus*) into the seminiferous tubules, through the urogenital papilla of full sexually mature Nile tilapia (*Oreochromis niloticus*). Host spermatogenesis had been previously depleted using a similar method to that of Majhi et al. (2009b), in which busulfan and high water temperature (35°C) were employed. The authors observed the presence of PKH26-labeled cells two weeks after transplantation, and in four weeks, donor germ cell cysts from *C. monoculus* were observed in the seminiferous tubules of the recipient animals (not published data).

In 2014, Majhi and collaborators performed another surrogate broodstock experiment using the pejerrey. At that time, however, the approach included transplanting female stem germ cells. Thus, both spermatogonial and oogonial stem germ cells were harvested from the gonads of *O. banariensis* juveniles and, after labeling using PKH26 cell membrane dying, were transplanted intra-papillary into both ovaries and testicles from the cytoablated host specimens of *O. hatcheri.* The experimental depletion of host gonads was different from the previous work (Majhi et al. 2009b) since the water temperature was warmer (26°C) (aiming to promote faster endogenous gametogenesis depletion), and four busulfan doses were used instead of two (Majhi et al. 2009b), containing 30 mg/kg body weight for female and 40 mg/kg body weight for male. Seven months after transplantation, the transplanted spermatogonial and oogonial stem cells could colonize the host gonads and produce functional donor-origin eggs (5% of female recipients) and sperm (17% of the recipients). After that, the authors crossed male specimens of the host (*O. hatcheri*) with *O. bonariensis* females, obtaining 12.6 to 39.7% of pure *O. bonariensis* offspring and female hosts (*O. hatcheri*) with *O. bonariensis* males. The results were still better, and 52.2% of the offspring were pure *O. bonariensis.* This study showed the surrogate broodstock technique using adult animals as hosts could be a promising strategy in the conservation of endangered species because it can rapidly propagate species (only seven months in this study).

Silva et al. (2016) evaluated the viability of adult Nile tilapia (*O. niloticus*) testes to receive spermatogonial stem cells removed from the testicles of sexually mature Jundia catfish (*Rhamdia quelen*) and produce sperm with the donor characteristics. They confirmed the presence of the Jundia catfish genetic material in the tilapia testes, demonstrating for the first time that the adult testes of the Nile tilapia possess functional conditions to promote the development of spermatogenesis, as well as, the production of sperm (120 days after spermatogonial transplantation) from a fish species belonging to a different order, thus enabling a new and important point for the advancement of surrogate broodstock approach for Neotropical species.

Lópes (2018) performed a similar technique, using two species of Neotropical Siluriformes, the mandi (*Pimelodus maculatus*) as host species and the frog catfish (*Pseudopimelodus mangurus*) as donor species, aiming to reconstitute the second species, threatened with extinction. For this purpose, a protocol of isolation, separation, transplantation, and cryopreservation of stem germ cells was established, through which stem germ cells from *P. mangurus* were transplanted, via the urogenital papilla, into triploid recipients of *P. maculatus*. Donor PKH26 labeled germ stem cells were found in recipient gonads 30 days later. This work was a pioneer in the transplantation of stem germ cells in Neotropical Siluriformes species, using adult triploid recipients (not published data).

In the following year (2019), Siqueira-Silva's group standardized the first interspecific spermatogonial transplantation between two Neotropical species from Characiformes order, using the same method as Majhi et al. (2009b) and Lacerda et al. (2010b). Spermatogonial transplantation was performed between two phylogenetically related species, the piracanjuba *Brycon orbignyanus* (donor species) and the yellow-tail-tetra *Astyanax altiparanae* (host species). The association between heated water temperature and the chemotherapeutic agent busulfan also suppressed the endogenous spermatogenesis of the host species. The authors also obtained sterile animals by incubating sexually mature males in heated water (35°C) only (Silva 2015). After transplantation, the authors described that approximately 67% of the individuals treated with busulfan and 85% of the specimens incubated at 35°C heated water showed at least one PKH26 labeled stem cell in the host germinal epithelium, and in some animals, the presence of PKH26 labeled spermatozoa was observed into the tubular lumen (Siqueira-Silva et al. 2019).

Aiming to further contribute to the knowledge about surrogate broodstock in Neotropical fishes, López (2022) performed spermatogonial stem cell transplantation from piracanjuba (*Brycon orbignyanus*) to both busulfan-treated triploid juvenile males of the yellowtail tetra (*A. altiparanae*), and adult males of the triploid hybrids between the species *Astyanax altiparanae* × *Astyanax fasciatus* (Piva et al. 2018). Triploid juvenile recipients received freshly isolated stem spermatogonia, while the adult triploid hybrid hosts received both freshly isolated and cryopreserved stem spermatogonia. In both of the transplantations, cells were injected intra-papillary. Up to 80 days after transplantation, PKH26-positive spermatogonia was visualized in the testes of the triploid juvenile hosts. In the adult triploid hybrid recipients, spermatogonial proliferation was observed on the 20th day after transplantation, independently of whether the transplanted cells were fresh or cryopreserved. These results are promising for the conservation of genetic material and reconstitution of the piracanjuba (*B. orbignyanus*) in case of extinction (not published data).

Using another method (spermatogonial transplantation into larvae as hosts), Lemos (2018) performed the surrogate broodstock approach with the pirarucu (*Arapaima gigas*), a large (three-meter-length) native species of the Amazon. *Arapaima gigas* is a high-market-value fish with difficult reproduction in captivity. In addition, the specimens only reach sexual maturity around five years of age. Therefore, Lemos (2018) aimed to establish tools for both the conservation of its genetic resources and understanding of the species' reproductive biology through a protocol standardization for stem spermatogonia cryopreservation, followed by their transplantation into the coelomic cavity of Nile tilapia larvae. The obtained results showed both cryopreserved and fresh cells remained biologically active and could migrate and colonize the recipient Nile tilapia gonads. Thus, spermatogonial stem cells' cryopreservation, followed by xenotransplantation, are highly promising and viable techniques for the genetic material maintenance and conservation of this important species, which is threatened with extinction.

Carvalho et al. (2022) performed primordial germ cell (PGC) transplantation in Neotropical species for the first time. Initially, the authors experimented the best condition to culture the host embryos. Thus, dechorionated and intact embryos were cultivated in three incubation systems. The control group was incubated in characin solution that was replaced every hour; the second group was cultivated in a close environment containing high oxygen pressure and manual replacement of the characin solution and; in a recirculation system of the characin solution. The biggest hatching percent (54.178%) and normal morphology (50.06%) were achieved with dechorionated embryos incubated in the recirculation system. Afterward, the authors evaluated the best anesthetic concentration of 2-phenoxyethanol to perform the transplantation procedures in *Prochilodus lineatus* larvae and observed the doses of 29.3 mM resulted in a shorter induction time and a recovery time between 5 and 10 minutes. Enough time to perform the transplantation. Finally, Green fluorescent protein-positive PGCs were isolated from *Astyanax altiparanae* and transplanted into the coelomic cavity in the gonadal ridge region from 1-day post-hatching *P. lineatus* larvae. Labeled PGCs were seen in two recipients, but the cells did not proliferate. In addition to this study, the same group tested different water temperatures for incubation of *Piaractus mesopotamicus* during embryo development and described *in vivo* PGC migration by using artificial GFP-nos1 3'UTR mRNA (Coelho et al. 2022). They observed that the best incubation temperature for the species was at room temperature (~ 26°C), in which approximately 52.9% of the embryos hatched. In relation to PGCs traceability, their first visualization occurred in the segmentation period when the embryos presented between 6 and 16 somites, and the number of GFP-positive cells varied from 7 to 18 per embryo. After hatching, the GFP-positive cells ranged from 7 to 22 per embryo and their visualization was possible until the 9th day after hatching. These preliminary studies are an important starting point for the Surrogate broodstock approach in Neotropical specie and they should be optimized to be employed in other endangered native species or in species of aquaculture importance.

5. Perspectives for the Use of South American Neotropical Species in Surrogate Broodstock Approach

In spite of the advances in the surrogate broodstock approach around the world, its development with Neotropical species is still incipient and new (Figure 2). Maybe the huge fish species diversity in the region and establishing one model make it difficult for the technique to advance. Yasui et al. (2022) group took an important step in its direction, proposing the yellowtail tetra *Astyanax altiparanae* as a model laboratory species for laboratory and aquaculture purposes. Moreover, as already mentioned, the short embryo development of Neotropical fishes hinders the performance of the methods involving germ cell transplantation on embryos. Finally, many species must still be studied on the basic aspects of their reproductive biology and physiology to submit them to such an approach.

Undoubtedly, the improvement of the surrogate broodstock method, as well as its association with other reproductive biotechniques, such as cells and tissue

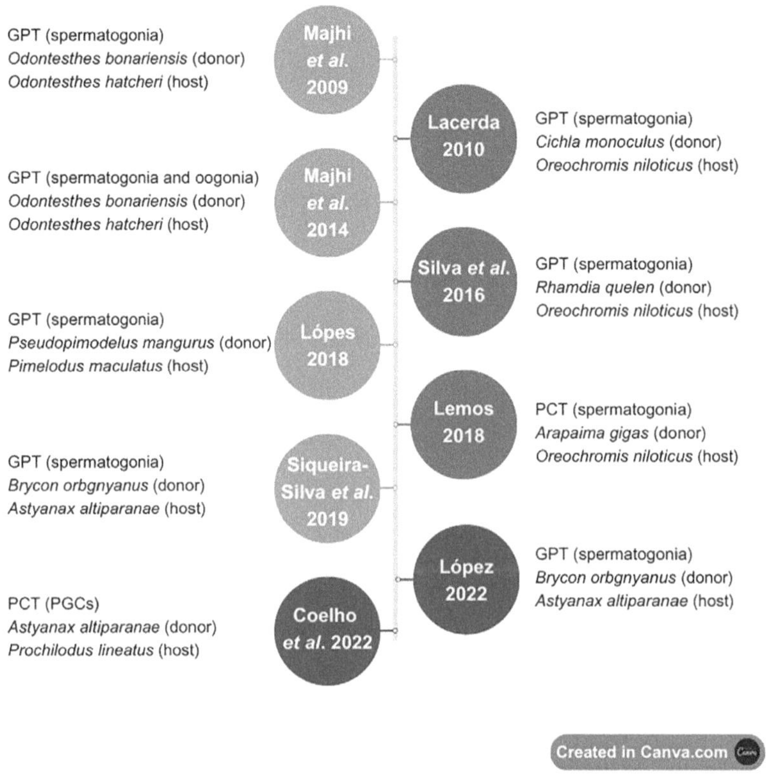

Figure 2. Schematic graphic resuming chronologically the surrogate broodstock studies performed in Neotropical fish species. GPT: genital pore transplantation; PCT: peritoneal cavity transplantation.

cryopreservation and the consequent establishment of transplantable stem cell cryobanks as well as their culture *in vitro* (Xie et al. 2020) will be essential for the preservation and even reconstitution of Neotropical fish species in a soon future.

References

Adamov, N.S.D.M., N.F.D. Nascimento, E.C.S. Maciel, M. Pereira-Santos, J.A. Senhorini, L.L. Calado et al. 2017. Triploid induction in the yellowtail tetra, *Astyanax altiparanae*, using temperature shock: tools for conservation and aquaculture. J. World Aquacult. Soc. 48: 741–750.

Agostinho, A.A., S.M. Thomaz and L.C. Gomes. 2005. Conservação da biodiversidade em águas continentais do Brasil. Megadiversidade 1: 70–78.

Alves, A.L., E.S. Varela, G.V. Moro and L.N. KirschniK. 2014. Riscos genéticos da produção de híbridos de peixes nativos. Embrapa Pesca e Aquicultura, Palmas.

Andrade, E.S., E.A. Andrade, V.O. Felizardo, D.A.J. Paula, G.C. Veras and L.D.S. Murgas. 2015. Biologia reprodutiva de peixes de água doce. Rev. Bras. Reprod. Anim. 39: 195–201.

Arashiro, D.R., G.S. Yasui, L.L. Calado, N.F.D. Nascimento, M.P. Dos Santos, S.C. Dos Santos et al. 2018. Synchronizing developmental stages in Neotropical catfishes for application in germ cell transplantation. Zygote 26: 135–48.

Baloch, A.R., R. Franěk, T. Tichopád, M. Fučíková, M. Rodina and M. Pšenička. 2019. Dnd1 knockout in sturgeons By CRISPR/Cas9 generates germ cell free host for surrogate production. Animals 9: 174.

Boldajipour, B., H. Mahabaleshwar, E. Kardash, M. Reichman-Fried, H. Blaser et al. 2008. Control of chemokine-guided cell migration by ligand sequestration. Cell 132: 463–473.

Braat, A.K., J.E. Speksnijder and D. Zivkovic. 1999. Germ line development in fishes. Int. J. Dev. Biol. 760: 745–760.

Brinster, R.L. and J.W. Zimmermann. 1994. Spermatogenesis following male germ-cell transplantation. Proc. Natl. Acad. Sci. 91: 11298–302.

Brinster, C.J., B.Y. Ryu, M.R. Avarbock, L. Karagenc, R.L. Brinster and K.E. Orwig. 2003 Restoration of fertility by germ cell transplantation requires effective recipient preparation. Biol. Reprod. 69: 412–20.

Carvalho, G.B., G.C.Z. Coelho, A.C. Alves, A.P.D.S. Silva, P. Sérgio Monzani, J.A. Senhorini et al. 2022. Embryo manipulation in neotropical characiform fish: incubation system, anaesthetic, and PGC transplantation in *Prochilodus lineatus*. Zygote 30: 773–780.

Castro, R.M.C.A. and R.P. Vari. 2003. Family Prochilodontidae. pp. 65–70. *In*: Reis, R.E., S.O. Kullander and C.J. Ferraris-JR [eds.]. Check List of the Freshwater Fishes of South and Central America. EDIPUCRS, Porto Alegre, Brazil.

Ciruna, B., G. Weidinger, H. Knaut, B. Thisse, C. Thisse, E. Raz et al. 2002. Production of maternal-zygotic mutant zebrafish by germ-line replacement. Proc. Natl. Acad. Sci. 99: 14919–14924.

Coelho, G.C., I.S. Yo, T.M. Mira-Lopez, P.S. Monzani, D.R. Arashiro, T. Fujimoto et al. 2019. Preparing a fish embryo (*Prochilodus lineatus*) for staging, chorion removal and PGC traceability. Int. J. Dev. Biol. 1: 57–65.

Coelho, G.C.Z. 2022. Caracterização e transplante de células germinativas primordiais em peixes neotropicais. doutorado Tese, Universidade Estadual Paulista (Unesp), Instituto de Biociências, Botucatu, Brasil.

Doitsidou, M., M. Reichman-Fried, J. Stebler, M. Köprunner, J. Dörries, D. Meyer et al. 2002. Guidance of primordial germ cell migration by the chemokine SDF-1. Cell 111: 647–659.

Franěk, R., Z. Marinović, J. Lujić, B. Urbányi, M. Fučíková, V. Kašpar et al. 2019a. Cryopreservation and transplantation of common carp spermatogonia. PLoS ONE 14: 1–17.

Franěk, R., T. Tichopád, M. Fučíková, C. Steinbach and M. Pšenička. 2019b. Production and use of triploid zebrafish for surrogate reproduction. Theriogenology 140: 33–43.

Franěk, R., T. Tichopád, C. Steinbach, X. Xie, J. Lujić, Z. Marinović et al. 2019c. Preservation of female genetic resources of common carp through oogonial stem cell manipulation. Cryobiology 87: 78–85.

Fricke, R., W.N. Eschmeyer and R. van der Laan. 2020. Eschmeyer's Catalog Of Fishes: Genera, Species, References. Disponível em: http://researcharchive.calacademy.org/research/ichthyology/c atalog/ fishcatmain.asp.

Froese, R. and D. Pauly. 2015. FishBase. World Wide Web Electronic Publication. Disponível em: www. fishbase.org.

Fujimoto, T., T. Nishimura, R. Goto, Y. Kawakami, E. Yamaha and K. Arai. 2010. Sexual dimorphism of gonadal structure and gene expression in germ cell-deficient loach, a teleost fish. Proc. Natl. Acad. Sci. USA 107: 17211–17216.

Golpour, A., M.A.M. Siddique, D.H. Siqueira-Silva and M. Pšenička. 2016. Induced sterility in fish and its potential and challenges for aquaculture and germ cell transplantation technology: a review. Biologia 71: 853–864.

Goto, R. and T. Saito. 2019. A state-of-the-art review of surrogate propagation in fish. Theriogenology 133: 216–227.

Grier, H.J. 2002. The germinal epithelium: its dual role in establishing male reproductive classes and understanding the basis for indeterminate egg production in female fishes. Gulf and Caribbean Fisheries Institute Proceedings. pp. 537–552.

Grier, H.J., M.C. Uribe, F.L. Lo Nostro, S.D. Mims and L.R. Parenti. 2016. Conserved form and function of the germinal epithelium through 500 million years of vertebrate evolution. J. Morphol. 277: 1014–1044.

Grimaldi, C. and E. Raz. 2019. Germ cell migration - Evolutionary issues and current understanding. Semin. Cell Dev. Biol. 100: 152–159.

Güralp, H., K. Pocherniaieva, M. Blecha, T. Policar, M. Pšenička and T. Saito. 2016. Early embryonic development in pikeperch (*Sander lucioperca*) related to micromanipulation. Czech J. Anim. Sci. 61: 273–80.

Hashimoto, D.T., J.A. Senhorini, F. Foresti and F. Porto-Foresti. 2012. Interspecific fish hybrids in Brazil: management of genetic resources for sustainable use. Rev. Aquac. 4: 108–18.

Hattori, R.S., T.T. Yoshinaga, N. Katayama, S. Hattori-Ihara, R.Y. Tsukamoto, N.S. Takahashi et al. 2019. Surrogate production of *Salmo salar* oocytes and sperm in triploid *Oncorhynchus mykiss* by germ cell transplantation technology. Aquaculture 506: 238–45.

Higaki, S., Y. Eto, Y. Kawakami, E. Yamaha, N. Kagawa, M. Kuwayama et al. 2010. Production of fertile zebrafish (*Danio rerio*) possessing germ cells (gametes) originated from primordial germ cells recovered from vitrified embryos. Reproduction 139: 733–740.

Higaki, S., T. Todo, R. Teshima, I. Tooyama, Y. Fujioka, N. Sakai et al. 2018. Cryopreservation of male and female gonial cells by vitrification in the critically endangered cyprinid honmoroko *Gnathopogon caerulescens*. Fish Physiol. Biochem. 44: 503–513.

Higuchi, K., Y. Takeuchi, M. Miwa, Y. Yamamoto, K. Tsunemoto and G. Yoshizaki. 2011. Colonization, proliferation, and survival of intraperitoneally transplanted yellowtail *Seriola quinqueradiata* spermatogonia in nibe croaker *Nibea mitsukurii* recipient. Fish Sci. 77: 69–77.

ICMBIO/MMA. 2018. Livro Vermelho da Fauna Brasileira Ameaçada de Extinção. ICMBio/MMA, Brasília.

Izadyar, F., L.B. Creemers, M.F. Dissel-Emiliani, A.M.M. Pelt and D.G. Rooij. 2000. Spermatogonial stem cell transplantation. Mol. Cell. Endocrinol. 169: 21–26.

Jin, Y.H., D. Robledo, J.M. Hickey, M.J. McGrew and R.D. Houston. 2021. Surrogate broodstock to enhance biotechnology research and applications in aquaculture. Biotechnol. Adv. 49: 107756.

Johnstone, R., R.M. Knott, A.G. Macdonald and M.V. Walsingham. 1989. Triploidy induction in recently fertilized Atlantic salmon ova using anaesthetics. Aquaculture 78: 229–36.

Kawakami, Y., R. Goto-Kazeto, T. Saito, T. Fujimoto, S. Higaki, Y. Takahashi et al. 2010. Generation of germ-line chimera zebrafish using primordial germ cells isolated from cultured blastomeres and cryopreserved embryoids. Int. J. Dev. Biol. 54: 1493–1501.

Kawamura, W., R. Tani, H. Yahagi, S. Kamio, T. Morita, Y. Takeuchi et al. 2020. Suitability of hybrid mackerel (*Scomber australasicus* × *S. japonicus*) with germ cell-less sterile gonads as a recipient for transplantation of bluefin tuna germ cells. Gen. Comp. Endocrinol. 295: 113525.

Kayo, D., S. Kanda and K. Okubo. 2022. Allogeneic testes transplanted into partially castrated adult medaka (*Oryzias latipes*) can produce donor-derived offspring by natural mating over a prolonged period. Zool. Lett. 8: 1–12.

Kobayashi, T., Y. Takeuchi, T. Takeuchi and G. Yoshizaki. 2007. Generation of viable fish from cryopreserved primordial germ cells. Mol. Reprod. Dev. 74: 207–213.

Lacerda, S.M.S.N. 2006a. Transplante de espermatogônias: A Tilápia-Nilótica (*Oreochromis niloticus*) como modelo experimental. dissertação Mestrado, Universidade Federal de Minas Gerais, Belo Horizonte, Brasil.

Lacerda, S.M.S.N., S.R. Batlouni, S.B.G. Silva, C.S.P. Homem and L.R. França. 2006b. Germ cells transplantation in fish: the Nile-tilapia model. Anim. Reprod. 3: 146–159.

Lacerda, S.M.S.N., S.R. Batlouni, L.H. Assis, F.M. Resende, S.M. Campos-Silva, R. Campos-Silva et al. 2008. Germ cell transplantation in tilapia (*Oreochromis niloticus*). Cybium 32: 115–118.

Lacerda, S.M.S.N., S.R. Batlouni, G.M. Costa, T.M. Segatelli, B.R. Quirino, B.M. Queiroz et al. 2010a. A new and fast technique to generate offspring after germ cells transplantation in adult fish: the Nile tilápia (*Oreochromis niloticus*) model. PLoS One 5: e10740.

Lacerda, S.M.S.N. 2010b. Transplante singênico e xenogênico de espermatogônias-tronco em tilápias-nilóticas (*Oreochromis niloticus*) adultas. tese Doutorado, Universidade Federal de Minas Gerais, Belo Horizonte, Brasil.

Lacerda, S.M.S.N., G.M. Costa, P.H. Campos-Junior, T.M. Segatelli, R. Yazawa, Y. Takeuchi et al. 2012. Germ cell transplantation as a potential biotechnological approach to fish reproduction. Fish Physiol. Biochem. 39: 3–11.

Lacerda, S.M.S.N., G.M. Costa and L.R. de França. 2014. Biology and identity of fish spermatogonial stem cell. Gen. Comp. Endocrinol. 207: 56–65.

Leal, M.C., E.R. Cardoso, R.H. Nóbrega, S.R. Batlouni, J. Bogerd, L.R. França et al. 2009. Histological and stereological evaluation of zebrafish (*Danio rerio*) spermatogenesis with an emphasis on spermatogonial generations. Biol. Reprod. 81: 177–87.

Lee, S., Y. Iwasaki, S. Shikina and G. Yoshizaki. 2013. Generation of functional eggs and sperm from cryopreserved whole testes. Proc. Natl. Acad. Sci. 110: 1640–45.

Lee, S., S. Seki, N. Katayama and G. Yoshizaki. 2015. Production of viable trout offspring derived from frozen whole fish. Sci. Rep. 5: 16045.

Lee, S. and G. Yoshizaki. 2016a. Successful cryopreservation of spermatogonia in critically endangered Manchurian trout (*Brachymystax lenok*). Cryobiology 72: 165–168.

Lee, S., N. Katayama and G. Yoshizaki. 2016b. Generation of juvenile rainbow trout derived from cryopreserved whole ovaries by intraperitoneal transplantation of ovarian germ cells. Biochem. Biophys. Res. Commun. 478: 1478–83.

Lemos, M.S. 2018. Caracterização do cDNA vasa, criopreservação e transplante xenogênico de espermatogônias de pirarucus (*Arapaima gigas*) em larvas de tilápias (*Oreochromis niloticus*). tese Doutorado, Universidade Federal de Minas Gerais. Belo Horizonte. Brasil.

Li, M., N. Hong, H. Xu, J. Song and Y. Hong. 2016. Germline replacement by blastula cell transplantation in the fish medaka. Sci. Rep. 6: 29658.

Lin, S., W. Long, J. Chen and N. Hopkins. 1992. Production of germ-line chimeras in zebrafish by cell transplants from genetically pigmented to albino embryos. Proc. Natl. Acad. Sci. 89: 4519–23.

Linhartova, Z. and T. Saito. 2014. Embryogenesis, visualization and migration of primordial germ cells in tench (*Tinca tinca*). J. Appl. Ichthyol. 30: 29–39.

Linhartová, Z., T. Saito, V. Kašpar, M. Rodina, E. Prášková, S. Hagihara et al. 2015. Sterilization of sterlet *Acipenser ruthenus* by using knockdown agent, antisense morpholino oligonucleotide, against dead end gene. Theriogenology 84: 1246–1255.

López, L.S. 2022. Transplante interespecífico de espermatogônia tronco de piracanjuba *Brycon orbignyanus*. tese Doutorado, Universidade Estadual Paulista (Unesp), Botucatu, Brasil.

Lonnie, D.R. and M.D. Griswold. 2000. Spermatogonial transplantation—an update for the millennium, Mol. Cell. Endocrinol. 161: 117–120.

Majhi, S.K., R.S. Hattori, M. Yokota, S. Watanabe and C.A. Strüssmann. 2009a. Germ cell transplantation using sexually competent fish: an approach for rapid propagation of endangered and valuable germlines. PLoS One 4: e6132.

Majhi, S.K., R.S. Hattori, S.M. Rahman, T. Suzuki and C.A. Strüssmann. 2009b. Experimentally induced depletion of germ cells in sub-adult Patagonian pejerrey (*Odontesthes hatcheri*). Theriogenology 71: 1162–72.

Majhi, S.K., R.S. Hattori, S.M. Rahman and C.A. Strussmann. 2014. Surrogate production of eggs and sperm by intrapapillary transplantation of germ cells in cyto ablated adult fish. PLoS One 9: e95294.

Majhi, S.K., A.R. Rasal, B. Kushwaha and S. Raizada. 2017. Heat and chemical treatments in adult *Cyprinus carpio* (Pisces cypriniformes) rapidly produce sterile gonads. Anim. Reprod. Sci. 183: 77–85.

Martino, G. 2002. Retrocruce de hembras hibridos (F1) (*Colossoma macropomum* x *Piaractus brachypomus*) con machos de las espécies parentales. Mem. Soc. Cienc. Nat. La Salle 53: 688–693.

Mazzoni, T.S., H.J. Grier and I. Quagio-Grassiotto. 2015. The basement membrane and the sex establishment in the juvenile hermaphroditism during gonadal differentiation of the *Gymnocorymbus ternetzi* (Teleostei: Characiformes: Characidae). Anat. Rec. 298: 1984–2010.

MClean, D.J. 2005. Spermatogonial stem cell transplantation and testicular function. Cell Tissue Res. 322: 21–31.

Miranda, J.C. 2012. Ameaças aos peixes de riachos da Mata Atlântica. Natureza on line 10: 136–139.

Morita, T., N. Kumakura, K. Morishima, T. Mitsuboshi, M. Ishida, T. Hara et al. 2012. Production of donor-derived offspring by allogeneic transplantation of spermatogonia in the yellowtail (*Seriola quinqueradiata*). Biol. Reprod. 86: 176.

Nagai, T., E. Yamaha and K. Arai. 2001. Histological differentiation of primordial germ cells in zebrafish. Zool. Sci. 18: 215–223.

Nagler, J.J., J.G. Cloud, P.A. Wheeler and G.H. Thorgaard. 2001. Testis transplantation in male rainbow trout (*Oncorhynchus mykiss*). Biol. Reprod. 64: 644–646.

Nakagawa, M., T. Kobayashi and K. Ueno. 2002. Production of germline chimera in loach (*Misgurnus Anguillicaudatus*) and proposal of new method for preservation of endangered fish species. J. Exp. Zool. 293: 624–631.

Nakamura, S., K. Kobayashi, T. Nishimura, S. Higashijima and M. Tanaka. 2010. Identification of germline stem cells in the ovary of the teleost. Science 328: 1561–1563.

Nakamura, S., K. Kobayashi, T. Nishimura and M. Tanaka. 2011. Ovarian germline stem cells in the teleost fish, Medaka (*Oryzias latipes*). Int. J. Biol. Sci. 7: 403–409.

Nakatani, K., A.A. Agostinho, G. Baumgartner, A. Bialetzki, P.V. Sanches, M.C. Makrakis et al. 2001. Ovos e larvas de peixes de água doce: desenvolvimento e manual de identificação. Eduem, Maringá.

Nascimento, N., D. De Siqueira-Silva, M. Pereira-Santos, T. Fujimoto, J. Senhorini, L. Nakaghi et al. 2017. Stereological analysis of gonads from diploid and triploid fish yellowtail tetra *Astyanax altiparanae* (Garutti & Britski) in laboratory conditions. Zygote 25: 537–544.

Nelson, J.S. 2006. Fishes of the World. John Wiley & Sons, Hoboken.

Nóbrega, R.H., S.R. Batlouni and L.R. França. 2009. An overview of functional and stereological evaluation of spermatogenesis and germ cell transplantation in fish. Fish Physiol. Biochem. 35: 197–206.

Nóbrega, R.H., C.D. Greeb, H.V. Kant, J. Bogerd, L.R. França and R.W. Schulz. 2010. Spermatogonial stem cell niche and spermatogonial stem cell transplantation in zebrafish. PLoS ONE 5: e12808.

Octavera, A. and G. Yoshizaki. 2022. Production of Chinese rosy bitterling offspring derived from frozen and vitrified whole testis by spermatogonial transplantation. Fish Physiol. Biochem. 46: 1431–1442.

Okutsu, T., A. Yano, K. Nagasawa, S. Shikina, T. Kobayashi, Y. Takeuchi et al. 2006a. Manipulation of fish germ cell: visualization, cryopreservation and transplantation. J. Reprod. Dev. 52: 685–93.

Okutsu, T., K. Suzuki, Y. Takeuchi, T. Takeuchi and G. Yoshizaki. 2006b. Testicular germ cells can colonize sexually undifferentiated embryonic gonad and produce functional eggs in fish. Proc. Natl. Acad. Sci. U.S.A. 103: 2725–29.

Okutsu, T., S. Shikina, M. Kanno, Y. Takeuchi and G. Yoshizaki. 2007. Production of trout offspring from triploid salmon parents. Science 317: 1517.

Orwig, K.E. and S. Schlatt. 2005. Cryopreservation and transplantation of spermatogonia and testicular tissue preservation of male fertility. J. Natl. Cancer Inst. Monogr. 34: 51–56.

Piferrer, F., R.M. Cal, C. Gómez, C. Bouza and P. Martínez. 2003. Induction of triploidy in the turbot (*Scophthalmus maximus*): II. Effects of cold shock timing and induction of triploidy in a large volume of eggs. Aquaculture 220: 821–31.

Piva, L.H., D.H. De Siqueira-Silva, C.A.G. Goes, T. Fujimoto, T. Saito, L.V. Dragone et al. 2018. Triploid or hybrid tetra: which is the ideal sterile host for surrogate technology? Theriogenology 108: 239–244.

Pšenička, M., T. Saito, Z. Linhartová and L. Gazo. 2015. Isolation and transplantation of sturgeon early-stage germ cells. Theriogenology 83: 1085–1092.

Pšenička, M., T. Saito, M. Rodina and B. Dzyuba. 2016. Cryopreservation of early stage siberian sturgeon *Acipenser baerii* germ cells, comparison of whole tissue and dissociated cells. Cryobiology 72: 119–22.

Quagio-grassiotto, I., H.J. Grier, T.S. Mazzoni, R.H. Nóbrega and J.P. Amorim. 2011. Activity of the ovarian germinal epithelium on the follicle formation and the oocyte development in the freshwater catfish *Pimelodus maculatus* (Teleostei: Ostariophysi: Siluriformes). J. Morphol. 8: 1290–1306.

Quagio-grassiotto, I., D.D. Wildner and R. Ishiba. 2013. Gametogênese em peixes: aspectos relevantes para o manejo reprodutivo. Rev. Bras. Reprod. Anim. 37: 181–191.

Raz, E. 2003. Primordial germ-cell development: The zebrafish perspective. Nat. Rev. Genet. 4: 690–700.

Raz, E., J. Stebler, D. Spieler, K. Slanchev, K.A. Molyneaux, U. Richter et al. 2004. Primordial germ cell migration in the chick and mouse embryo: the role of the chemokine SDF-1/CXCL12. Dev. Biol. 272: 351–361.

Raz, E. and M. Reichman-Frie. 2006. Attraction rules: germ cell migration in zebrafish. Curr. Opin. Genet. Dev. 16: 355–359.

Reis, R.E., J.S. Albert, F. Di Dario, M.M. Mincarone, P. Petry and L.A. Rocha. 2016. Fish biodiversity and conservation in South America. J. Fish Biol. 89: 12–47.

Robles, V., M.F. Riesco, M. Psenicka, T. Saito, D.G. Valcarce, E. Cabrita et al. 2017. Biology of teleost primordial germ cells (PGCs) and spermatogonia: Biotechnological applications. Aquaculture 472: 4–20.

Rodrigues, M.D., D.H. Siqueira-Silva, P.P. Quirino, A. Ninhaus-Silveira and R. Verissimo-Silveira. 2015. Spermatogenesis in the yellowtail tetra *Astyanax altiparanae*: a histological analyses with emphasis to spermatogonial and spermatid types. Bol. Inst. Pesca 41: 697–705.

Rushton, D.N. 2003. Functional electrical stimulation and rehabilitation—an hypothesis. Med. Eng. Phys. 25: 75–8.

Saito, T., R. Goto-Kazeto, K. Arai and E. Yamaha. 2008. Xenogenesis in teleost fish through generation of germ-line chimeras by single primordial germ cell transplantation. Biol. Reprod. 78: 159–66.

Saito, T., R. Goto-Kazeto, T. Fujimoto, Y. Kawakami, K. Arai and E. Yamaha. 2010. Inter-species transplantation and migration of primordial germ cells in cyprinid fish. Int. J. Dev. Biol. 54: 1481–86.

Saito, T., R. Goto-Kazeto, Y. Kawakami, K. Nomura, H. Tanaka, S. Adachi et al. 2011. The mechanism for primordial germ-cell migration is conserved between japanese eel and zebrafish. PLoS ONE 6: e24460.

Saito, T., M. Pšenička, R. Goto, S. Adachi, K. Inoue, K. Arai et al. 2014. The origin and migration of primordial germ cells in sturgeons. PLoS ONE 9: e86861.

Saito, T. and M. Psenicka. 2015. Novel technique for visualizing primordial germ cells in sturgeons (*Acipenser ruthenus, A. gueldenstaedtii, A. baerii,* and *Huso huso*). Biol. Reprod. 93: 1–7.

Saito, T., H. Güralp, V. Iegorova, M. Rodina and M. Pšenicka. 2018. Elimination of primordial germ cells in sturgeon embryos by ultraviolet irradiation. Biol. Reprod. 99: 556–564.

Sato, L.S., P.H. Jorge, V.A. Mastrochirico-Filho, M.E. Hata, S.R. Batlouni, R.Y. Kuradomi et al. 2020. Triploidy in tambaqui *Colossoma macropomum* identified by chromosomes of fish larvae. J. Aquac. Mar. Biol. 9: 65–69.

Schulz, R.W., L.R. De França, J.J. Lareyre, F. Le Gac, H. Chiarini-Garcia, R.H. Nobrega et al. 2010. Spermatogenesis in fish. Gen. Comp. Endocrinol. 165: 390–411.

Shinohara, T., K.E. Orwig, M.R. Avarbock and R.L. Brinster. 2002. Germ line stem cell competition in postnatal mouse testes. Biol. Reprod. 66: 1491–97.

Silva, D.H.S. 2015. Transplante interespecífico de espermatogônias-tronco de *Brycon* orbignyanus em *Astyanax altiparanae* (Teleostei: Characidae). tese Doutorado, Universidade Estadual Paulista Julio de Mesquita Filho, Piracicaba.

Silva, M.A., G.M.J. Costa, S.M.S.N. Lacerda, P.F.P. Brandão-Dias, E. Kalapothakis, A.F. Silva Júnior et al. 2016. Successful xenogeneic germ cell transplantation from Jundiá catfish (*Rhamdia Quelen*) into adult Nile Tilapia (*Oreochromis Niloticus*) testes. Gen. Comp. Endocrinol. 230: 48–56.

Siqueira-Silva, D.H., C.A. Vicentini, A. Ninhaus-Silveira and R. Veríssimo-Silveira. 2013. Reproductive cycle of the Neotropical cichlid yellow peacock bass *Cichla kelberi*: A novel pattern of testicular development. Neotrop. Ichthyol. 11: 5587–596.

Siqueira-Silva, D.H., A.P.S. Silva, A. Ninhaus-Silveira and R. Veríssimo-Silveira. 2015. The effects of temperature and busulfan (Myleran) on the yellowtail tetra *Astyanax altiparanae* (Pisces, Characiformes) spermatogenesis. Theriogenology 84: 1033–1042.

Siqueira-Silva, D.H., T. Saito, A.P. Dos Santos-Silva, R. Da Silva Costa, M. Psenicka and G.S. Yasui. 2018. Biotechnology applied to fish reproduction: tools for conservation. Fish Physiol. Biochem. 44: 1469–1485.

Siqueira-Silva, D.H., A.P. Dos Santos Silva, R. Da Silva Costa, J.A. Senhorini, A. Ninhaus-Silveira and R. Veríssimo-Silveira. 2019. Preliminary study on testicular germ cell isolation and transplantation in an endangered endemic species *Brycon orbignyanus* (Characiformes: Characidae). Fish Physiol. 47: 767–776.

Suarez López, L. 2018. Transplante interespecífico de células germinativas-tronco em Siluriformes Neotropicais. dissertação Mestrado, Universidade Estadual Paulista "Júlio de Mesquita Filho", Botucatu, Brasil.

Takeuchi, Y., G. Yoshizaki and T. Takeuchi. 2001. Production of germ-line chimeras in rainbow trout by blastomere transplantation. Mol. Reprod. Dev. 59: 380–89.

Takeuchi, Y., G. Yoshizaki and T. Takeuchi. 2003. Generation of live fry from intraperitoneally transplanted primordial germ cells in rainbow trout. Biol. Reprod. 69: 1142–1149.

Takeuchi, Y., G. Yoshizaki and T. Takeuchi. 2004. Surrogate broodstock produces salmonids. Nature 430: 629–30.

Tolussi, C.E., A.D.O. Gomes, A. Kumar, C.S. Ribeiro, F.L.L. Nostro, P.A. Bain et al. 2018. Environmental pollution affects molecular and biochemical responses during gonadal maturation of *Astyanax fasciatus* (Teleostei: Characiformes: Characidae). Ecotoxicol. Environ. Saf. 147: 926–934.

Vilela, D.A., S.G. Silva, M.T. Peixoto, H.P. Godinho and L.R. França. 2003. Spermatogenesis in teleost: insights from the Nile tilapia (*Oreochromis niloticus*) model. Fish Physiol. Biochem. 28: 187–190.

Wakamatsu, Y., K. Ozato, H. Hashimoto, M. Kinoshita, M. Sakaguchi, T. Iwamatsu et al. 1993. Generation of germ-line chimeras in medaka (*Oryzias latipes*). Mol. Mar. Biol. Biotechnol. 2: 325–332.

Wong, T.T., T. Saito, J. Crodian and P. Collodi. 2011. Zebrafish germline chimeras produced by transplantation of ovarian germ cells into sterile host larvae. Biol. Reprod. 84: 1190–1197.

Xie, X., R. Nóbrega and M. Pšenička. 2020. Spermatogonial stem cells in fish: characterization, isolation, enrichment, and recent advances of *in vitro* culture systems. Biomolecules 10: 644.

Xu, J., F. You, X. Wu, P. Zhang, Y. Lin, H. Jiang et al. 2008. Induction of triploidy in large yellow crocker *Pseudosciaena crocea* (Richardson, 1846): effects of pressure shocks and growth performance in the first rearing year. Aquac. Res. 39: 1369–1376.

Xu, D., T. Yoshino, M.B. Cioffi, H. Yoshikawa, Y. Ino, R. Yazawa et al. 2020. Production of donor-derived eggs after ovarian germ cell transplantation into the gonads of adult, germ cell-less, triploid hybrid fish. Biol. Reprod. 103: 1289–99.

Yamaha, E., M. Kazama-Wakabayashi, S. Otani, T. Fujimoto and K. Arai. 2001. Germ-line chimera by lower-part blastoderm transplantation between diploid goldfish and triploid crucian carp. Genetica 111: 227.

Yamaha, E., M. Murakami, K. Hada, S. Otani, T. Fujimoto, M. Tanaka et al. 2003. Recovery of fertility in male hybrids of a cross between goldfish and common carp by transplantation of PGC (Primordial Germ Cell)-containing graft. Genetica 119: 121–31.

Yamaha, E., T. Saito, R. Goto-Kazeto and K. Arai. 2007. Developmental biotechnology for aquaculture, with special reference to surrogate production in teleost fishes. J. Sea Res. 58: 8–22.

Yamaha, E., R. Goto-Kazeto, T. Saito, Y. Kawakami, T. Fujimoto, S. Adachi et al. 2010. Primordial germ cell in teleost fish with special references to its specification and migration. J. Appl. Ichthyol. 26: 816822.

Yang, F., K. Ichida and G. Yoshizaki. 2022. Gametogenesis commencement in recipient gonads using germ cells retrieved from dead fish. Aquaculture 552: 737952.

Yasui, G.S., T. Fujimoto, S. Sakao, E. Yamaha and K. Arai. 2011. Production of loach (*Misgurnus anguillicaudatus*) germ-line chimera using transplantation of primordial germ cells isolated from cryopreserved blastomeres. J. Anim. Sci. 89: 2380–2388.

Yasui, G.S., N.F. Do Nascimento, M. Pereira-Santos, A.P. Dos Santos Silva, G.C. Coelho, J.A. Visintin et al. 2022. Establishing a model fish for the Neotropical region: The case of the yellowtail tetra *Astyanax altiparanae* in advanced biotechnology. Front. Genet. 13: 903990.

Yazawa, R., Y. Takeuchi, T. Morita, M. Ishida and G. Yoshizaki. 2013. The Pacific bluefin tuna (*Thunnus orientalis*) dead end gene is suitable as a specific molecular marker of type A spermatogonia. Mol. Reprod. Dev. 80: 871–880.

Yön, N.D. and C. Akbulut. 2015. Identification of primordial germ cells: Cytological, histological and immunohistochemical aspects. Braz. Arch. Biol. Technol 58: 222–228.

Yoon, C., K. Kawakami and N. Hopkins. 1997. Zebrafish vasa homologue RNA is localized to the cleavage planes of 2- and 4-cell-stage embryos and is expressed in the primordial germ cells. Development 124: 3157–3165.

Yoshizaki, G., Y. Takeuchi, T. Kobayashi and T. Takeuchi. 2003. Primordial germ cell: A novel tool for fish bioengineering. Fish Physol. 14: 453457.

Yoshizaki, G., M. Ichikawa, M. Hayashi, Y. Iwasaki, M. Miwa, S. Shikina et al. 2010. Sexual plasticity of ovarian germ cells in rainbow trout. Development 137: 1227–1230.

Yoshizaki, G., K. Fujinuma, Y. Iwasaki, T. Okutsu, S. Shikina, R. Yazawa et al. 2011. Spermatogonial transplantation in fish: A novel method for the preservation of genetic resources. Comp. Biochem. Physiol. 6: 55–61.

Yoshizaki, G., T. Okutsu, T. Morita, M. Terasawa, R. Yazawa and Y. Takeuchi. 2012. Biological characteristics of fish germ cells and their application to developmental biotechnology. Reprod. Domest. Anim. 47: 187–192.

Yoshizaki, G., K. Takashiba, S. Shimamori, K. Fujinuma, S. Shikina, T. Okutsu et al. 2016. Production of germ cell-deficient salmonids by dead end gene knockdown, and their use as recipients for germ cell transplantation. Mol. Reprod. Dev. 83: 298–311.

Yoshizaki, G. and R. Yzawa. 2019. Application of surrogate broodstock technology in aquaculture. Fish. Sci. 85: 429–437.

Zhou, L., X. Wang, Q. Liu, J. Yang, S. Xu, Z. Wu et al. 2021. Successful spermatogonial stem cells transplantation within pleuronectiformes: first breakthrough at inter-family level in marine fish. Int. J. Biol. Sci. 17: 4426–4441.

CHAPTER 12

Potential Use of Germplasm Banks of South American Neotropical Species in Environmental Conservation in Brazil

*Evoy Zaniboni-Filho** and *Josiane Ribolli*

INTRODUCTION

The premises of conservation focus not only on preserving species but also on preserving their genetic variability. However, the development model adopted by humanity has caused an acceleration in the disappearance of fish species, notably migratory species from freshwater environments. Using the World Wildlife Fund for Nature's Living Planet Index for migratory freshwater fish populations, this group declined by 76% from 1970 to 2016, mainly caused by habitat loss and damming of rivers (W. W. F. 2022).

In Brazil, more than half of the electricity consumed comes from hydroelectric power plants (EPE 2022), making it considered one of the countries that build the most dams in the world and the second-greatest generator of hydroelectric power after China (Oliveira 2018). Water pollution is another crucial factor in habitat alteration and is considered the most sensitive indicator of ecological stress in structuring fish communities (Sarkar and Islam 2020). Despite efforts to collect and treat sewage in Brazil, the volume of effectively treated sewage, compared to what is generated, still

Laboratório de Biologia e Cultivo de Peixes de Água Doce. Universidade Federal de Santa Catarina, SC, Brazil, 88066-260.
Email: josianeribolli@gmail.com
* Corresponding author: evoyzaniboni@gmail.com

needs to be higher in the country, reaching only 50.3% (Brasil 2022). Additionally, new threats are emerging along with technological development, whose effects on habitat and biodiversity still need to be thoroughly evaluated. According to (Reid et al. 2019), 12 emerging and entirely new threats have emerged since 2006 or have since intensified, namely: (i) changing climates; (ii) e-commerce and invasions; (iii) infectious diseases; (iv) harmful algal blooms; (v) expanding hydropower; (vi) emerging contaminants; (vii) engineered nanomaterials; (viii) microplastic pollution; (ix) light and noise; (x) freshwater salinization; (xi) declining calcium; and (xii) cumulative stressors. Thus, the conservation of fish species and the maintenance of the genetic variability of fish populations become increasingly dependent on specific actions, such as biobanks. However, it seems insufficient to guarantee this conservation with the restricted use of *in situ* conservation techniques, especially when considering the distribution of fish biodiversity on the South American continent and the extensive areas of rivers directly affected by dams and other threats. Article 9 of the Convention on Biological Diversity recommends *ex situ* genetic conservation to support *in situ* conservation actions. It is an example of using the precautionary approach to conservation (United Nations 2018).

In addition to conservationist interests, the approval by the United Nations in 2015 of the 2030 Agenda for the Sustainable Development of the Planet gave even greater importance to genetic biobanks when, among the recommended actions to guarantee the food security of humanity, it suggested the conservation of wild lineages of populations used as food. Some South American endemic fish species, including *Colossoma macropomum*, *Brycon amazonicus,* and *Pseudoplatystoma corruscans*, are already important for commercial fish farming.

1. Characterization of the Brazilian and South American Ichthyofauna: A Look at Conservation

When analyzing the historical biogeography of South American freshwater fishes, Hubert and Renno (2006) suggest the areas of endemism identified seem to be deeply influenced by the distribution of the emerged land during the 100-m marine highest that occurred during the late Miocene, but the establishment of the modern South American freshwater fish biota is the result of an interaction between marine incursions, uplift of the palaeoarches, and historical connections allowing cross-drainage dispersal.

According to Albert et al. (2020), the freshwater fish fauna of South America is the most diverse on Earth, with a current species richness of about 5,160 freshwater fish species, and the estimate points to a final diversity between 8,000 and 9,000 species (Reis et al. 2016). To assess the degree of knowledge of the ichthyofauna in the region, it is worth looking at the best-known and most-studied watershed in Brazil, the upper Paraná basin. The species richness of the basin in the first decade of the 21st century was still expanding, with 15% of the total described being recognized as new and under the description (Langeani et al. 2007). Therefore, to plan and define species conservation actions, it is recommended to have information on the biology of the fish and the genetic variability of the populations, under penalty of wasting energy without reaching the desired goals.

2. Influence of Fish Biology and Ecology on Conservation Actions

The planning of conservation actions is closely linked to the biology and ecology of the fish under management. Therefore, it is essential to recognize these traits and the strategies adopted by the species in the reproductive process when the gene flow to the new generations effectively occurs. It is easy to assume that different actions should be taken for species homogeneously distributed in a basin or several genetically distinct populations, either by homing behavior or another characteristic of reproductive behavior. Below, some of the leading life traits of South American fish and their influence on the definition of conservation measures to be adopted will be discussed.

2.1 Species That Perform Long Reproductive Migrations

Several species of importance for fishing in South America carry out reproductive migrations, including the genera *Salminus*, *Leporinus*, *Prochilodus*, *Brycon*, and *Semaprochilodus*. This group of fish has a wide home range and can populate extensive stretches of the watersheds in which they occur. They are also among the groups most affected by the construction of dams along rivers. The simple installation of a physical barrier between breeding and feeding homes can lead to the gradual local disappearance of an entire fish population. There are records of local extinction of migratory species due to the installation of hydroelectric dams, such as *Salminus brasiliensis* in the upper Uruguay River (Bem et al. 2021) or the collapse of the reproductive process of *Prochilodus argenteus* and *P. costatus* in São Francisco River (Pereira Arantes et al. 2011). Installing fish transposition structures has not adequately solved this problem (for more information, see Agostinho et al. 2002, 2011; Oldani et al. 2007; Pelicice and Agostinho 2008; Pompeu et al. 2012; Pelicice et al. 2015; Lira et al. 2017). As the home range of these species is extensive, multiple anthropic actions have reproductive success failures to guarantee the periodic recruitment of new generations. These stocks with limited recruitment are threatened with disappearing, taking with them genetic information that characterizes this population, and that is important for conservation and sustainability. Specific actions must be adopted if there is an interest in maintaining the genetic diversity that characterizes this population at risk. (See Item 3.1. Biobanks as environmental conservation tools: use in fish stocking actions.)

2.2 Species That Form Populations Through Homing Behavior, IBT, and Reduced Reproductive Migration

Among the species that carry out reproductive migration, there are records of some peculiarities in the reproductive behavior that can strongly interfere with the actions of conservation of the genetic diversity of a particular species. The homing behavior for large migratory Amazonian catfish is described by Duponchelle et al. (2016)

for *Brachyplatystoma rousseauxii*. It can use extensive areas of the lower Amazon River as a feeding area and carry out a reproductive migration that can exceed 8,000 km to the reproduction areas in a homing behavior, with spawning areas mapped in tributaries from the upper part of the Amazon basin, located in Bolivia, Peru.

Yet *Brachyplatystoma platynemum*, despite not having recorded natal homing behavior, performs a migration of several thousand kilometers in the Amazon basin, between estuary in Brazil to at least Bolivia and Peru (Hauser et al. 2019). The current and planned hydroelectric development for the region will severely disrupt migrations to access feeding areas and breeding grounds, affecting recruitment and population dynamics (Hauser et al. 2019). Alternatively, even the *Brachyplatystoma vaillantii*, which forms a panmictic population along the Solimões-Amazonas River and has greater genetic variability than other species Brachyplatystoma genus available so far, is possibly associated with the tributary's role (Formiga et al. 2021). Different strategies for *B. rousseauxii*, *B. platynemum,* and *B. vaillantii*, when *in situ* conservation, seem insufficient to guarantee stocks' genetic conservation.

Another reproductive behavior that interferes with the genetic diversity of a population that cohabits a given watershed is the sharing of spawning areas at different times (IBT), as recorded for *Salminus brasiliensis* in the Uruguay River (Ribolli et al. 2016), a condition that allows the formation of populations with distinct genetic characteristics and therefore should be separately treated when it comes to conservation.

A group of migratory fish also perform small migratory displacements, such as species of the genera Pseudoplatystoma and Steindachneridium. In both genera, there was a record of variation in the genetic structure of stocks located in the same basin, such as *S. scriptum* in the upper Uruguay River (Paixão et al. 2018) and *P. reticulatum* and *P. corruscans* in the basins of Paraguay and upper Paraná River (Prado et al. 2017) as well as among *P. corruscans* populations in the upper and middle Paraná River (Rueda et al. 2023). Thus, actions that aim to conserve genetic variability and the sustainability of stocks must consider these particularities listed above, with biobanks being valuable tools.

2.3 Species Widely Distributed Across the Continent

Considering that there are already cases of genetic structure between species that occupy the same hydrographic basin, it seems logical to assume that genetic differences are even greater between populations that inhabit different hydrographic basins. Some fish species are widely distributed across the American continent, occurring in basins from Central America to southern Argentina, such as *Rhamdia quelen*, proven a tremendous genetic differentiation between the different lineages (Ribolli et al. 2021a).

Other species of importance for fishing and aquaculture have a smaller distribution area. However, they are also distributed in different watersheds, such as *Colossoma macropomum* and *Brycon amazonicus*, which occur in the Amazon and Orinoco basins. In agreement with species-typical behavior of migratory movements driven by dispersal for feeding and reproduction during its life cycle,

C. macropomum forms a panmictic population along the Solimões-Amazon River channel (Santos et al. 2007), and *B. amazonicus* forms just one population in the central Amazon basin (de Oliveira et al. 2018). Considering the lack of gene flow between the isolated populations in these two watersheds, it is expected that there are genetic differences between the stocks. Therefore, they should be treated as separate stocks when considering conservation actions.

2.4 Species That Do Not Perform Reproductive Migration

Among species with sedentary reproductive behavior, greater spatial genetic variability is expected due to population genetic structure even within the same hydrographic basin. This situation is even more marked when the species is widely distributed. An example of this situation is seen in *Hoplias malabaricus*, widespread in the South American hydrographic basins. It is frequently recorded along the Orinoco, Amazonas, São Francisco, Paraná, Paraguay, and Uruguay River basins. This species is structured in six subpopulations genetically differentiated by high Fst values and restricts gene flow (Guimarães et al. 2022). From the point of view of conservation of genetic variability, the use of biobanks applies very well to the preservation of these distinct populations, being an adequate complement to *in situ* conservation actions.

2.5 Species and Populations Delimitation

Molecular identification of individuals is essential for species reproductive management, broodstock assembly, and correct gamete allocation in semen banks. Broodstock characterization allows targeted crossing, minimizing the loss of genetic variability resulting from the crossings of related individuals (An et al. 2013; Ferreira et al. 2019; Gonçalves et al. 2019; Urrea-Rojas et al. 2020). Furthermore, molecular identification of species prevents the unintentional intersection of morphologically similar individuals but constitutes distinct species (Melo et al. 2018; Ribolli et al. 2021a), thus avoiding introgressions into the natural environment, as in the case of Pseudoplatystoma genus in the Paraguay River (Vaini et al. 2014), both by repopulation programs and by accidental escapes from cultivation. Individual characterization can be done through microsatellite markers and individual and related attribution software. At the same time, molecular identification of species is widely carried out through DNA barcoding and genetic databases such as BOLD and GeneBank.

Characterizing populations within and between basins is the basis for the genetic conservation of natural fish stocks. Unstructured populations can be managed as a single panmictic population, permitting the collection of genetic material without restrictions as to place or time of year. However, genetically structured populations should have their structures considered, aiming to attain the evolutionary potential of each of them, such as environmental adaptations. The exchange of individuals between hydrographic basins, without consideration of these aspects, both of species identification and demarcation of populations, is the fundamental basis for conservation actions, the collection of breeders for cryopreservation of gametes, and their subsequent use in restoking programs.

3. Influence of the Geopolitical Configuration of Watersheds on Conservation Actions

The main river basins in South America are multinational, such as the Amazon, Prata, and Orinoco. Some species have a home range that goes beyond the territorial limits of Brazil, such as the large migratory catfish of the genus Brachyplatystoma in the Amazon (Formiga et al. 2021) and *Salminus brasiliensis* in the Prata basin (Rosso et al. 2018). For species whose life cycle of the species uses habitats in several countries, their management and conservation depend heavily on international efforts, and the results can be expanded with joint actions.

3.1 Biobanks as Environmental Conservation Tools: Use in Fish Stocking Actions

The current state of scientific knowledge has allowed for identifying the fish stocks' diversity and genetic variability and keeping viable samples in biobanks for future use. This condition allows us to use specific conservation-oriented actions, like managed relocation of species (Reid et al. 2019).

For stocking fish species programs, Krueger et al. (1981) recommend making separate collections of fish from each of several genetically different populations, performing all possible crosses between the different sources, and only then stocking the progeny. This strategy maximizes genetic variability and then relies on the environment to select the most appropriate genotypes. They also advise sampling populations from waters that are environmentally similar to those being rehabilitated that may contain preadapted genotypes. Finally, they recommend that this strategy consider using gene banks or natural refugia as a source of preadapted genetic variability (Krueger et al. 1981).

There are cases where the collapse of isolated fish populations is predicted over the years, mainly due to the reduction in the number of alleles, like *S. brasiliensis* in the upper Uruguay basin (Ribolli et al. 2021b), when genetic structure preserves populations partially isolated by an ancient natural barrier, but now completely wholly isolated by a dam. In this case, a specific action could allow the entry of genes from the adjacent population to increase genetic variability and avoid collapse. Here, both using biobanks and/or through the translocation of specimens between populations.

The research group on biology and freshwater fish culture from the Federal University of Santa Catarina, of which the authors of this chapter are part, worked for decades with conservation actions involving endemic species from the Uruguay River basin and biobanks. Below are some of these activities that exemplify the potential use of biobanks for this purpose. First, four action strategies followed almost simultaneously: (i) complementing studies on the biology and ecology of fish with an assessment of the genetic variability of stocks; (ii) the assembly of biobanks (*in vivo* and *in vitro*) that represented, when possible, the genetic variability observed in the natural environment; (iii) development or improvement of technologies for storing samples in biobanks and their use in assisted reproduction; (iv) the development or improvement of assisted reproduction and larviculture techniques. Moreover, later, (v) stocking fish species programs of the natural environment in specific conservation-oriented actions and guiding the management of areas of ecological importance relevant to fish.

The group's greatest dedication was working with species of migratory fish, among the groups most threatened by anthropic actions. The details of these action steps are as follows:

I) *Study of Fish Biology and Ecology*: This action allows the recognition of the distribution of the species and its genetic structure, as well as the home range involved in its life cycle, a condition that favors the following decision-making, such as the capture of samples for assembly of biobanks or the definition of areas of ecological importance and relevant for the specie. These studies made it possible to confirm the disappearance of fish in large stretches of the basin, such as *Brycon orbignyanus*, previously abundant in the upper part of the basin and which later occupied a small stretch on the border with the middle Uruguay River (Zaniboni-Filho and Schulz 2003). In this remaining stretch of occurrence of the species, areas of reproduction and recruitment of the species in adjacent areas were identified (Reynalte-Tataje et al. 2020), thus allowing the delimit of the area of vital importance for maintaining the stock. Additionally, strategies were set up to assemble biobanks for the species. Other studies carried out along the upper Uruguay River, a stretch of about 938 km that currently has four dams built on the main river channel, made it possible to map spawning and rearing areas of various migratory fish species, including *Salminus brasiliensis*, *Megaleporinus obtusidens*, *Steindachneridion scriptum*, and *Prochilodus lineatus* (Hermes-Silva et al. 2009; Corrêa et al. 2011). More punctual studies continuously developed for almost a decade showed that *S. brasiliensis*, *P. lineatus*, and *S. scriptum* can spawn in small lotic river stretches within two reservoirs, but only under very specific conditions and not common environmental conditions (Reynalte-Tataje et al. 2012b).

II) *Assembly of Biobanks* (*In Vivo* and *In Vitro*): The identification of reproduction areas and specific spawning conditions favored actions for the collection of semen destined for the creation of the *in vitro* bank, with the removal of the semen and the subsequent release of the fish. In other cases, immature and even young specimens were collected and kept in captivity until gamete production, enabling the collection and storage of gametes. For species with greater sensitivity to handling and transport, such as *S. brasiliensis* and *B. orbignyanus*, capturing juveniles proved quite efficient for setting up the biobank.

III) *Development or Improvement of Technologies for Storing Samples in Biobanks and Their Use*: An important partnership took place with the NGO World Fisheries Trust from Canada, allowing that in a few years, there was a protocol for storing South American semen samples (Carolsfeld et al. 2003)—a condition that allowed the starting of the assembly of the regional fish semen bank of the upper Uruguay River. This line of research was maintained, seeking to improve the results of the ideal inseminating dose for fertilization (Weingartner et al.

2015) or semen cryopreservation with new substances available on the market, like powder coconut water (Zanandrea et al. 2014, 2016).

IV) *Development and/or Improvement of Assisted Reproduction and Larviculture Techniques*: The indefinite storage of cryopreserved viable cells and tissues, followed by assisted or advanced assisted reproductive technology, maybe the only hope for species' long-term survival (Bolton et al. 2022). However, species-specific protocol optimization and increased knowledge of basic biology taxa are required (Bolton et al. 2022). In the case of *S. brasiliensis,* reproductive success through hormonal induction had already been obtained; however, always with a reduced fertilization rate (~ 10%). After years of investigation, this fertilization rate was increased by more than 500% simply by increasing the volume of water used in gamete activation (Weingartner et al. 2015). The larviculture of this species is equally challenging, with larvae practicing first-feeding cannibalism that results in very low survival rates. The use of live food and the definition of the daily consumption rates of the larvae (Weingartner et al. 2008), as well as the identification of the enzymatic activity of the larvae in the first days of life, determined when there was a physiological capacity to digest food inert (Vega-Orellana 2006), allowed a revolution in survival rates during larviculture.

In reproduction work with *B. orbignyanus*, research was even more challenging. A species with low stocks in the Uruguay River basin and listed in the red book of endangered Brazilian fauna (Agostinho et al. 2008), the strategy was to develop the entire protocol from juveniles collected in the Paraná River basin. While the research was being carried out, a huge effort was necessary to capture wild specimens in the Uruguay River, whose presence was restricted to the Turvo State Park area between the Brazilian state of Rio Grande do Sul and Argentina. Duly authorized for this work, the team took several years to obtain a small number of wild specimens. At that time, induced reproduction and larviculture techniques had already been mastered, making it possible to obtain the first production of juveniles in captivity descended from the Uruguay River (Weingartner et al. 2008). Part of this production (3,000 juveniles) was used in a widely publicized environmental education program, with marked fish being released in an isolated stretch between dams (in the Itá Reservoir, on the upper Uruguay River) where the species had been extinct for over 20 years. Although, at that moment, a publicity campaign encouraged the return of the marks of the fish captured with the capture data for a small reward, in addition to providing essential data for the research, it stimulated the discussion of the motivations for this disappearance of the species in the region, as well as build future measures to revert this situation and avoid the repetition of other extinctions. Additionally, some of these animals released into the reservoir in that campaign appeared in the region's artisanal fishing statistics a few years later (Schork and Zaniboni-Filho 2017).

V) *Stocking Fish Species Programs in Specific Conservation-Oriented Actions*: In addition to the environmental education program carried out with *B. orbignyanus*, as described above, other stocking actions with a similar purpose were developed with other species of migratory fish in stretches of the upper Uruguay River restricted between dams. Only two storage projects carried out in the region were different in terms of objectives:

- Recognizing that some migratory species did not reproduce annually in that stretch, as they depended on specific environmental conditions for this, but in the environment altered by the dams, this happened only sporadically (Reynalte-Tataje et al. 2012a); there were records of failures in the annual recruitment and consequent reduction in the population, also felt by data on artisanal fisheries in the region (Beux and Zaniboni-Filho 2008; Schork et al. 2013; Schork and Zaniboni-Filho 2017, 2018). At that moment, having a breeding stock originating from the region and with mastered assisted reproduction technology, it was possible to implement a program for the release of these species, which were also marked and the target of a campaign to return the marks through reward, so that they would fulfill their ecological role in the environment and also increase in fisheries.
- By identifying *S. scriptum* reproduction areas in a small lotic stretch of the Uruguay River located between two dams (Reynalte-Tataje et al. 2012b), adult specimens were stored, born in captivity, and originating from parents of the region, to increase the reproductive potential of the species in that stretch. In this region a genetic study with the species revealed low nucleotide diversity, spatial population structure, and the reduction of effective population size (Paixão et al. 2018).

In addition, special areas for spawning and recruitment of the species were mapped along the stretch, which were subsequently protected by the environmental agency. The possibility of using biobanks to support stocking actions favors that other objectives, in addition to simply increasing fish production, can be achieved more easily through careful planning of the action. Both are actions aimed at releasing fish to fulfill their ecological role in structuring the fish community, regardless of whether acting in a bottom-up direction, as observed with *Prochilodus mariae* (Taylor et al. 2006) or top-down, as observed with *Salminus brasiliensis* (Bem et al. 2021), stocking can also be used to increase population genetic variability and the effective number of fish in the region. Or, as suggested by Pullin et al. (1998), transferring representative groups from populations risk elsewhere to aquatic protected areas provided that health and biosafety requirements are met.

4. Conditions Provided for in Brazilian Legislation

Since 1934, Brazil has had regulations designed to guarantee fish stocks against the impacts caused by the construction of dams (Decree nº 24.643, of 1934), both for food security and for the conservation of stocks and the free movement of fish (Brasil 1934). Since then, this legislation has discussed and changed much to seek an evolution of obligations and reconcile developmental and conservationist interests (vide Section 2 in Vidal and Batlouni 2022). Nowadays, despite several laws generally dealing with the environmental licensing of HPPs and the protection of the environment, none of them necessarily requires specific ichthyofauna conservation programs. Thus, it is up to the licensing agency to establish such regulations as a condition for operating licenses for these projects (Vidal and Batlouni 2022). Considering the particularities of each enterprise and each hydrographic basin where it will be installed, it seems correct to assume that a detailed analysis of each enterprise by the environmental agency can favor more assertive decision-making on the actions to be adopted instead of national legislation that unifies these procedures.

For decades, sheltered by Brazilian legislation, the release of fish has become mandatory and is seen as one of the main ways to increase fish populations and preserve the ichthyofauna. The efficiency of this method has been questioned by the Brazilian scientific community due to the lack of monitoring based on scientific criteria (Agostinho et al. 2005; Casimiro et al. 2022).

5. Possibilities for Improving the Results of Current Actions

Any management action adopted after technically well-founded planning needs to be monitored, evaluated, and revised, thus making it possible to improve and redirect efforts to achieve the desired objectives. In the case of management actions involving the storage of fish produced in captivity, the fish farming stations intended to produce fish must have special production procedures different from those used to produce fish intended for aquaculture. Hatcheries must be designed to produce fish intended for storage and restricted to a watershed. Thus, breeding stocks can be set up for specific actions in certain regions, maintaining the breeding stock's genetic quality compatible with the natural population's genetic composition under management action. Additionally, broodstock must be chipped and genetically characterized to avoid crosses between genetically related individuals and performed directed crosses can be made to maximize the genetic variability of the fish batch to be stored, favoring the expansion of the genetic variability of the population under management action. Special care for the health of the fish must be present, thus preventing this action of storage from being a vector for the dissemination of diseases and pathogens. Some publications suggest a protocol to standardize and guide new conservationist policies of hatchery fish stocking (Casimiro et al. 2022).

The evolution of genetic diversity assessment techniques has allowed the identification and separation of genetically distinct fish populations, even within the same basin, as observed for *S. brasiliensis* in the Uruguay River (Ribolli et al. 2021b) and *P. corruscans* in the Paraná River (Rueda et al. 2023). This knowledge favors the adoption of assertive conservation management strategies.

Another important technological advance for conservation purposes is the development of surrogate broodstock via germ cell transplantation, which produces millions of fish gametes with desirable qualities. In this way, spawning the threatened species in captivity is unnecessary when gonadal tissue is available in biobanks. This technique offers a unique and adaptable method for storing and subsequently reintroducing genetic material into a population, which could be hugely beneficial to fish conservation (Mitra et al. 2023). There are already successful studies with primordial germ cell transplantation with South American fish, such as *P. lineatus* (Coelho et al. 2019), bringing new insights into gene banking and subsequent reconstitution through a germinal germline chimera in Neotropical fish species.

Considering the technological advances and the importance of actions to guarantee the conservation of the genetic diversity of South American fish, it is necessary to create a database of cryopreserved material (e.g., gametes, embryos) that can be shared, considering the peculiarities of the population structure of the species in the different hydrographic basins, can be a viable alternative to be implemented for Neotropical fish, such as cryobanks, which were created in 1981 in Europe. In addition, forming a single genetic bank parallel to the banks of research institutions could be an alternative, given the need for long-term maintenance, which could be used to recompose populations after disasters, such as after the tailing dam collapse on the Doce Basin, which happened in Brazil in 2015. Additionally, a database on the genetic diversity of fish in the Neotropics basins, like a sequence database such as BOLD, would allow the possibility of inserting and exchanging information on the genetic structure of species in the different hydrographic basins in an enlarged way, with the ability to group data, such as the study performed on *Pseudoplatystoma corruscans* from the Upper and Lower Paraná River (Rueda et al. 2023) and *S. scriptum* from upper Paraná and upper Uruguay rivers (Paixão et al. 2018). A database of genotypes and their locations would support gamete cryopreservation measures to preserve genetic variability while respecting aspects of the species' population structure.

Recognizing that 10% of all fish species in South America face some extinction risk, mainly due to habitat loss and degradation (Reis et al. 2016). Added to the knowledge that South American fish species living in tropical areas are more susceptible to global warming since they are already living close to their maximum habitat temperature (Campos et al. 2021), the use of these techniques may be increasingly necessary to avoid the disappearance of species and/or the reduction of their genetic variability.

Summary

Genetic diversity mapping techniques have gradually made it possible to identify and map this biodiversity richness in South American fish. Together with this knowledge, in addition to providing information about the life history of several species of fish, weaknesses that put the sustainability of these populations at risk have been pointed out. This prior knowledge allows for the adoption of specific measures to reduce these risks, helping to conserve this biodiversity. Biobanks have enormous potential to be used as germplasm sources to act punctually. In extreme cases of local extinctions, tissues stored in biobanks can be associated with assisted reproduction techniques with surrogate broodstock to rehabilitate missing populations. According to (Mitra et al. 2023), conservation programs often require that large numbers of new individuals be produced over many years to sustain declining populations successfully; therefore, the application of germ cell surrogacy is arguably better suited to support large-scale fish conservation efforts than other methods.

References

Agostinho, A.A., L.C. Gomes, D.R. Fernandez and H.I. Suzuki. 2002. Efficiency of fish ladders for Neotropical ichthyofauna. River Res. Appl. 299–306.

Agostinho, A.A., S.M. Thomaz and L.C. Gomes. 2005. Conservation of the biodiversity of Brazil's inland waters. Conserv. Biol. 19: 646–652.

Agostinho, A.A., F.M. Pelicice and L.C. Gomes. 2008. Dams and the fish fauna of the Neotropical region: impacts and management related to diversity and fisheries. Brazilian J. Biol. 68: 1119–1132.

Agostinho, C.S., F.M. Pelicice, E.E. Marques, A.B. Soares and D.A.A. Almeida. 2011. All that goes up must come down? Absence of downstream passage through a fish ladder in a large Amazonian river. Hydrobiologia 675: 1–12.

Albert, J.S., V.A. Tagliacollo and F. Dagosta. 2020. Diversification of neotropical freshwater fishes. Annu Rev. Ecol. Evol. Syst. 51: 27–53.

An, H.S., E.H. Shin, J.W. Lee, M.M. Nam, J.I. Meyong and C.M. An. 2013. Comparative genetic variability between broodstock and offspring populations of Korean starry flounder used for stock enhancement in a hatchery by using microsatellite DNA analyses. Genet. Mol. Res. 12: 6319–6330.

Bem, J.D., J. Ribolli, C. Röpke, K.O. Winemiller and E. Zaniboni-Filho. 2021. A cascade of dams affects fish spatial distributions and functional groups of local assemblages in a subtropical river. Neotrop. Ichthyol. 19: e200133.

Beux, L.F. and E. Zaniboni-Filho. 2008. Produção pesqueira do reservatório de Itá, Alto Rio Uruguai. pp. 87–108. *In*: Zaniboni Filho, E. and A.P.O.N. Nuñer [eds.]. Reservatório de Itá: estudos ambientais, desenvolvimento de tecnologias de cultivo e conservação da ictiofauna. UFSC, Florianópolis, Brasil.

Bolton, R.L., A. Mooney, M.T. Pettit, A. Bolton, L. Morgan, G.J. Drake et al. 2022. Resurrecting biodiversity: advanced assisted reproductive technologies and biobanking. Reprod. Fertil. 3: R121.

Brasil. 1934. Decreto nº 24.643; de 10 de julho de 1934. Decreta o Código de Águas. Diário Oficial da República Federativa do Brasil. Seção 1, p. 14738, 20 jun. 1934, Brasil.

Brasil. 2022. Estudo do Governo Federal mostra que Brasil tem 753,2 mil km de redes de distribuição de água e 365 mil km de redes de esgoto. Brasilia, Brasil.

Campos, D.F., R.D. Amanajás, V.M.F. Almeida-Val and A.L. VaL. 2021. Climate vulnerability of South American freshwater fish: Thermal tolerance and acclimation. J. Exp. Zool. Part. A. 335: 723–734.

Carolsfeld, J., B. Harvey, C. Ross and A. Baer. 2003. Migratory Fishes of South America. World Fisheries Trust, Victoria, Canada.

Casimiro, A.C.R., A.C. Vizintim Marques, A. Claro-Garcia, D.A.Z. Garcia, F.S. de Almeida and M.L. Orsi. 2022. Hatchery fish stocking: case study, current Brazilian state, and suggestions for improvement. Aquac. Int. 30: 2213–2230.

Coelho, G.C., I.S. Yo, T.M. Mira-López, P.S. Monzani, D.R. Arashiro, T. Fujimoto et al. 2019. Preparing a fish embryo (*Prochilodus lineatus*) for staging, chorion removal and PGC traceability. Int. J. Dev. Biol. 63: 57–65.

Corrêa, R.N., S. Hermes-Silva, D. Reynalte-Tataje and E. Zaniboni-Filho. 2011. Distribution and abundance of fish eggs and larvae in three tributaries of the Upper Uruguay River (Brazil). Environ. Biol. Fishes 91: 51–61.

de Oliveira, R.C., M.D.C.F. Santos, G. Bernardino, T. Hrbek and I.P. Farias. 2018. From river to farm: an evaluation of genetic diversity in wild and aquaculture stocks of *Brycon amazonicus* (Spix & Agassiz, 1829), Characidae, Bryconinae. Hydrobiologia 805: 75–88.

Duponchelle, F., M. Pouilly, C. Pécheyran, M. Hauser, J.F. Renno, J. Panfili et al. 2016. Trans-Amazonian natal homing in giant catfish. J. Appl. Ecol. 53: 1511–1520.

EPE. 2022. Brazilian Energy Research Company. *In*: http://www.epe.gov.br. Acessed in April 20, 2023.

Ferreira, L.D.A., P.F. Fazzi-Gomes, S. Guerreiro, M.D.N. Rodrigues, A.K. Ribeiro-dos-Santos, S. Santos et al. 2019. Genetic variability of tambaqui broodstocks in the Brazilian state of Para. Rev. Bras. Zootec. 48: e20180106.

Formiga, K.M., J.S. Batista and J.A. Alves-Gomes. 2021. The most important fishery resource in the Amazon, the migratory catfish *Brachyplatystoma vaillantii* (Siluriformes: Pimelodidae), is composed by an unique and genetically diverse population in the Solimões-Amazonas River System. Neotrop. Ichthyol. 19: e200082.

Gonçalves, R.A., C.H.D.A.D. Santos, C.S. de Sá Leitão, E.M.S. de Souza and V.M.F. de Almeida-Val. 2019. Genetic basis of *Colossoma macropomum* broodstock: Perspectives for an improvement program. J. World Aquac. Soc. 50: 633–644.

Guimarães, K.L.A., M.P. Lima, D.J. Santana, M.B.B. de Souza, R.S. Barbosa and L.R.R. Rodrigues. 2022. DNA barcoding and phylogeography of the *Hoplias malabaricus* species complex. Sci. Rep. 12: 5288.

Hauser, M., C.R. Doria, R.V. Santos, A. García-Vasquez, M. Pouilly and C. Pécheyran. 2019. Shedding light on the migratory patterns of the Amazonian goliath catfish, *Brachyplatystoma platynemum*, using otolith 87 Sr/86 Sr analyses. Aquat. Conserv. Mar. Freshw. Ecosyst. 29: 397–408.

Hermes-Silva, S., D. Reynalte-Tataje and E. Zaniboni-Filho. 2009. Spatial and temporal distribution of ichthyoplankton in the Upper Uruguay river, Brazil. Brazilian Arch. Biol. Technol. 52: 933–944.

Hubert, N. and J.F. Renno. 2006. Historical biogeography of South American freshwater fishes. J. Biogeogr. 33: 1414–1436.

Krueger, C.C., A.J. Gharrett, T.R. Dehring and F.W. Allendorf. 1981. Genetic aspects of fisheries rehabilitation programs. Can. J. Fish. Aquat. Sci. 38: 1877–1881.

Langeani, F., R.M.C. Castro, O.T. Oyakawa, O.A. Shibatta, C.S. Pavanelli and L. Casatti. 2007. Diversidade da ictiofauna do Alto Rio Paraná : **composição** atual e perspectivas **futuras Introdução** O Alto Paraná é área complexa devido às atividades tectônicas. Biota Neotrop. 7: 1–18.

Lira, N.A., P.S. Pompeu, C.S. Agostinho, A.A. Agostinho, M.S. Arcifa and F.M. Pelicice. 2017. Fish passages in South America: An overview of studied facilities and research effort. Neotrop. Ichthyol. 15: 1–14.

Melo, B.F., B.F. Dorini, F. Foresti and C. Oliveira. 2018. Little divergence among mitochondrial lineages of *Prochilodus* (Teleostei, Characiformes). Front. Genet. 9: 107.

Mitra, A., F.K. Abdel-Gawad, S. Bassem, P. Barua, L. Assisi, C. Parisi et al. 2023. Climate change and reproductive biocomplexity in fishes: innovative management approaches towards sustainability of fisheries and aquaculture. Water 15: 725.

Oldani, N.O., C.R.M. Baigún, J.M. Nestler and R.A. Goodwin. 2007. Is fish passage technology saving fish resources in the lower La Plata River basin? Neotrop. Ichthyol. 5: 89–102.

Oliveira, N.C.C. 2018. The great acceleration and hydroelectric dam building in Brazil. Varia. Hist. 34: 315–346.

Paixão, R.V., J. Ribolli and E. Zaniboni-Filho. 2018. Genetic variation of the endangered neotropical catfish *Steindachneridion scriptum* (siluriformes: Pimelodidae). Front. Genet. 9: 48.

Pelicice, F.M. and A.A. Agostinho. 2008. Fish-passage facilities as ecological traps in large neotropical rivers. Conserv. Biol. 22: 180–188.

Pelicice, F.M., P.S. Pompeu and A.A. Agostinho. 2015. Large reservoirs as ecological barriers to downstream movements of Neotropical migratory fish. Fish Fish. 16: 697–715.

Pereira Arantes, F., H. Batista dos Santos, E. Rizzo, Y. Sato and N. Bazzoli. 2011. Collapse of the reproductive process of two migratory fish (*Prochilodus argenteus* and *Prochilodus costatus*) in the Três Marias Reservoir, São Francisco River, Brazil. J. Appl. Ichthyol. 27: 847–853.

Pompeu, P.S., A.A. Agostinho and F.M. Pelicice. 2012. Existing and future challenges: The concept of successful fish passage in South America. River Res. Appl. 28: 504–512.

Do Prado, F.D., R. Fernandez-Cebrián, D.T. Hashimoto, J.A. Senhorini, F. Foresti, P. Martínez et al. 2017. Hybridization and genetic introgression patterns between two South American catfish along their sympatric distribution range. Hydrobiologia 788: 319–343.

Pullin, R.S.V., J. Bell, J.C. Danting and F. Longalong. 1998. Gene banking for fish and other aquatic organisms: ICLARMs perspectives and experiences. pp. 31–43. *In*: Brian, H., C. Ross, D. Greer and J. Carolsfled [eds.]. Action Before Extinction. World Fish Trust, Victoria, Canada.

Reid, A.J., A.K. Carlson, I.F. Creed, E.J. Eliason, P.A. Gell, P.T. Johnson et al. 2019. Emerging threats and persistent conservation challenges for freshwater biodiversity. Biol. Rev. 94: 849–873.

Reis, R.E., J.S. Albert, F. Di Dario, M.M. Mincarone, P. Petry and L.A. Rocha. 2016. Fish biodiversity and conservation in South America. J. Fish Biol. 89: 12–47.

Reynalte-Tataje, D.A., A.A. Agostinho, A. Bialetzki, S. Hermes-Silva, R. Fernandes and E. Zaniboni-Filho. 2012a. Spatial and temporal variation of the ichthyoplankton in a subtropical river in Brazil. Environ. Biol. Fishes 94: 403–419.

Reynalte-Tataje, D.A., A.P. Nuñer, M.C. Nunes, V. Garcia, C.A. Lopes and E. Zaniboni-Filho. 2012b. Spawning of migratory fish species between two reservoirs of the upper Uruguay river, Brazil. Neotrop. Ichthyol. 10: 829–835.

Reynalte-Tataje, D.A., M.D.L. Soares, M.V. Massaro, R. Bastian and F.M. Pelicice. 2020. First evidence of a spawning site of the endangered fish *Brycon orbignyanus* (Valenciennes, 1850) (characiformes, bryconidae) in the middle Uruguay River, Brazil. Acta Limnol. Bras. 32: 1–5.

Ribolli, J., D.J. Hoeinghaus, J.A. Johnson, E. Zaniboni-Filho, P.D. de Freitas and P.M. Galetti Jr. 2016. Isolation-by-time population structure in potamodromous Dourado *Salminus brasiliensis* in southern Brazil. Conserv. Genet. 18: 67–76.

Ribolli, J., E. Zaniboni Filho, B.M.S. Scaranto, O.A. Shibatta and C.B. Machado. 2021a. Cryptic diversity and diversification processes in three cis-Andean *Rhamdia* species (Siluriformes : Heptapteridae) revealed by DNA barcoding. Genet. Mol. Biol. 44: e20200470.

Ribolli, J., E. Zaniboni-Filho, C.B. Machado, T.C.D.S. Guerreiro, P.D.D. Freitas and P.M. Galetti Jr. 2021b. Anthropogenic river fragmentation reduces the long-term viability of the migratory fish *Salminus brasiliensis* (Characiformes: Bryconidae) populations. Neotrop. Ichthyol. 19: 1–17.

Rosso, J.J., E.C. Rueda, S. Sanchez, M.C. Bruno, J. Casciotta, G. Aguiler et al. 2018. Basin-scale distribution and haplotype partitioning in different genetic lineages of the Neotropical migratory fish *Salminus brasiliensis*. Aquat. Conserv. 28: 444–456.

Rueda, E.C., C.B. Machado, V. Castro, A. Braga-Silva, G. Ojeda, F. Vargas et al. 2023. Genetic population structure of *Pseudoplatystoma corruscans* (Siluriformes: Pimelodidae) and evidence of temporal variation in structure. J. Fish Biol. 102: 1040–1048.

Santos, M.C.F., M.L. Ruffino and I.P. Farias. 2007. High levels of genetic variability and panmixia of the tambaqui *Colossoma macropomum* (Cuvier, 1816) in the main channel of the Amazon River. J. Fish Biol. 71: 1–12.

Sarkar, B. and A. Islam. 2020. Drivers of water pollution and evaluating its ecological stress with special reference to macrovertebrates (fish community structure): a case of Churni River, India. Environ. Monit. Assess. 192: 45.

Schork, G., S. Hermes-Silva and E. Zaniboni-Filho. 2013. Analysis of fishing activity in the Itá reservoir, Upper Uruguay River, in the period 2004–2009. Braz. J. Biol. 73: 559–571.

Schork, G. and E Zaniboni-Filho. 2017. Structure dynamics of a fish community over ten years of formation in the reservoir of the hydroelectric power plant in upper Uruguay River. Braz. J. Biol. 77: 710–723.

Schork, G. and E. Zaniboni-Filho. 2018. Influence of spatial gradient caused by a large dam on fish assemblage in a subtropical reservoir – Upper Uruguay river. Bol. Inst. Pesca 44: e236.

Taylor, B.W., A.S. Flecker and R.O. Hall. 2006. Loss of a harvested fish species disrupts carbon flow in a diverse tropical river. Science 313: 833–836.

United Nations. 2018. World Youth Report: Youth and the 2030 Agenda for Sustainable Development. United Nations Publications, New York.

Urrea-Rojas, A.M., F.P. de Souza, E.C.S. de Lima, A.L. Yamachita, V.C.F. Pandolfi and S.M. de Godoy. 2020. Genetic diversity of Tambaqui (Teleostei - Characidae) broodstocks from Northern region of Brazil using microsatellite markers. Semin. Cienc. Agrar. 41: 3249–3258.

Vaini, J.O., A.B. Grisolia, F.D.D. Prado and F. Porto-Foresti. 2014. Genetic identification of interspecific hybrid of neotropical catfish species (*Pseudoplatystoma corruscans* vs. *Pseudoplatystoma reticulatum*) in rivers of mato grosso do sul state, Brazil. Neotrop. Ichthyol. 12: 635–641.

Vega-Orellana, O.M., D.M. Fracalossi and J.K. Sugai. 2006. Dourado (*Salminus brasiliensis*) larviculture : Weaning and ontogenetic development of digestive proteinases. Aquaculture 252: 484–493.

Vidal, M.V. and S.R. Batlouni. 2022. The environmental licensing of hydroelectrics and the interface with migratory fish and aquaculture in Brazil. Bol. Inst. Pesca 48: e696.

Weingartner, M., D.M. Freacalossi, L.F. Beux and A.P.O.N. Nuñer. 2008. Desenvolvimento de tecnologias de cultivo para peixes nativos do alto rio Uruguai. pp. 257–306. *In*: Zaniboni Filho, E. and A.P.O.N. Nuñer [eds.]. Reservatório de Itá: estudos ambientais, desenvolvimento de tecnologias de cultivo e conservação da ictiofauna. UFSC, Florianópolis, Brasil.

Weingartner, M., A.C.V. Zanandrea and E. Zaniboni-Filho. 2015. Cryopreserved sperm for oocyte fertilization of dourado *Salminus brasiliensis*. Cienc. Rural 45: 892–897.

W. W. F. 2022. Relatório Planeta Vivo 2022 – Construindo uma sociedade positiva para a natureza. WWF, Gland, Suíça.

Zanandrea, A.C.V., M. Weingartner and E. Zaniboni-Filho. 2014. Use of ACP-104 at different dilutions for sperm cryopreservation of dourado, *Salminus Brasiliensis*. J. World Aquac. Soc. 45: 82–87.

Zanandrea, A.C.V., M. Weingartner and E. Zaniboni-Filho. 2016. Induced reproduction of dourado (*Salminus brasiliensis*): fertilization with sperm cryopreserved in ACP®-104. Acta Sci. 38: 17–22.

Zaniboni-Filho, E. and H.U. Schulz. 2003. Migratory fishes of the Uruguay River. pp. 157–194. *In*: Baer, A. and C. Ross [eds.]. Migratory Fishes of the South America: Biology, Social Importance and Conservation Status. World Fisheries Trust, Victoria, Canada.

CHAPTER 13

Potential Use of Germplasm Banks of Neotropical South American Species for Fish Farming in Brazil

Danilo P. Streit Jr.,[1,2,*] *Jayme A. Povh,*[3] *Darci C. Fornari,*[1] *Jonhy L. Benato,*[1] *Fernanda De Mello,*[3] *Eric A. B. Routledge,*[4] *Estefania Paredes*[5] and *Tiantian Zhang*[6]

INTRODUCTION

Brazil is one of the leading players in global food production (both plant and animal). Regarding animal protein production, about the three terrestrial species considered domestic—cattle, swine, and poultry—Brazil is respectively the second, fourth, and second-largest producer (USDA 2023). However, the country is not yet among the top ten aquaculture producers. Brazilian fish production has been gradually increasing its significance as a source of animal protein in the national agricultural scenario.

[1] Aquam Research Group, Animal Science Research Program, Federal University of Rio Grande do Sul, Porto Alegre, RS, Brazil.

[2] Veterinary Science Program, Federal University of Rio Grande do Sul, Porto Alegre, RS, Brazil.

[3] Department of Animal Science, Federal University of Mato Grosso do Sul, Campo Grande, Mato Grosso do Sul, MS, Brazil.

[4] Department of Physiology, Bioscience Institute, University of São Paulo (IB/USP), São Paulo, SP, Brazil.

[5] Centro de Investigación Mariña (CIM)-Universidade de Vigo, Ecoloxía e Bioloxía Animal, Facultade de Ciencias do Mar, Vigo (Pontevedra), Spain.

[6] Faculty of Science and Technology, Bournemouth University, Poole, United Kingdom.

Email: darci.peixegen@gmail.com, jhony.benato@ufrgs.br, jaymepovh@gmail.com, fernandade.mello@gmail.com, eric.routledge@embrapa.br, eparedes@uvigo.gal, tzhang@bournemouth.ac.uk

* Corresponding author: danilo.streit@ufrgs.br

Over the past ten years, there has been an average growth of 5% in fish production, reaching the current 560,000 tons (IBGE 2022). Several factors have contributed to this favorable scenario, including the Brazilian population's increased demand for fish as a source of animal protein and the intensification of technological processes in tilapia production.

The importance of genetic banks gained significant attention with the approval of the United Nations' 2030 Agenda for Sustainable Development in 2015. The FAO (Food and Agriculture Organization) established the goals for sustainable development indicators regarding food production. The hands of Target 2.5.1b, published on the FAO website (2023), address the crucial importance of animal germplasm banks in contributing to the future subsistence of the global population through sufficient, diversified, and nutritious diets. Another critical consideration in this document relates to genetic reserve (germplasm banks) and their significance as the most reliable long-term strategy for conserving genetic resources worldwide.

The sustainability future of aquaculture is consistently linked to the diversity of species that should be produced, according to a document by the FAO (2019). Germplasm banks are essential in achieving the goals of the leading organization for eradicating hunger worldwide. According to the same FAO document (2019), three factors are indispensable for attaining aquaculture sustainability: (I) increasing the number of species produced, (II) enhancing the uniformity of the species produced, and (III) promoting diversity among the currently produced species through the development of new varieties. In this context, we can see how germplasm banks can and are crucial in developing and consolidating different South American species with significant potential from a zootechnical perspective.

1. The Number of Species Produced Versus Germplasm Banks

The diversity of South American freshwater fish species is known to be the richest among all continents on the planet (Reis et al. 2016; Dagosta and de Pina 2018). This richness is reflected in the production of freshwater species, as consistently shown in the data published regarding Brazilian production. In the latest official document (IBGE 2022), eight of every ten species produced in the country are native (or hybrids). However, native species account for just over 30% of the total fish production in Brazil. Therefore, an important question arises: Is there a germplasm bank that houses the diversity of South American fish specimens? Ideally, there should be one. However, such a bank does not exist, and it would be practically unfeasible from a technical and operational standpoint, given the rich diversity of South American fish species.

One solution to establish and operate a functional germplasm bank, housing a good representation of species diversity and ensuring significant conservation of the target basin's fish fauna, would be to identify priority species. However, selecting these species for gene conservation in a germplasm bank is not a simple task; on the contrary, it is highly complex and involves numerous factors that must be considered in the decision-making process. In this chapter, we should consider the approach of germplasm banks as potential safeguards for their genes, following Geist's (2015) theory and practice of conserving priority species. This approach can serve as a pathway for assembling a germplasm bank with a good representation of South American fish fauna. According to Geist, flagship species are often the focus of conservation efforts. These are typically iconic or charismatic species that attract significant public attention but, in contrast to keystone species, are not necessarily of great importance for ecosystem resilience or functioning. The same author had previously emphasized this understanding in another study. In conserving freshwater fish species, the most significant effort should typically be devoted to specific target species, considered flagships, keystones, indicators, and umbrellas (Geist 2010).

Considering that the diversity of South American freshwater fish fauna is regarded as the most mega-diverse, surpassing 5,500 species (Reis et al. 2016; Dagosta and de Pina, 2018), how can we determine or evaluate Neotropical South American species as "key" and prioritize them as targets for conservation through a macro germplasm bank? Geist (2015) gives us a possible answer, which uses reproductive migratory species (such as salmon) as an example. In this case, different species of salmon were considered emblematic (for recreational/commercial fishing and, in this context, aquaculture) but also potential indicators as keystone species in numerous eco-biological flows. Fortunately, this model resembles some Neotropical migratory species found in South America. By applying Geist's (2015) proposed model and using *Colossoma macropomum* (known as tambaqui) as an equivalent, it is possible to consider it emblematic in the South American context as a keystone species for gene conservation through germplasm banks. In the Amazonian rivers, it is a crucial species in the floodplain dynamics of the Amazon basin, migrating extensive distances (Araújo-Lima et al. 2004). It also survives for several years in flooded forests (igapós) and contributes significantly to the dispersal of seeds of riparian and spring trees in the Amazon (Araújo-Lima et al. 1997; Lucas 2008).

While *C. macropomum* was once abundant in the Amazonian rivers, it is now the second most produced species in Brazilian fish farming (IBGE 2022). Intrinsic zootechnical characteristics of the species have been instrumental in making it a strong candidate for aquaculture in Brazil, Colombia, Peru, and Ecuador. High prolificacy is a characteristic of *C. macropomum*, and many medium and large-bodied native South American species are known to migrate during the reproductive period. According to the Brazilian Institute of Geography and Statistics (IBGE 2022), over 98% of the Brazilian production of South American species consists of migratory species. Coincidentally, nearly all the research on seminal cryopreservation of South American species focuses on medium to large-sized reproductive migrants. Among the 131 articles published in South America over the past 40 years, only two studies involving *Austrolebias minuano* (Fernandes et al. 2019) and *Astyanax lacustris* (Carneiro-Leite et al. 2020) do not fit the criteria of medium to large-sized reproductive migrants. Among the six families that have studied semen cryopreservation protocols, all possess two measures that can be considered fundamental for being considered keystone species in a large-scale genetic conservation program: (1) strong appeal for recreational fishing and aquaculture and (2) essential for the complex bio-ecological flow of the major South American river basins.

2. National Center for Research in Genetic Resources and Biotechnology - CENARGEN

In Brazil, the National Center for Research in Genetic Resources and Biotechnology (CENARGEN) belongs to the Brazilian Agricultural Research Corporation (EMBRAPA). It is responsible for maintaining the collection of animal and plant genetic resources in the country. In its access portal called Sistema ALELO, they have records of cryopreserved seminal samples from ten native South American species. The maintenance and expansion of CENARGEN's collection of Brazilian ichthyofauna samples are fundamentally linked to the approval of projects commissioned by EMBRAPA's management for genetic resource conservation. In this regard, one of the company's objectives is to increase its germplasm bank, both *in situ* and *ex situ*. An example of this model is cited by Villela et al. (2020), based on the cryopreservation of tissue and semen samples from three native fish species: *C. macropomum*, *Pseudoplatystoma* sp., and *Piaractus brachypomus* (known as Caranha or Pirapitinga). In particular, the *in situ* germplasm bank (GBA) has been structured and maintained by EMBRAPA Fisheries and Aquaculture, one of EMBRAPA's research centers, in collaboration with partner fish farms through cooperation agreements (Satellite Centers). The germplasm bank has two conservation nuclei: Tambaqui (*C. macropomum*), composed of 200 broodstock (100 animals of each sex), and Pirarucu (*Arapaima gigas*) with 50 broodstock (25 animals of each sex). This EMBRAPA center also has a biological collection of caranha/pirapitinga (*P. brachypomus*) with 63 broodstock (30 males and 33 females). Maintenance activities, phenotypic characterization, germplasm enrichment, and exchange are periodically carried out in the germplasm bank, in addition to the cryopreservation of tambaqui and caranha semen. DNA and tissues of these species are sent to CENARGEN for conservation in the genetic bank for characterization (genetic structure and diversity; identification of gene introgression; consanguinity and parentage relationships) and for future use. The information collected in the germplasm bank is registered in the Sistema ALELO, allowing the cataloging and understanding of these genetic resources (Villela et al. 2020).

Therefore, it is crucial that this initiative is permanent and strengthened to increase the number of species significantly conserved in CENARGEN and thus ensure the first prerequisite of aquaculture sustainability recommended by FAO (2019). Notably, out of the species with cryopreserved seminal samples in CENARGEN, only seven native South Americans are reproductive migrators.

3. Increased Uniformity of Species Produced Versus Germplasm Bank

When analyzing production data in Brazilian fish farming regarding species diversity, we can consider that a relatively high number of species have been produced in Brazil, according to IBGE (2022). This is undoubtedly related to the diversity of South American ichthyofauna. On the other hand, the contribution of these species to Brazilian production is not directly proportional, as it has decreased from 52.6% (2014) to only 31.6% (2020) in the past eight years (IBGE 2022) (Figure 1). It is

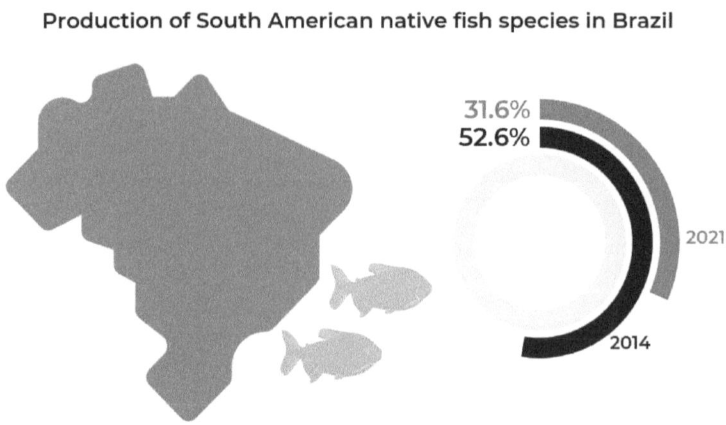

Figure 1. Production of South American native fish species in Brazil in 2014 ■ and 2021 ■, according to IBGE (2022).

important to note that approximately 26% of this volume of South American fish produced in the country is represented by *C. macropomum* and its hybrids. In this sense, it is crucial to understand what happened in the past for South American species (especially *C. macropomum*) to transition from a promising future in national aquaculture to an uncertain present and future regarding their representation in national production.

The alarming decline in the production of native South American species in Brazil, in our view, is directly related to a fundamental technical factor for the zootechnical development of a species: Genetic Improvement Program. To achieve optimal statistical accuracy indices, the animals that will be tested must necessarily have a response pattern as identical as possible. Therefore, a classical Genetic Improvement Program for a species becomes the first task for the expected zootechnical development regarding diet quality, intensive management, and animal health.

The first fish breeding programs based on families were conducted in the 1970s by AKVAFORSK with *Salmo salar* (Gjedrem 2010). A little over 20 years ago, according to Neira (2010) and Rye et al. (2010), around 86 Genetic Improvement Programs were being carried out worldwide for different fish species, two in Brazil, both for tilapia. In 2008, EMBRAPA started the first Genetic Improvement Program for two South American species in Brazil, *C. macropomum*, and *Pseudoplatystoma reticulatum* (EMBRAPA 2009; Marcos et al. 2016). At that time, EMBRAPA Pantanal, led by researcher Dr. Emiko Resende, initiated a network project called AQUABRASIL to make the country one of the leading players in global aquaculture production. One initial objective identified by the AQUABRASIL technicians was to develop Genetic Improvement Programs (EMBRAPA 2009).

In this chapter, we consider it essential to exemplify how it was possible to build a Genetic Improvement Program using the germplasm bank as a tool. The operational advantages of the germplasm bank were fundamental in the initial process of creating the Genetic Improvement Program for the two target species of the AQUABRASIL project. One example of this plasticity was the ability to transport genetic material (sperm) between different locations to compose families. The Genetic Improvement Program of the AQUABRASIL project was designed on the composition of half-sibling families. Subsequently, these families were subjected to the same selection environment to select the best animals that responded to the zootechnical growth parameter.

Next, we summarize the Genetic Improvement Program of the AQUABRASIL project, highlighting in this chapter the essential importance of the germplasm bank of both species in the composition of the family structure:

1. Seminal cryopreservation of populations previously mapped for genetic distance;
2. Transport of cryopreserved semen for previously established targeted crosses to form half-sibling families, aiming to avoid the formation of inbred families;
3. Obtaining oocytes from a female, dividing them into two equal-weight portions, and fertilizing them with cryopreserved semen from two different animals;
4. Embryos and larvae are produced separately from each targeted parental combination;
5. All half-sibling families were transported to a single place for the growth of the individuals. The juveniles were tagged when they reached 8 grams using microchips and raised in the same location to undergo the same selection pressure;
6. After the formation of the initial or zero population, the best individuals from each family were selected for a new targeted crossbreeding to form new half-sibling families, resulting in F1;
7. New seminal cryopreservation (formation of a new germplasm bank) to be used in the composition of F1.

At this point, one crucial piece of information relates to the semen samples: males of these species release semen in their second year of life, while females reach maturity between three and four years. Therefore, for the formation of the F1 population in the AQUABRASIL project, the semen was frozen one year before the formation of half-sibling families.

Some considerations that we deem extremely important in the legacy left by this prematurely terminated project.

1. Operationality of the germplasm bank:
   a. Brazil is a country with continental dimensions, and transporting live breeders would be a costly and complex operation. Thus, transporting cryopreserved semen was strategic for the project's development.
   b. Both *C. macropomum* and *P. reticulatum* are annual migratory reproductive species. Therefore, the availability of cryopreserved semen (one year in advance) for use in targeted crossbreeding was not dependent on the timing of the female's oocyte spawning, enabling the composition of families for the Genetic Improvement Program. This process occurred in forming the initial

population and in the design of F1. It is worth noting that oocyte spawning was obtained from one female and divided into equal-weight parts to compose these half-sibling families and then fertilized with cryopreserved semen from two animals previously genetically characterized. This model was the basis for the family formation, isolating the female's effect and proving the operational zootechnical efficiency of a germplasm bank.

2. Zootechnical gain with the Genetic Improvement Program of the AQUABRASIL project, using the germplasm bank as a tool:
    a. During the zootechnical evaluation of the Initial Population, there were observed gains of over 20% in average weight gain in some families compared to a non-improved group of animals. After the close of the AQUABRASIL project, F1 families were maintained by some collaborating private producers. Some of these F1 families were evaluated by Marcos et al. (2016), confirming a gain of over 14.8% compared to non-improved animals. This value corroborates the observation by Gjedrem and Rye (2018) that genetic gains for growth rate average 13% for fish.

Unfortunately, the Genetic Improvement Program of the AQUABRASIL Project was finalized upon obtaining F1 (except for a few F2 individuals obtained for research purposes). At that time, the potential for zootechnical competitiveness of the species in the future market could be envisioned, but, unfortunately, a part was lost. EMBRAPA, through the transfer of animals from former partners of the AQUABRASIL project and cryopreserved semen samples maintained at EMBRAPA Pantanal, preserved a portion of these improved families, even though the Improvement Program did not have the expected development afterward. We understand that this fact contributes to the loss of competitiveness of native South American, especially of *C. macropomum* in the Brazilian market, reflected in the fall in the volume produced by over 30% in seven years, according to IBGE data (2022). Here, we emphasize that the germplasm bank worked as an operational tool for developing the zootechnical gain process for one species of two South American species. This pioneering work on using a germplasm bank of native species as a tool for building a Genetic Improvement Program can be found in the article by Streit Jr. et al. (2015).

4. Increased Diversity Within Currently Cultivated Species Through the Development of New Strains Versus Germplasm Banks (GBA)

In this chapter, we discuss the use of the germplasm bank to support fish production in Brazil, using the aquaculture sustainability premises of the FAO to demonstrate the necessity of a germplasm bank, which is inseparable from the previous assumption (2). The acquisition of new strains would be a consequence for a species through its Genetic Improvement Program. In this case, as the Genetic Improvement Program progresses, producing new generations, new genes with desired zootechnical characteristics can be fixed from animals stored in germplasm banks. Therefore, if there is no established Genetic Improvement Program with generations being produced for native South American species, there are no lineages for these species. EMBRAPA, through the Aquaculture and Fishing Research Center, is promoting initiatives to establish a foundation for a new Genetic Improvement Program for native Neotropical South American species, with the initial focus on the species *C. macropomum*.

Even without improved populations for native South American species, it is possible to critically analyze and anticipate future problems if a Genetic Improvement Program, like the AQUABRASIL project, is established at some point. As mentioned in this chapter, the Brazilian Government is responsible for maintaining the CENARGEN (EMBRAPA 2023), which conserves samples of about 25% of Brazil's most cultivated fish species, according to those reported by IBGE (2022). Let us use *C. macropomum* as an example due to it being the most produced native South American species throughout the South American continent and also because there may have been a Genetic Improvement Program at some point.

The germplasm bank of CENARGEN currently has 2,843 seminal samples of *C. macropomum*, which is a significantly high number. However, due to temporary operational issues, the ALELO system needs to provide information on the genotype of these samples. Although there are no significant technical difficulties in extracting DNA and genetically characterizing individuals from cryopreserved spermatozoa, there is a considerable risk that these samples may exhibit low genetic variability. In this case, the available samples of *C. macropomum* maintained by the Brazilian Government would have limited use in ensuring the species' genetic conservation. Aguiar et al. (2018) reinforced the low genetic variability, where the authors report a dramatic loss of genetic variability in *C. macropomum* breeders, mainly due to the founder effect due to the absence of adequate genetic management for these animals.

A solution to this problem would be to capture *C. macropomum* breeders wild (non-domesticated) and, with this strategy, ensure an increase in the genetic variability of this species in future Genetic Improvement Programs. However, it is essential to remember that fattening animals from fish farms in the Amazon escape to wild populations, with introgression of genes fixed by selection, as already pointed out by Aguiar et al. (2013). Many consequences can occur due to the gene flow between farm animals (domesticated) and wild populations. According to Bekkervold et al. (2006), salmon is seen as a model for understanding direct genetic interactions between domesticated and wild fish stocks and to understand the problem from this point of view. In this species, a change in life cycle characteristics (Glover et al. 2017), a reduction in egg size due to body size selection (Hagen et al. 2019), and a reduction in the survival rate in juveniles (Sylvester et al. 2019; Wacker et al. 2021) has been observed. However, it is essential to emphasize that studies to characterize the degree of genetic introgression caused by escape for wild populations of salmon showed that there is no direct response model (Besnier et al. 2011; Karlsson et al. 2011; Glover et al. 2013). On the other hand, it was evident in studies with salmon that the most important thing is the density of the wild population. In this case, population size and good genetic diversity would be able to absorb a possible genetic introgression from farmed salmon (Glover et al. 2012), supported by modeling by Heino et al. (2015).

Another critical point to ensure a future Genetic Improvement Program, which consequently can guarantee new lineages, is ensuring that the cryopreserved semen samples are functional. Chapter 10 contains all the articles produced with seminal cryopreservation of South American migratory species until the publication of this book. It is a fact that *C. macropomum* is the second South American species with the highest number of scientific articles on seminal cryopreservation, with 20 articles in total. However, no defined protocol has been efficiently and exhaustively repeated, which allows for ensuring the production of animals from the germplasm bank. When we contrast, for example, with *Rhamdia quelem*, another South American species, the protocol is defined, efficient, safe, and repeated in ten of the 11 articles published with the species. However, unfortunately, this species is not representative in the aquaculture scenario Brazilian, being classified as "other species" by IBGE (2022).

Another example of "genetic reserve insecurity" in South American species is the case of *Leiarius marmoratus*. To better understand the importance of this species in the Brazilian scenario, a brief history of South American catfish production is necessary, which began with the genus *Pseudoplatystoma*. In Brazil, a significant portion of the population that consumes fish meat values the taste of South American catfish, especially those belonging to the genus *Pseudoplatystoma*, as Almeida Filho (2015) stated. About 30 years ago, production began with a high financial investment (Kubitza et al. 1998), using hybrids, specifically *Pseudoplatystoma corruscans* x *P. fasciatum*. The catfish carried the hybrid vigor of their parents. Still, there were complex zootechnical issues to overcome, such as high production costs resulting from cannibalism in the early stages and high nutritional requirements. To solve these problems and enable the productive sector to use hybrids of the genus *Pseudoplatystoma*, EMBRAPA included the species *P. reticulatum* in the Genetic Improvement Program through the AQUABRASIL project. However, as described in this chapter, the AQUABRASIL project was prematurely finished, and the use of hybrids for this genus continued to be the production reality for the sector over the years.

The Brazilian production sector overcame the zootechnical efficiency difficulties of catfish hybrids by crossing the genus *Pseudoplatystoma* with a species from another genus, specifically *L. marmoratus*. This new hybrid (*Pseudoplatystoma* sp. x *L. marmoratus*), popularly known as "Pintado da Amazônia," provided numerous zootechnical advantages compared to pure specimens of the *Pseudoplatystoma* genus or hybrids, especially in terms of omnivorous feeding habits (Fornari and Ribeiro 2017). Consequently, the "Pintado da Amazônia" dominated and virtually ended the cycle of hybrids derived from specimens of the *Pseudoplatystoma* genus. It is important to note that currently, the official data from IBGE (2022) do not differentiate the "Pintado da Amazônia" hybrid from other catfish produced in the country; out of the current 11,000 tons, are produced only a tiny volume of *Pseudoplatystoma* hybrids.

With the context of Brazilian catfishes, critical reflections become relevant for future actions related to the production of strains for South American siluriform species.

a. Hybrids continue to be the alternative found by the production sector to produce South American catfish;
b. Fish farmers strongly consider genetic introgression to be a problem;
c. Low reliability or efficiency of cryopreservation protocols for South American catfishes that make up the hybrid production in Brazil.

Regarding the first observation (a), it is evident that it will not be possible to produce strains if the production of South American catfish continues to rely on hybrids. As for the second issue (b), the problem of genetic introgression in wild populations has already been discussed in this chapter. Specifically, regarding the genus *Pseudoplatystoma*, some studies report issues such as the massive production of hybrids in fish farms (Campos 2010), and inadequate management likely facilitates escapes and introductions of these hybrids into the natural environment (Porto-Foresti et al. 2010; Hashimoto et al. 2014). In the study by Prado et al. (2017), the authors suggested that South American catfish hybrids may be reproducing, potentially causing genetic introgression of selectively fixed genes into wild populations of the *Pseudoplatystoma* genus.

Concerns about the genetic reserve from constructing germplasm banks for catfish species used in hybrid production exist. Once again, by accessing the data from CENARGEN, it is possible to identify that the three main catfish species involved in hybrid production have some deposits: *L. marmoratus*, *P. corruscans*, and *P. fasciatum*. While there are seminal sample deposits for species of the *Pseudoplatystoma* genus, for *L. marmoratus*, only purified DNA is available. There is no record of genetic variability characterization for the donor populations of cryopreserved samples from *Pseudoplatystoma* catfish species. However, it is likely that the samples already have genetic identification based on the variability characterization of individuals generated from thawed semen.

Nonetheless, as previously discussed for *C. macropomum*, the germplasm bank will only be applicable when this information becomes available in the ALELO system again. The recommendation by Vaini et al. (2014) to first identify the genetic variability of the animal and then determine the population genetic variability is crucial to avoid the risk of misinterpretation, and this identification can perfectly apply to establishing a germplasm bank for native South American species. The appropriate model for this process considers the following steps: (1) genetic identification of the species that are providing the semen to ensure the exclusion of hybrids; (2) mapping the genetic variability of populations that would provide seminal samples for cryopreservation, aiming to generate a population with high heterozygosity; (3) properly constructing the germplasm bank for the selected species. Therefore, the genetic purity of the animals providing seminal samples must be prioritized, and this issue is detailed in Chapter 4 of this book, emphasizing the need for adequate management to avoid future problems.

The other issue concerns the existing protocols to cryopreserve seminal samples of the *Pseudoplatystoma* and *Leiarius* genera, which need to be more reliable to guarantee future use. There are only five studies published to date on *Pseudoplatystoma* species, including *P. fasciatum* (Arciniegas et al. 2005), *P. metaense* (Ramírez-Merlano et al. 2011), *P. magdaleniatum* (Herrera-Cruz et al. 2019), and *P. corruscans* (Carolsfeld et al. 2003; Velarde et al. 2023). Even the most recent article on the *Pseudoplatystoma* genus did not provide a definitive protocol. Since the samples at CENARGEN were deposited before the current year, there is doubt about the actual quality of this preserved material. For the *Leiarius* genus, the studies are still more limited, with only two studies conducted by Gheller et al. (2019) and Borges et al. (2020), and there are no cryopreserved seminal samples of this species at CENARGEN.

5. Use of the Germplasm Bank of South American Native Species as a Solution in the Juvenile Production Market in Brazil – Model Developed by AQUAM

Numerous publications recurrently mention the benefits of using a germplasm bank directly in a fish farm (Asturiano et al. 2017), among which the following stand out: (1) providing semen for species with reproductive asynchrony during the breeding period (Egger et al. 2021); (2) working with a semen pool to maximize genetic variability (Caffey and Tiersch 2000; Streit Jr. et al. 2015); (3) ensuring the potential genetic reserve of breeding stock (Carolsfeld et al. 2003; Taitson et al. 2008; Torres et al. 2016; Martinez-Páramo et al. 2021; Galo et al. 2019). Operationalizing the germplasm bank and making it practical is considered the most complex aspect, with few solid records in the literature. The situation is similar for South American species, as discussed in this and other chapters, involving a series of technical difficulties, such as consistent cryopreservation protocols and the high cost of maintaining the germplasm bank.

Brazil has an ongoing economic discussion regarding establishing a central facility that distributes cryopreserved semen to farms specializing in the juvenile production of *C. macropomum*. Reducing the number of male breeders on juvenile production farms is the most evident advantage of having access to cryopreserved semen provided by a distribution center. However, in this case, the economic factor needs to be considered as an aspect, if not the most important one, that could render this operation unfeasible. In summary, published for Streit Jr. et al. (2019), we conducted an economic analysis considering the low productive efficiency of *C. macropomum* breeders, which, coupled with the low value of juveniles, practically makes a distribution center for cryopreserved semen unviable for farms producing juvenile forms of these species. It is essential to consider that this situation could be changed if these produced juveniles had added genetic value, meaning that the semen used to produce them originated from genetically improved animals. In this case, the unit value of the juvenile would be higher, offsetting the use of cryopreserved semen.

A promising scenario for using cryopreserved semen directly in producing juvenile South American species is the ongoing experience we are developing.

Two years ago, we started a project based on demand from the productive sector. Farm-producing juveniles of the "Pintado da Amazônia" hybrid (*P. corruscans* x *L. marmoratus*) requested that we establish a germplasm bank that could be operated during the breeding period of the species that give rise to the hybrid. It is possible to find mature *P. corruscans* females up to two months before *L. marmoratus* males begin seminal release. It is worth noting that the seminal quality of *L. marmoratus* at the beginning of the breeding season is low (kinetics and concentration), and there is also a tiny semen volume. This condition complicates the management of hybrid juvenile production when there is a high demand for this fish in the Brazilian market.

Initially, we successfully made cryopreserved semen of *L. marmoratus* available using the protocol developed by Gheller et al. (2019) as a pilot for the study that motivated the development of a germplasm bank application project. We cannot confirm the motility of 31.4% reported by Gheller et al. (2019) in our tests. To improve the freezability efficiency of *L. marmoratus* semen and to use a carbohydrate molecule with a lower cost than trehalose (used in the Gheller et al. 2019 protocol), we tested numerous combinations of permeable and non-permeable cryoprotectants. In this case, we applied an experimental process flowchart developed at AQUAM to construct a cryopreservation protocol. The newly obtained protocol proved to be more efficient and cost-effective than what is available in the literature, consisting mainly of methanol and skimmed milk powder (data submitted for publication). With this adapted protocol, combined with information on the productive efficiency of cryopreserved semen (fertilization rates, hatching rates, and the number of normal larvae produced), the farm-producing juveniles will be able to rely on the germplasm bank to produce the "Pintado da Amazônia" hybrid whenever deemed necessary. As a market strategy, the use of the germplasm bank in this case for hybrid production serves two points in the production chain: (1) the fry producer, who anticipates juvenile production by approximately two months and, consequently, gains faster access to the market compared to competitors, and (2) the producer, who can stock the "Pintado da Amazônia" earlier, accelerating the production cycle of this fish.

6. Conclusions

When formed and utilized appropriately the germplasm bank of South American species has played a leading role in developing a Genetic Improvement Program for two South American species. However, some points must be further reflected upon from a governance perspective.

1. South American species must have priority in conservation programs by the Brazilian Government. The country's readily available material and technical infrastructure guarantees this objective's physical and technical capacity. EMBRAPA has the structural ability to expand the number of cryopreserved samples and increase the number of species focused on production. Moreover, in Brazil, at least seven research groups in public universities, in addition to EMBRAPA, are fully dedicated to developing qualified scientific and technological knowledge for the genetic conservation of native South American species through cryobiology. Thus, it is possible to create a comprehensive,

integrated plan for the genetic security of native species with production potential based on germplasm banks by uniting these institutions through a joint effort by the Brazilian Federal Government;

2. As discussed at the beginning of this chapter, the South American ichthyofauna is richly diverse. Therefore, it is imperative to establish a classificatory criterion for the representativeness of different native species in the formation of future germplasm banks;
3. The use of germplasm bank in establishing the Genetic Improvement Program for two native South American species was one of the few worldwide. Suppose there is an option to build a new Improvement Program for South American species. In that case, the formation of a germplasm bank will be crucial for its operation, as has been pursued by EMBRAPA in recent years through collaboration with Brazilian producers and universities. The knowledge is documented and must be mandatorily used;
4. Brazil's most produced catfish is currently a hybrid animal derived from *L. marmoratus* semen, but no cryopreserved sample of it at CENARGEN. We understand that it is a priority for the Government to pay greater attention to this vital institution, CENARGEN. There is a need for more significant effort from the Brazilian Government, especially in collaboration with university research groups and EMBRAPA, to advance the development of a broad and efficient genetic reserve security plan. The research should be networked to build a truly efficient and representative germplasm bank, ensuring a more comprehensive understanding of the isolated components and their interactions.

Based on the FAO guidelines (2019) for the aquaculture sustainability of countries, it was possible to observe in this chapter that Brazil has a high diversity of fish species being produced. However, the Brazilian Government needs to pay more attention to the conservation of genes from these species. It is essential that, for renewed growth in the production of native South American species, there is a considerable qualitative and quantitative genetic reserve available for zootechnical use.

References

Almeida, F.J.E.P. 2015. Tópicos Especiais em Piscicultura de Água Doce Tropical Intensiva. Editora TecArt, São Paulo, Brasil.

Araújo-Lima, C.A.R.M. and M.L. Ruffino. 2004. Migratory fishes of the Brazilian Amazon. pp. 233–306. *In*: Carolsfeld, J., H. Brian, C. Ross and A. Baer [eds.]. Migratory Fishes of South America: Biology, Fisheries and Conservation Status. World Fisheries Trust - World Bank. IDRC, Victoria, Canada.

Arciniegas, S.P., J.M. Rodríguez and P.C. Casallas. 2005. Ensayos preliminares sobre crioconservación de semen de bagre rayado (*Pseudoplatystoma fasciatum* Linnaeus, 1766). Orinoquía 9: 28–37.

Asturiano, J.F., E. Cabrita and A. Horváth. 2017. Progress, challenges and perspectives on fish gamete cryopreservation: A mini-review. Gen. Comp. Endocrinol. 245: 69–76.

Bekkevold, D., M.M. Hansen and E.E. Nielsen. 2006. Genetic impact of gadoid culture on wild fish populations: predictions, lessons from salmonids, and possibilities for minimizing adverse effects. ICES J. Mar. Sci. 63: 198–208.

Besnier, F., K.A. Glover and O. Skaala. 2011. Atlantic salmon populations invaded by farmed escapees: quantifying genetic introgression with a Bayesian approach and SNPs. Aquac. Environ. Interact. 2: 75–86.

Caffey, R.H. and T.R. Tiersch. 2000. Economics and marketing of cryopreserved fish sperm. pp. 388–408. *In*: Tiersch, T.R. and P.M. Mazik [eds.]. Cryopreservation in Aquatic Species. Advances in World Aquaculture. World Aquaculture Society, Baton Rouge, USA.

Campos, J.L. 2010. O cultivo do pintado (*Pseudoplatystoma corruscans*, Spix; Agassiz, 1829), outras espécies do gênero *Pseudoplatystoma* e seus híbridos. pp. 335–361. *In*: Baldisserotto, B. and L.C. Gomes [eds.]. Espécies Nativas Para a Piscicultura no Brasil. Editora UFSM, Santa Maria, Brasil.

Carolsfeld, J., J. Harvey, H.P. Godinho and E. Zaniboni. 2003. Cryopreservation of sperm in Brazilian migratory fish conservation. J. Fish Biol. 63: 472–481.

Carneiro-Leite, L., C. Bashiyo-Silva, Y.A.A. Oliveira, L.P. Borges, M.P. Sanchez, L.G. Silva et al. 2020. Seminal characteristics and sensitivity of *Astyanax lacustris* (Characiformes: Characidae) sperm to cryoprotective solutions based on dimethylsufoxide and methylglicol. Neotrop. Ichthyol. 18: e200039.

Dagosta, F.C.P. and M.C.C. Pinna. 2018. A history of the biogeography of Amazonian fishes. Neotrop. Ichthyol. 16: 1–19.

Egger, R.C., N.C. Motta, T.S. França, A.V. Oliveira, W. Fernandes-Braga, J.I. Alvarez-leite et al. 2021. Sperm cryopreservation of *Prochilodus lineatus* throughout the same reproductive season. Aquac. Res. 52: 6453–6463.

EMBRAPA - Empresa Brasileira de Pesquisa Agropecuária. 2009. Projeto AquaBrasil: Bases tecnológicas para o desenvolvimento sustentável da aqüicultura no Brasil. Embrapa Meio-Norte, Parnaíba, Brasil.

EMBRAPA - Empresa Brasileira de Pesquisa Agropecuária. 2023. REGEN - Gestão de Recursos Genéticos para Alimentação, Agricultura e Bioindústria. Acessed 1 June 2023. https://agrin.ars.usda.gov/genomic_handling_dev?language=PT&record_source=BR&only_genomic=Y.

FAO - Food and Agriculture Organization of the United Nations. 2019. The State of the World's Aquatic Genetic Resources for Food and Agriculture. FAO Commission on Genetic Resources for Food and Agriculture assessments. Rome.

FAO - Food and Agriculture Organization of the United Nations. 2023. SDG Indicators Data Portal. Accessed 1 June 2023. https://www.fao.org/sustainable-development-goals/data/en.

Fernandes, M.D.O., D.K. Garcez, I.B. Acosta, S.M.M. Gheller, C.D. Corcini, L.J. Robe et al. 2020. Cryopreservation of sperm in annual fish *Austrolebias minuano* (Cyprinodontiformes; Rivulidae). Aquac. Res. 51: 147–154.

Fornari, D.C. and R.P. Ribeiro. 2017. A produção do pintado amazônico em Mato Grosso. Aquac. Bras. 6: 34–40.

Galo, J.M., D.P. Streit-Junior, S. Oliveira, J.P. Povh, D.C. Fornari, M. Digmayer et al. 2019. Quality of fresh and cryopreserved semen and their influence on the rates of fertilization, hatching and quality of the larvae of *Piaractus mesopotamicus*. Braz. J. Biol. 79: 438–445.

Geist, J. 2010. Strategies for the conservation of endangered freshwater pearl mussels (*Margaritifera margaritifera* L.): a synthesis of Conservation Genetics and Ecology. Hydrobiologia 644: 69–88.

Geist, J. 2015. Seven steps towards improving freshwater conservation. Aquatic Conserv: Mar. Freshw. Ecosyst. 25: 447–453.

Gheller, S.M.M., C.D. Corcini, C.R.C. de Brito, I.B. Acosta, G.C. Tavares, S.L. Soares et al. 2019. Use of trehalose in the semen cryopreservation of Amazonian catfish *Leiarius marmoratus*. Cryobiology 87: 74–77.

Gjedrem, T. 2010. The first family-based breeding program in aquaculture. Rev. Aquac. 2: 2–15.

Gjedrem, T. and M. Rye. 2018. Selection response in fish and shellfish: a review. Rev. Aquac. 10: 168–179.

Glover, K.A., M. Quintela, V. Wennevik, F. Besnier, A.G.E. Sørvik and Ø. Skaala. 2012. Three decades of farmed escapees in the wild:a spatio-temporal analysis of population genetic structure throughout Norway. PLoS ONE 7: e43129.

Glover, K.A., C. Pertoldi, F. Besnier, V. Wennevik, M. Kent and Ø. Skaala. 2013. Atlantic salmon populations invaded by farmed escapees: quantifying genetic introgression with a Bayesian approach and SNPs. BMC Genet 14: 74.

Glover, K.A., M.F. Solberg, P. McGinnity, K. Hindar, E. Verspoor, M.W. Coulson et al. 2017. Half a century of genetic interaction between farmed and wild Atlantic salmon: Status of knowledge and unanswered questions. Fish Fish. 18: 890–927.

Hagen, I.J., A.J. Jensen, G.H. Bolstad, O.H. Diserud, K. Hindar, H. Lo et al. 2019. Supplementary stocking selects for domesticated genotypes. Nat. Comum. 10: 199.

Hashimoto, D.T., F.D. Prado, J.A. Senhorini, F. Foresti and F. Porto-Foresti. 2014. Aquaculture of neotropical catfish hybrids: genetic strategies for conservation and management. pp. 1–10. *In*: Bradley, R. [ed.]. Carp and Catfish: Biology, Behavior and Conservation Strategies. Nova Science Publishers, New York, USA.

Heino, M., T. Svåsand, V. Wennevik and K.A. Glover. 2015. Genetic introgression of farmed salmon in native populations: quantifying the relative influence of population size and frequency of scapees. Aquac. Environ. Interact. 6: 185–190.

Herrera-Cruz, E., J. Aristizabal-Regino, J. Yepes-Blandón, A. Estrada-Posada, J. Espinosa-Araujo and V. Atencio-Garcia. 2019. Evaluation of three cryoprotectants to preserve striped catfish (*Pseudoplatystoma magdaleniatum*) semen. Rev. Colomb. Biotechnol. 21: 55–62.

IBGE – Instituto Brasileiro de Geografia e Estatística. 2022. SIDRA – Pesquisa da Pecuária Municipal. Acessed 9 June 2023. https://sidra.ibge.gov.br/Tabela/3940#.

Jézéquel, C., P.A. Tedesco, R. Bigorne, J.A. Maldonado-Ocampo, H. Ortega, M. Hidalgo et al. 2020. A database of freshwater fish species of the Amazon Basin. Sci Data. 7: 96.

Karlsson, S., T. Moen, S. Lien, K.A. Glover and K. Hindar. 2011. Generic genetic differences between farmed and wild Atlantic salmon identified from a 7K SNP-chip. Mol. Ecol. Resour. 11: 247–253.

Kubitza, F., J.L. Campos and J.A. Brum. 1998. Produção Intensiva no Projeto Pacu Ltda. e Agropeixe Ltda. Panor. Aqüi. 8: 41–49.

Lucas, C.M. 2008. Within flood season variation in fruit consumption and seed dispersal by two characin fishes of the Amazon. Biotropica 40: 581–589.

Marcos, R., J.A. Povh, D.C. Fornari, C.A.L. De Oliveira, R.P. Ribeiro, N.M. Lopera-Barrero et al. 2016. Ganho de peso e crescimento morfométrico de tambaqui (*Colossoma macropomum*) melhorado geneticamente. Semin. Cienc. Agrar. 37: 2521–2527.

Martínez-Páramo, S., A. Horváth, C. Labbé, T. Zhang, V. Robles, P. Herráez et al. 2017. Cryobanking of aquatic species. Aquaculture 472: 156–177.

Neira, R. 2010. Breeding in aquaculture species: genetic improvement programs in developing countries. 9th World Congress on Genetics Applied to Livestock Production, Germany 1: 8.

Pedroza Filho, M.X., R.M.V. Flores, H.S. Rocha, H.J.T. da Silva, D.Y. Sonoda, V.B. de Carvalho et al. 2020. The fish farming market in Brazil: study of supermarket segment. Embrapa Pesca e Aquicultura, Palmas, Brasil.

Porto-Foresti, F., D.T. Hashimoto, J.A. Senhorini and F. Foresti. 2010. Hibridação em piscicultura: monitoramento e perspectivas. pp. 589–606. *In*: Baldisserotto, B. and L.C. Gomes [eds.]. Espécies Nativas Para Piscicultura no Brasil. Editora UFSM, Santa Maria, Brasil.

Prado, F.D., R. Fernandez-Cebrián, D.T. Hashimoto, J.A. Senhorini, F. Foresti, M. Paulino et al. 2017. Hybridization and genetic introgression patterns between two South American catfish along their sympatric distribution range. Hydrobiologia 788: 319–343.

Ramírez-Merlano, J.A., V.M. Medina-Robles and P.E. Cruz-Casallas. 2011. Crioconservación seminal de bagre rayado *Pseudoplatystoma metaense* (Teleostei, Pimelodidae), bajo diferentes protocolos de congelación. Arch. Med. Vet. 43: 135–144.

Reis, R.E., J.S. Albert, F. Di Dario, M.M. Mincarone, P. Petry and L.A. Rocha. 2016. Fish biodiversity and conservation in South America. J. Fish Biol. 89: 12–47.

Rye, M., B. Gjerde and T. Gjedrem. 2010. Genetic development programs for aquaculture species in developed countries. 9th World Congress on Genetics Applied to Livestock Production, Leipzig, Germany.

Spica, L.N., E.A. Sanches, D.P. Streit Junior, R.A.C. Corrêa Filho, L.O. Brasileiro, R.B. Rodrigues et al. 2021. Successive seminal collections from *Leiarius marmoratus* during the reproductive period. Anim. Reprod. Sci. 234: 106852.

Streit Jr., D.P., D.P. Fornari, J.A. Povh, L.C. Godoy, F. Mello, C.A.L. Oliveira et al. 2015. Germplasm banking and its role in the development of the fish genetic improvement programme in Brazil. CryoLetters 36: 399–404.

Streit Jr., D.P., M.A. Rotta, U.S. Morales, J.A. Povh, R.B. Rodrigues and L.C. Godoy. 2019. Estimate of brazil's breeding stock and young "Round Fish". Proc. LAQUA 2019.

Sylvester, E.V.A., B.F. Wringe, S.J. Duffy, L.C. Hamilton, I.A. Fleming, M. Castellani et al. 2019. Estimating the relative fitness of escaped farmed salmon offspring in the wild and modelling the consequences of invasion for wild populations. Evol. Appl. 12: 705–717.

Taitson, P.F., E. Chami and H.P. Godinho. 2008. Gene banking of the neotropical fish *Leporinus obtusidens* (Valenciennes, 1836): a protocol to freeze its sperm in the field. Anim. Reprod. Sci. 105: 283–291.

Torres, L., E. Hu and T.R. Tiersch. 2016. Cryopreservation in fish: Current status and pathways to quality assurance and quality control in repository development. Reprod. Fertil. Dev. 28: 1105–1115.

USDA - United States Department of Agriculture Foreign Agricultural Service. 2023. Livestock and Poultry: World Markets and Trade. Acessed 21 april 2023. https://apps.fas.usda.gov/psdonline/app/index.html#/app/home.

Velarde, J.M.C., N.M. Bastos, L.C. Leite, L.P. Borges, E.G. Vieira and R. Veríssimo-Silveira. 2023. Dimethyl acetamide and dimethyl sulfoxide associated at glucose and egg yolk for cryopreservation of *Pseudoplatystoma corruscans* semen. Neotrop. Ichthyol. 21: e220071.

Villela, L.C.V., L.N. Ganeco-Kirschnik, E.S. Varela, E.B. Routledge, F.P. Rezende, L.S. Torati et al. 2020. Programa de Conservação de Recursos Genéticos de Peixes Nativos de Interesse Econômico da EMBRAPA. Embrapa Pesca e Aquicultura, Palmas.

Wacker, S., T. Aronsen, S. Karlsson, O. Ugedal, O.H. Diserud, E.M. Ulvan et al. 2021. Selection against individuals from genetic introgression of escaped farmed salmon in a natural population of Atlantic salmon. Evol. Appl. 14: 1450–1460.

Index

A

Aggregate Collection 6, 7
Aquabrasil 260–264
Aquaculture 173, 244, 250
Assay 185–187

B

Betta splendens 70, 86, 87
Biobanks 242–247, 249–251
Biodiversity 54–56, 58, 59, 64
Breeding Conservation Programs 58–60
Broodstock Characterization 245
Budget 185–188

C

Carassius auratus 70, 86
Cells 183–191
CENARGEN 259, 263, 265, 266, 268
Central Processing 6, 9, 14
Chilean Salmon Farming 47
Colossoma macropomum 258
Conservation 177
Cryobank 58, 59
Cryobanking 233
Cryopreservation 54, 58–63, 70–88, 171–178
Cryopreservation Medium 28, 32, 34, 35, 48
Cryopreservation Protocol 193, 196, 200–203, 206–211
Cryopreserved Semen Genetic Improvement Program 262
Cryoprotectants 28, 34, 35, 40–44, 73, 75, 78–83, 85, 87, 196–211
Curimba 104, 110, 112, 113, 128, 134, 136, 138, 168

D

Danio rerio 70, 74, 79
Density 121, 122, 124, 126, 127
Dimethylsulfoxide 35
DMSO 196–198, 200–210
DNA 58–61
Dourado 106, 139, 141, 254, 255

E

Ecology 243, 246, 247
Egg 177
Embryo 60–62, 70, 71, 73, 79, 82–88, 171–173

F

Female 184–186, 188
Female Milt 31
Fish 171–178
Fish Biodiversity 242
Fish Chimera 222–224
Fish Conservation 54, 57, 58
Fish Management 243
Fish Reproduction 118, 120, 132, 153, 157, 158
Fish Sterilization 224–226
Fish Sustainability 251
Fish-Induced Reproduction 145, 147
Fish-Sexual Maturity 150
Freshwater 241, 242, 246

G

Gametes 118, 119, 121, 126, 245, 247, 248, 250, 251
Genetic Conservation 143, 144
Genetic Resources 1, 3, 5, 15, 21, 23, 24
Genetic Variability 241–246, 249–251
Genome Cryodamage 63

Germ Cells 97, 100, 102, 103, 220–227, 229–232
Germinal Epithelium 97–102
Germplasm 59, 227
Germplasm Banks 87
Germplasm Repository 1–3, 5–7, 11, 13, 14, 20, 21
Germplasm Sources 251
Gonadal Differentiation 143, 148, 149
Gonadal Maturation 97

H

Host Specimen 229, 230

K

Kinetics 127, 128, 131, 132

L

Leiarius marmoratus 264

M

Male 184–188, 190
Maternal Genome 174, 178
Methanol 196–198, 200, 201, 203–206, 209, 210
Methods 183, 185–191
Mobile Facilities 6, 15, 21–23
Motilidade 134
Motility 126–133

N

Native 184
Neomale 31, 33, 34
Neotropical Fish 225, 226, 229, 231–233
Neotropical South American Species 256, 258, 263

O

Oocyte 71, 73, 74, 79–83, 88, 97, 98, 100–104, 107–109, 172, 174–178
Oogonia 174, 177, 178
Organizational Principles 3
Oryzias latipes 70, 82
Ovarian 97, 100–102
Ovarian Tissue 171, 172, 174–177

P

Pintado da Amazônia 264, 267
Piracanjuba 106, 112, 213–216, 231, 236
Population Restoration 58
Primordial Germ Cells 55, 62, 71, 83–85, 87, 221, 225, 227, 232
Process Mapping 3
Pseudoplatystoma reticulatum 260

Q

Quality Assurance 7
Quality Control 4, 7, 11, 13, 14
Quality Evaluation 7, 8
Quality Management 4–8, 14, 19
Quick Freezing 42–44

R

Repository Network 4, 20
Reproducibility 13, 18, 19, 21, 22
Reproductive Behavior 243–245
Reproductive Systems 158

S

Salmonids 27, 28, 30–33, 35, 42–45, 47, 48
Sample Processing 4, 6, 7, 13, 14, 16, 19
Slow Freezing 41–43, 80, 83, 84
South American Catfish Hybrids 265
Species 183, 184, 186, 189–191
Species and Population Delimitation 245
Sperm 71–79, 87, 88, 97, 103, 104, 106, 107, 118–122, 125–129, 131–133, 193, 194, 196, 199, 201–206, 208, 209, 211
Sperm Cryopreservation 27, 28, 32, 35, 47
Spermatogenesis 100
Spermatozoa 60, 61
Standardization 13, 16, 18, 19, 21, 22
Straws and Intratesticular Sperm 32, 33
Study 184, 185, 187, 191
Surrogate Broodstock 221–224, 227, 229–233, 250, 251

T

Tambaqui 203
Testis 98, 99
Tissue 183–185, 187, 188
Transplantation 221–224, 226–232

U

Ultra-Quick Freezing 43, 44

V

Velocity 119, 127–129, 132
Vitrification 77, 80–86
Volume 120–122, 124, 125